Electronic Measurements and Instrumentation

J.G. JOSHI

Head, Department of Industrial Electronics Engineering
Shri Sant Gadge Baba Polytechnic,
Z.T.C. Bhusawal (425 203) M.S.

KHANNA BOOK PUBLISHING CO. (P) LTD.

Publisher of Engineering and Computer Books

4C/4344, Ansari Road, Darya Ganj, New Delhi-110002
Phone : 011-23244447-48 **Mobile:** +91-9910909320
E-mail : contact@khannabooks.com
Website : www.khannabooks.com

Dear Readers,

To prevent the piracy, this book is secured with HIGH SECURITY HOLOGRAM on the front title cover. In case you don't find the hologram on the front cover title, please write us to at contact@khannabooks.com or whatsapp us at +91-99109 09320 and avail special gift voucher for yourself.

Specimen of Hologram on front Cover title:

Moreover, there is a SPECIAL DISCOUNT COUPON for you with EVERY HOLOGRAM.

How to avail this SPECIAL DISCOUNT:

Step 1: Scratch the hologram
Step 2: Under the scratch area, your "coupon code" is available
Step 3: Logon to www.khannabooks.com
Step 4: Use your "coupon code" in the shopping cart and get your copy at a special discount
Step 5: Enjoy your reading!

Price: ₹ 599.00

ISBN: 978-93-86173-62-1

Electronic Measurements and Instrumentation
by J.G. Joshi

First Edition: 2000
Reprint: 2022, 2023, 2024, **2025**

To view complete list of books, Please scan the QR Code:

Published by:
Khanna Book Publishing Co. (P) Ltd.
CIN: U22110DL1998PTC095547

Visit us at: **www.khannabooks.com**
Write us at: **contact@khannabooks.com**

Printed in Bharat.

PREFACE

The industrial scenario is changing very rapidly due to the fact that automatic, advanced, sophisticated measurement and instrumentation systems are used in research fields, industries, laboratories etc. In this book various aspects and methodologies used for measurement, recording and controlling of various electrical and non-electrical quantities is discussed.

Chapter 1 covers the objectives of measurement and instrumentation systems. In this chapter general aspects of measurement system, characteristics, types of errors involved in measurement methods is discussed. In last section of this chapter introduction and developments in instrumentation field is highlighted.

Chapter 2 covers the basic ideas and preliminary concepts of instruments, performance characteristics of instruments used for measurement of different electrical quantities like voltage, current, power etc. Types of instruments such as deflection type, null type, analog and digital type etc. are also covered.

In **Chapter 3** various transducers and sensing elements involved in instrumentation system are discussed. This chapter defines type of transducers, selection parameters of transducers. Transducers used for measurement of parameters like temperature, pressure, displacement, strain etc. are explained.

For measurement methods and systems of non-electrical quantities **Chapter 4** and **5** are devoted. These chapters are divided into two parts *i.e.* measurement systems part I and II. These chapter covers measurement methods of non-electrical quantities such as **displacement, flow, pressure, velocity, thickness, vibrations** etc. Important features of chapter 5 are Biomedical Instrumentation and non-destructive testings (NDT).

Detailed applications and significance of **signal conditioning devices and** need of these devices such as **amplifiers, V to I and I to V converters, filters** etc. are discussed in **Chapter 6**. It covers different characteristics and various applications of operational amplifier (OP-AMP).

Analog and Digital data acquisition systems and their components are explained in **Chapter 7**. Microprocessor and microcomputer based data acquisition systems are also discussed. These systems are used for continuous recording and monitoring the data or information accurately and precisely in processes industries. Detailed operation of real time clock (RTC) is also elaborated in this chapter.

Chapter 8 is "**Recorders and Display Devices**" in this chapter graphical recorders are discussed which provides continuous graphical recording or representations of the parameters involved in the processes. Working and Applications of Display devices such as LED, LCD, light indicators etc. are explained respectively.

Chapter 9 is "**Controllers**" in this chapter different **control actions, type of controllers** are explained. Controllers such as **Pneumatic, Hydraulic, Electronics** etc. are covered in detailed. Feature of this chapter is **Fuzzy logic Controllers (FLC).** Various applications of FLC in instrumentation field are explained by assuming that the reader is well familiarize with ideas and concepts regarding FLC.

This book is useful for **Degree** and **Undergraduate** students in **Electronics** and **Instrumentation** fields.

This book is written by keeping the views of academic requirements regarding instrumentation systems. Important feature of this book is that the text is lucid and simple. All the measurement methods and different measurement techniques are discussed in detailed manner. At the end of every chapter a set of questions *i.e.* Evaluation component is provided for understanding the text.

I hope this book will be useful for students of degree and undergraduate level. My sincere request to the readers to point out any shortcomings in the book which will help to improve and update the book.

J.G. JOSHI.

CONTENTS

Chapter 1

QUALITIES OF MEASUREMENTS

Learning Objectives

- To define significance of measurement.
- To study generalized block diagram of instrumentation system.
- To study dynamic characteristics.
- To study static characteristics.
- To define terms accuracy, precision, sensitivity, linearity.
- To discuss resolution, threshold, dead zone, dead time.
- To define errors and types of errors.
- To study statistical analysis.
- To calculate arithmetic mean, deviation, average deviation, standard deviation.
- To define sinusoidal transfer function.
- To study various types of input signals.
- To study frequency domain analysis.
- To discuss importance of instrumentation system.
- To study applications of instrumentation system.
- To explain recent development in instrumentation system.

1.1 Introduction

Measurement is a phenomenon mostly used in Science and Engineering branches, laboratories, industries for monitoring, recording and controlling different operations. Generally, measurement provides the information and clear-cut idea of the quantity or parameter under measurement. For measuring purpose instruments as well as monitoring devices are used.

Faithful, accurate and precise instrument provides exact measurement. For measuring purposes relative instrument is selected by considering quantity or parameter under measurement, disturbance in the system (if any), possible errors and performance characteristics of particular instrument etc.

Measurement generally involves application of an instrument by physical means for determining the unknown parameter or variable. In measurement process the physical parameters get converted into meaningful numbers by electrical means.

In industries, research laboratories, medicine etc. instrumentation is used for continuous measurement and controlling of parameters like temperature, pressure, liquid level, flow, acceleration, torque etc. For meeting todays need of fast, intelligent and accurate measurement systems, improvements and developments are essential.

In this chapter general concept of measurement systems and instruments, their performance characteristics are discussed in detail.

1.2 Generalized Block Diagram of a Measurement System

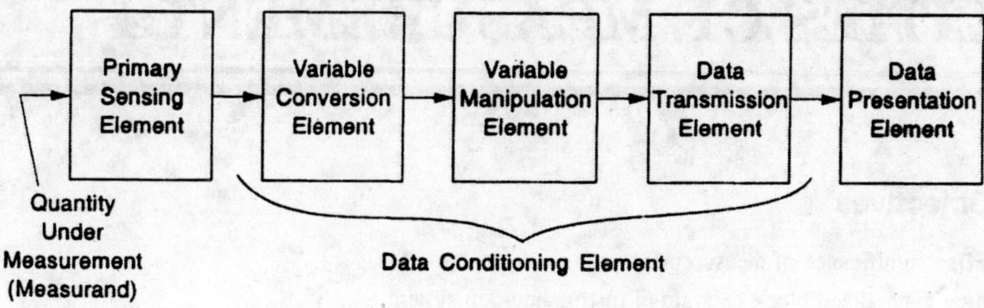

Fig. 1.1 Block Diagram of Measurement System

Fig. 1.1 shows the generalized block diagram of measuring system. The operation of any measuring instrument and the performance of any measuring system without resource to specific physical hardware can be analyzed. The operation can be analyzed by the functional elements of the system but the performance can be described by static and dynamic performance characteristics which shall be discussed in the next sections of this chapter.

Important functional elements of generalized measurement system are discussed below. The functional element consists of various components which perform their respective task.

(A) **Primary Sensing Element** : Initially the quantity under measurement or input signal (measurand) is received by the primary sensing element *i.e.* primary sensor. The function of this element is to convert the measurand into an analogous electrical signal. This particular function is carried out by transducer.

The function of transducer is to convert energy from one form to another. By electrical means transducer converts the physical quantities into an electrical signals.

In some cases the output signal may be non-electrical signal this output signal is coverted into electrical signal by transducer–detector stage. During the conversion process some energy is required which is extracted from the measurand medium, this may lead to loading effect. Due to this, the quantity under measurement gets disturbed so it is necessary to minimize this effect or design such system which needs external energy for conversion.

(B) **Variable Conversion Element** : As discussed the output of primary sensing element is electrical signal *i.e.* it may be any electrical parameter or variable (voltage, current, frequency etc.). If this output is not compatible with the system or not suitable for the instruments, under such conditions it may be necessary to convert this output to desired form before preserving it or processing it. This conversion is performed by the variable conversion element. If the electrical parameter is compatible with the system or instrument then the variable conversion element is not required. To introduce the variable conversion element on the other side some systems require several stages of it. *e.g.*-Analog-to-Digital converter, Voltage-to- current converter etc.

(C) **Variable Manipulation Element** : Here manipulation is defined as change in numerical value of the output signal (variable) with respect to particular condition. The function of this element is to manipulate the signal so as to preserve it in its original form.

Most of the transducers produce low output signal in mV and μV such low signal gets affected by noise signals or distortions. It is not possible to transmit or record these signals. To overcome these problems various operations like amplification, attenuation, filtering, chopping etc. are carried out to get the signal in desired form. These operations are nothing but the conversion processes and they are known as signal conditioning elements.

(D) **Data Transmission Element** : When the elements of an instruments or measuring systems are located separately or are physically separated then it is essential to transmit the data *i.e.* output signals or variables from one place to another. This function is performed by data transmission element. Through wires, cables, leadwires etc. the data can be transmitted.

(E) **Data Presentation Element** : This element acts as a communicating media between instrument and the user *i.e.* human being by providing the information of quantity under measurement. This information is necessary to monitor or to record for analysis purpose. It is also applicable for controlling the processes. This function can be performed by instruments like ammeter, voltmeter, magnetic tape recorders, CRO, digital memories, printers, recorders, microprocessors, computers, display devices etc.

1.3 Performance Characteristics of Instruments and Measurement System

As stated in section 1.2 that the performance of measurement system is analyzed by considering its performance characteristics, these characteristics are divided into two types:

(i) Static characteristics, and
(ii) Dynamic characteristics.

This classification is on the basis of quantity under measurement. If the quantity under measurement is constant or varying very slowly, then it is possible to decide the performance criterion which provides the meaningful information of the quantity without considering the dynamic description which uses the differential equations. This phenomenon is known as static characteristics *i.e.* in static characteristics the quantities under measurement are varying very slowly with respect to time.

Another aspect of measurement system is that if the measurement quantity is varying or changing rapidly then under such circumstances dynamic relationship between the instrument input and output must be considered. This relationship is normally considered by using differential equations. This phenomenon is known as dynamic characteristics.

1.4 Static Characteristics

In this section we shall concentrate on general static characteristics of instruments while the specific characteristics will be discussed in section 1.7.

The static performance characteristics are studied by considering the term **Static Calibration**. Actually, static calibration is to check the instrument against known (reference) standard and to find out accuracy, sensitivity, errors etc. Calibration is the method of determination by comparing the measurand with a standard or correct value of each scale reading on a meter. For instrument calibration the important approach is to test the instrument construction, possible inputs to the instrument, find out significant input at which the instrument should be calibrated, develop static input-output characteristics by keeping one input constant and varying others and recording the output.

For every instrument static characteristics are related so now we shall concentrate upon general static characteristics in the further text.

Important static characteristics are :

Accuracy, Sensitivity, Precision, Linearity, Static errors, Resolution, Threshold, Dead space etc. These characteristics are discussed below.

(a) **Accuracy** : Accurate means exact. Accuracy of a measurement system is a degree of exactness or confirmation to the true value. Accuracy is the closeness with which a measured value is equal to the specified standard. Factors like drift, static and dynamic errors, temperature, vibrations, non-linearity, environmental parameters affect the accuracy of the measurement system.

(b) **Sensitivity** : It is defined as the ratio of change in the magnitude of instrument output or response to the corresponding change in the magnitude of the measurand or input signal. Always the instrument or measurement system should have high sensitivity.

The reciprocal of sensitivity is called as inverse sensitivity or deflection factor. The unit of sensitivity is defined by the type of input and output.

(c) **Significant figures** : Significant figures indicate the precision of measurement. They are the numbers in which the result is expressed. Significant figures provide the information of magnitude and measurement precision of quantity.

For example :

Case 1. : The reading of voltmeter is 27V, then its voltage should be closer to 27V than 26V or 28V.

Case 2. : Consider if the voltage recorded is 27·0V in this case the measured voltage is closer to 27·0V than 26·9V or 27·1V.

In this example the significant figures in Case (1) are two and in Case (2) there are three. If the significant figures are more, then the precision of measurement is greater.

(d) **Precision** : The term **precise** means clearly defined. It is defined as **the degree of exactness for which an instrument is designed to perform**. It indicates that the instrument output must be consistent for given value of the input. Always measurements are carried out for getting the magnitude of quantity under measurement, precision has two characteristics, conformity and the number of significant figures for which the measurements are carried out. Meaning that the significant figures provide the magnitude and the measurement precision of the quantity which is under measurement.

(e) **Linearity** : In measurement system it is desirable that the system should be linear *i.e.* output of a system should be linearly proportional to input.

(f) **Hysteresis** : HYSTERESIS effects may be possible in electrical, mechanical, chemical or in any physical phenomenon.

Instruments or various devices are used for the measurements and analysis of unknown quantities and parameters. If the input quantity (measurand) is increased from zero to full scale and after decreasing to zero, the two output readings obtained sometimes differs from each other because of internal or external friction, loading effects etc.

The maximum difference in any part of the output readings obtained during any one calibration cycle is the hysteresis of the device. Hysteresis effects are present due to internal and external friction losses, backlash of gears (in mechanical systems), loading effects, magnetic materials etc.

(g) **Threshold** : Instruments are used to measure and analyze the unknown quantities. These quantities are applied to the input of an instrument. If this input to the instrument is increased gradually from zero, there will be some minimum value below which no output change can be observed or detected. This minimum value of reading is called Threshold of an instrument.

This indicates that the output is changing with respect to change in the input quantity. So threshold can be defined as the definite numerical value in the output corresponding to change in input.

(h) **Resolution** : If the input signal to an instrument is increased from non zero input value the output does not changes until a certain fixed input increment is exceeded, This increment is called resolution. Resolution is increment in the input quantity which produces a fixed change in output.

Threshold is smallest measurable input and resolution is smallest measurable input change.

(i) **Drift** : Is unwanted parameter in the instruments and measurement systems. Drift is defined as variation in an instrument output which is not due to change of input. It may occur by internal temperature variations and circuit or component instability due to stray electric and magnetic fields. It is very essential to minimize the drift by using compensation or proper shielding techniques and methods.

e.g. : In op-amp the **drift** due to temperature is compensated by different techniques.

(j) **Dead zone** : If input signal is applied to an instrument and the instrument does not provide the output it is in dead zone. It is defined as the largest value of the input signal to which the instrument will not respond. This is due to HYSTERESIS and backlash.

(k) **Dead time** : It is defined as the time required for the measurement system to provide output with respect to change in input quantity.

1.5 Errors in Measurement System and Analysis of Errors

In measurement process the unknown quantity (parameter) is compared with reference quantity. In measurement process an instrument is used or interfaced with unknown quantity. The measured quantity should be accurate and precise. It is very much necessary and important to know the validity of measured quantity or data. But it is not possible to measure the quantity or parameter with high accuracy because various sources are present which introduce errors in the measurement system. These sources are incorrect design parameters, design limitations, improper maintenance, loading effect environmental conditions, mistakes caused by the user or operator etc.

Errors are classified as follows :.

(a) Gross errors (b) Systematic errors (c) Random errors (d) Interference errors, which are discussed as follows :

(a) **Gross errors** : These errors are mainly occurring due to user or human mistakes in reading the instruments, recording the readings and observations, analyzing the results etc.

For example, the reading of Voltmeter is actually 26·8 Volt but the operator or user records it as 28·6 Volt. The gross errors are present due to mistakes by human beings. As a matter of fact human being is involved in measurements it is not possible to eliminate gross errors completely. But these errors can be minimized by taking great care in reading and recording the data, observe two, three, or more readings then select final reading of the quantity under measurement.

(b) **Systematic errors** : These errors are also known as bias since they affect complete measurement system. These can be removed by calibration. These errors are classified as follows :

(i) Instrumental errors (ii) Environmental errors (iii) Observational errors

(i) **Instrumental errors** : These errors are occurring due to misuse of the instrument, loading effect, improper mechanical structure. Misuse means faults by the user because of wrong concepts, improper use of cables and leads with the instruments, wrong adjustment of zero, overloading of instruments, friction in mechanical parts etc.

Instrumental errors can be minimized by accurate calibration of the instruments, using proper mechanical parts, to avoid overloading and overheating of the instrument etc.

(ii) **Environmental errors** : These errors are due to external conditions where the instruments and measurement systems are installed or placed. The parameters which introduces these errors are temperature, pressure, humidity, dust, vibrations, moisture, stray electric and magnetic fields.

Such type of errors can be removed by using magnetic and electrostatic shields, hermetic sealing, air conditioning etc.

(iii) **Observational errors** : These errors are due to the observer or user. Example of this error is "PARALLAX" while observing the reading.

For minimizing the "PARALLAX" error accurate meter with mirrored scales should be used. The precaution necessary is that only one observer or user should read the scale or readings on the meter which provides the line of vision of the observer exactly above the pointer.

(c) **Residual or random errors** : These errors remain after removing gross and systematic errors substantially. These errors are not determinable and very small. Residual errors can be analyzed statistically by using mathematical approach.

(d) **Interference errors** : It becomes necessary sometimes to measure low level input signals under any process but the unwanted interfering signals like noise, humming sound, distortion, ripples, transients, thermal sources, switching and relay arc etc. gets introduced with input signals which produces errors known as interference errors. These errors can be eliminated by using isolation techniques.

1.6 Statistical Analysis

In measurement system statistical analysis is used to verify and determine the final results. In large number of measurements it is necessary that the data or information should be accurate with meaningful interpretations. This is possible by means of statistical analysis, it also helps to calculate the statistical errors which are small. For meaningful and accurate analysis large number of measurements are necessary.

(A) **Arithmetic Mean** : It is defined as the most probable or approximate value of measured variable or number of readings taken. When the number of readings of same quantity are large then arithmetic mean is considered.

Arithmetic mean is given by :

$$\overline{X} = \frac{x_1 + x_2 + x_3 + x_4 + \ldots + x_n}{n} = \frac{\sum_{n=1}^{n} x_n}{n} \quad \ldots 1.1$$

where ; $\overline{X}$ is the arithmetic mean,

$x_1, x_2, x_3, x_4, \ldots x_n$ are the readings or observations taken (samples)

n = total number of readings.

Example 1.1 : The following values were taken from the measurements of the values of voltages in Volt by using digital voltmeter as 22·03, 22·34, 22·45, 22·78, 22·00, 22·80 calculate the arithmetic mean.

Solution :

The arithmetic mean is given by :

$$\overline{X} = \frac{x_1 + x_2 + x_3 + x_4 + \ldots + x_n}{n}$$

Here, $x_1 = 22{\cdot}03,\ x_2 = 22{\cdot}34,\ x_3 = 22{\cdot}45,\ x_4 = 22{\cdot}78,\ x_5 = 22{\cdot}00,\ x_6 = 22{\cdot}80$

$$\overline{X} = \frac{22{\cdot}03 + 22{\cdot}34 + 22{\cdot}45 + 22{\cdot}78 + 22{\cdot}00 + 22{\cdot}80}{6}$$

Arithmetic mean $\overline{X} = 22{\cdot}40$ Volt

(B) **Deviation** : Is defined as the departure of recorded reading from the arithmetic mean of the group of readings. Means the change or difference between reading from the group of readings and the arithmetic mean of group.

Let us suppose for reading x_1 deviation is d_1 which is given by :

$$d_1 = x_1 - \overline{x} \qquad \ldots 1.2$$

For d_2 it is $d_2 = x_2 - \overline{x}$, ... for $d_n = x_n - \overline{x}$

The algebraic sum of deviation is $d_1 + d_2 + d_3 + \ldots + d_n$ Deviation from the mean may have positive value or negative value but the algebraic sum of the deviation is zero.

(C) **Average Deviation** : It is defined as the sum of the absolute values of the deviations divided by number of readings.

Average deviation is given by :

$$D_{av} = \frac{|d_1| + |d_2| + |d_3| + \ldots + |d_3|}{n} = \frac{\Sigma\,|d_n|}{n} \qquad \ldots 1.3$$

where ; D_{av} is Average deviation

$|d_1| + |d_2| + |d_3| + \ldots + |d_n|$ are absolute values of deviations.

$n =$ total number of readings

Arithmetic deviation provides the precision of instruments used in measurement system. The highly precise instruments yield a low average Deviation between readings.

(D) **Standard Deviation** : It is also known as root mean square deviation as it is given by :

$$S.D. = \sigma = \sqrt{\frac{d_1^2 + d_2^2 + d_3^2 + \ldots + d_n^2}{n}} = \sqrt{\frac{\Sigma\, d_n^2}{n}} \qquad \ldots 1.4$$

It is defined as the square root of sum of all the individually squared deviations to the number of readings and denoted by 'σ'.

where ; σ is standard deviation.

the number of observations are greater than '20' then it is denoted by 'σ' and if less than '20' it is given by ;

$$\therefore \qquad S.D. = \text{'}S\text{'} = \sqrt{\frac{d_1^2 + d_2^2 + d_3^2 + \ldots + d_n^2}{n-1}}$$

$$S = \sqrt{\frac{\Sigma d_n^2}{n-1}} \qquad \text{...1.5}$$

Example 1.2 : If the data recorded during the current measurement by five observers is as 9·8 mA 9·7 mA, 9·2 mA, 9·9 mA and 9·4 mA calculate (a) Arithmetic mean (b) Deviation (c) Average deviation (d) Standard deviation.

Solution :

(a) The arithmetic mean is given by ;

$$\overline{X} = \frac{9{\cdot}8 + 9{\cdot}7 + 9{\cdot}2 + 9{\cdot}9 + 9{\cdot}4}{5}$$

$$\overline{X} = 9{\cdot}6 \text{ mA}$$

(b) The deviation is calculated as,

$$d_1 = x_1 - \overline{X} = 9{\cdot}8 - 9{\cdot}6 = 0{\cdot}2 \text{ mA}$$

$$d_2 = x_2 - \overline{X} = 9{\cdot}7 - 9{\cdot}6 = 0{\cdot}1 \text{ mA}$$

$$d_3 = x_3 - \overline{X} = 9{\cdot}2 - 9{\cdot}6 = -\,0{\cdot}4 \text{ mA}$$

$$d_4 = x_4 - \overline{X} = 9{\cdot}9 - 9{\cdot}6 = 0{\cdot}3 \text{ mA}$$

$$d_5 = x_5 - \overline{X} = 9{\cdot}4 - 9{\cdot}6 = -\,0{\cdot}2 \text{ mA}$$

The algebraic sum of deviations $d_1 + d_2 + d_3 + d_4 + d_5 = 0$.

(c) The average deviation is,

$$D_{av} = \frac{0{\cdot}2 + 0{\cdot}1 + 0{\cdot}4 + 0{\cdot}3 + 0{\cdot}2}{5} = 0{\cdot}24 \text{ mA}; \ (n = 5)$$

(d) Standard deviation,

$$\therefore \quad S = \sqrt{\frac{(0{\cdot}2)^2 + (0{\cdot}1)^2 + (0{\cdot}4)^2 + (0{\cdot}3)^2 + (0{\cdot}2)^2}{5-1}} \qquad \because \ (n = 5)$$

$$\therefore \quad S = 0{\cdot}291 \text{ mA}$$

$\therefore$ Standard deviation is 0·291 mA

The algebraic sum of the deviation is ;

$$= 0{\cdot}2 + 0{\cdot}1 - 0{\cdot}4 + 0{\cdot}3 - 0{\cdot}2 = 0$$

1.7 Dynamic Characteristics of Instruments and Measurement System

(A) **Transfer Function** : The transfer function of a linear time-invariant system is defined as the ratio of the Laplace transform of the output variable to the Laplace transform of the input variable assuming all the initial conditions are zero. In measurement and instrumentation systems the transfer function is used for designing, determination of overall characteristics and performance analysis of the systems.

Therefore the transfer function is given by;

$$G(s) = \frac{q_0(s)}{q_i(s)} = \frac{b_m s^m + b_{m-1} s^{m-1} + \ldots + b_1 s + b_0}{a_n s^n + a_{n-1} s^{n-1} + \ldots + a_1 s + a_0} \; ; m < n \qquad \ldots 1.6$$

where ; $q_0(s)$ is Laplace transform of the output.

$q_i(s)$ is Laplace transform of the input.

Symbolic Representation of Transfer Function

$$\xrightarrow[\text{(Input)}]{q_i} \boxed{\frac{b_m s^m + b_{m-1} s^{m-1} + \ldots\ldots\ldots\ldots + b_1 s_0 + b_0}{a_n s^m + a_{n-1} s^{n-1} + \ldots\ldots\ldots + a_1 s_0 + a_0}} \xrightarrow[\text{(Output)}]{q_0}$$

(B) **Sinusoidal Transfer Function** : Under dynamic conditions the quality of measurement can be analyzed by applying "standard" inputs to the system.

In the linear system as the transients die out from the system it tries to become stable and in steady state. As the transients die out, the output quantity will be a sine wave as that of input angular frequencies. While in the process of dying out of the transients it may be possible that the amplitude of output signal gets changed to that of input and it may be possible to shift the phase. Due to same frequency the input and output relation is specified by considering the amplitude ratio and phase shift as a function of frequency.

The steady state response of a measurement system in sinusoidal input is obtained by replacing 's' with '$j\omega$' in the transfer function of system.

Therefore substitute '$j\omega$' for 's'

The sinusoidal transfer function is;

$$\frac{q_0(j\omega)}{q_i(j\omega)} = \frac{b_m (j\omega)^m + b_{m-1} (j\omega)^{m-1} + \ldots + b_1 (j\omega) + b_0}{a_n (j\omega)^n + a_{n-1} (j\omega)^{n-1} + \ldots + a_n (j\omega) + a_0} \qquad \ldots 1.7$$

The term $\frac{q_0(j\omega)}{q_i(j\omega)}$ is a complex number at any given frequency. The polar form of this equation is as;

$$\frac{q_0(j\omega)}{q_i(j\omega)} = \frac{|q_0|}{|q_i|} \angle \phi = M \angle \phi \qquad \ldots 1.8$$

where ; M is the amplitude ratio of output to the input, $\angle \phi$ is the phase angle by which output q_0 leads the input q_i.

(C) **Dynamic Performance of Measurement System** : It can be analyzed by (i) Time domain analysis (ii) Frequency domain analysis.

(i) **Time domain analysis** : For most of the measurement systems the input signal is applied as a function of time. The performance of a system is measured by computing different time response parameters and steady state accuracy. In dynamic behaviour different inputs or standard test signals like step, ramp, parabolic and impulse are applied to the system for time response analysis.

In time domain analysis following input signals are used :

(1) **Step signal** : In the step signal its value changes from one level to another level A in zero time.

The graphical representation of a step signal is shown in Fig. 1.2(a). The mathematical representation is given by :

$$r(t) = Au(t)$$
$$u(t) = 1 \; ; t > 0$$
$$= 0 \; ; t < 0$$

The laplace transform is $R(s) = A/s$ ...1.9

(2) **Ramp signal** : This signal starts at a value of zero and increases linearly with time.

Graphical representation of ramp signal is shown in Fig. 1.2(b). The mathematical representation is given by :

$$r(t) = At \; ; t > 0$$
$$= 0 \; ; t < 0$$

The laplace transform is $R(s) = A/s^2$ ...1.10

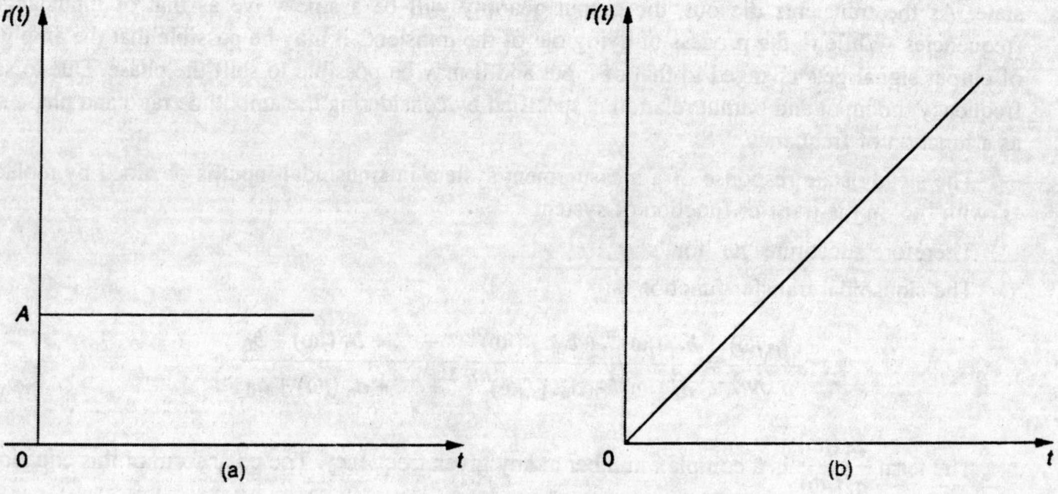

Fig. 1.2 (a) and (b)

(3) **Parabolic signal** : This signal is integral of a ramp signal.

Graphical representation of parabolic signal is shown in Fig. 1.2(c). The mathematical representation is given by :

$$r(t) = At^2/2 \; ; t > 0$$
$$= 0 \; ; t < 0$$

The laplace transform is $R(s) = A/s^3$ ...1.11

(4) **Impulse signal** : Is defined as a signal which has zero value everywhere except at $t = 0$, where its magnitude is infinite.

The graphical representation of impulse signal is shown in Fig. 1.2(d).

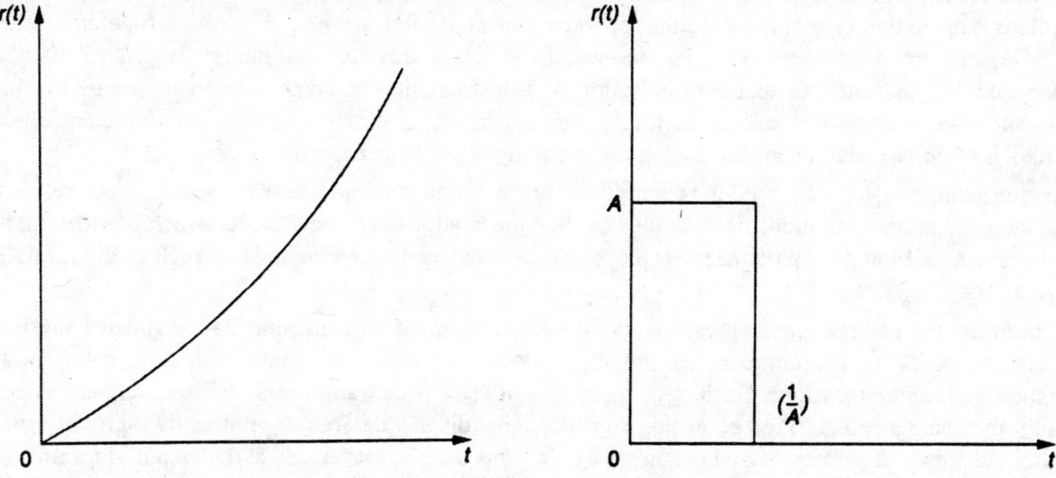

Fig. 1.2 (c) and (d).

The mathematical representation is given by :

$$\delta(t) = 0 \; ; \; t \neq 0$$

The laplace transform of impulse signal is 1.

$$R(s) = 1 \qquad ...1.12$$

According to the type of system and input signals, system performance aspects get changed.

(ii) **Frequency domain analysis** : The magnitude and phase relationship between the sinusoidal input and the steady state output is known as frequency response. In linear time-invariant system the frequency response of measurement system is independent of amplitude and phase of the input signal.

The frequency response can be obtained for the sinusoidal transfer function by replacing 's' with '$j\omega$' in the system transfer function as mentioned earlier. The transfer function $G(j\omega)$ obtained is a function of both magnitude and a phase angle. These parameters are represented by graphical plots in the frequency response analysis.

Consider a linear system with a sinusoidal input

$$r(t) = A \sin \omega t \qquad ...1.13$$

Under steady state condition the output is represented as ;

$$c(t) = B \sin (\omega t + \phi) \qquad ...1.14$$

where ; B is the amplitude, ϕ is phase shift of the output.

In the frequency response analysis 'B' and phase shift 'ϕ' is determined by keeping the amplitude 'A' fixed for suitable range of frequencies.

1.8 Importance of Instrumentation System

Now a days instrumentation system plays an important role in every aspect of human life. This field *i.e.*

instrumentation is involved in scientific and technological activities. In our day-to-day life we are observing different instrumentation system it *e.g.* automatic water tank controller, washing machine, refrigerator, water heater (Geyser), front panel of vehicle etc. which indicates different parameter associated with it (speedometer, fuel indicator, temperature indicator etc.). Instrumentation comes into existence by making attempts and efforts for measurement of various electrical and non-electrical quantities or parameters. These parameters may be physical, chemical, biological, metallurgical, environmental etc..

Instrumentation systems are used in power plants, process industries such as paper and sugar, oil refinery plants, automatic production industries. Biomedical instrumentation is the new aspect which is widely used for measurement of biological parameters or biopotentials associated with human body such as ECG, EMG, blood flow, body temperature etc.

Instruments are the backbone of measurement system. By using proper instrumentation systems physical parameters or variable like temperature, pressure, velocity, acceleration, strain, humidity, mass, flow, acceleration etc. can be measured. Such systems consist of sensors or transducers, different output devices (recorders and analog meters to record analog signals, digital display devices or printers for digital signals) to monitor the output of system. Signal conditioning elements may be introduced at the output of transducer to process their output signals.

1.9 Applications of Instrumentation System

(1) Measurement of Various Parameters

To measure or determine various parameters and information involved in the system or process. In addition to this the instruments can display, monitor or register the information or data using proper signal conditioning elements.

(2) Controlling of Parameters Involved in the Process

After measuring or getting the knowledge of unknown parameters it helps to control the process at set point (fixed value) or changes according to pre-set program. This phenomenon is used in automatic control system. In process industries and plants automatic control system is used to control different process variables such as temperature, pressure, velocity, flow rate, liquid level, etc. This system is also used in air crafts, radar systems etc.

(3) To Carry out Various Manipulating Functions

The elements or stages involved in instrumentation system performs various operations for conditioning the output signals of sensors or transducers. These operations are like signal addition, subtraction, multiplication, integration, differentiation, logarithmic and anti logarithmic functions, filtration, linearisation, ratio control, conversions like analog to digital and digital to analog, also current to Voltage and Voltage to current converters etc. These elements are discussed in the next chapter of this book.

(4) Designing and Developing New Products

Instrumentation system helps to design and develop new products or systems. Testing, modifications, trial and error methods can be performed using available data, information, designing procedure, calculations, standard equations, theory and operating principles. Various testing and experiments can be carried out on prototype model so that new product or system can be developed.

(5) Quality Control Testing of Products

Instrumentation system helps to perform different quality control tests on the products or manufactured industrial products. These testing helps to detect or trace the defects like cracks, flaws, blow holes, cavities, internal shrinkages etc. present in the newly manufactured product. These testings also helps to check hardness, internal structure, strength of the materials used in product fabrication. There are different testing like X-ray, ultrasonic, eddy current, radiography etc. which can be performed or carried out to detect these defects. Advantages of these testings are (1) non-destructive (2) continuous or on-line testing (3) no damage and wastage of products or samples under test (4) defective product or piece gets rejected.

Because of such type of testing good quality and reliable product or system can be developed which helps to satisfy the needs and demands of the customer or user.

1.10 Merits of Proper Instrumentation System

(1) It monitors or records the on line processes and operations continuously.

(2) It helps to control various processes and operations using proper control actions.

(3) Human errors are nearly eliminated due to which fast and accurate controlling of processes can be achieved.

(4) Accuracy and precision of system gets improved.

(5) Speed of operation is increased.

(6) Easy to store the information using memory, analog to digital converters and microprocessors.

(7) It performs the functions like linearisation, processing, calculations, etc. of the data.

(8) Useful in hazardous applications or environments.

1.11 Recent Developments in Instrumentation Systems

Present era is computer era, Microprocessor or computer systems are replacing the old and tedious instrumentation system. Most of the industries and power plants are introducing microprocessor based data acquisition systems and data loggers rapidly for monitoring, storing and processing the information or data (which is nothing but the outputs derived from various sensors or transducers used in the system). Different programming softwares such as C++. Turbo C etc. can be used for controlling various processes.

PC workstations with UNIX or MS-WINDOWS based systems can be used. It helps to carry out the functions like linearisation, plotting of graph, correlation of data, comparison of two or more data.

Due to progress and advancement in computer technology Laptop computer with giga-byte memory are being introduced in instrumentation fields.

(A) **Microelectronics Instrumentation** : Recently semiconductor transducers or sensors are used in numerous applications, biomedical instrumentations, meteorology, robotics, industrial process control etc. These sensors can be fabricated in the form of integrated circuit chip using different fabrication technologies such as very large scale integration (VLSI), very high speed integrated circuit (VHSIC), ultra large scale integration (ULSI).

Sensors or transducers produced by using these techniques are pressure transducers, micro-diaphgram pressure sensor, semiconductor strain gauges, photodiode etc.

e.g. : The integrated circuit pressure sensor chip has size of 3 mm square and thickness of 25 μm. It has output voltage range is about 1·2 to 3·4 V for a pressure range of 0 to − 550 mmHg, pressure sensitivity of 4 mV/mmHg.

(B) **Fibre Optic Sensors** : Recently fibre optic sensors are developed which are applicable for measurement of various parameters. Features include high operating speed, more accuracy, high sensitivity, more reliability, longer life, higher bandwidth can be achieved. Important advantage is better noise reduction is possible.

These type of sensors are used in digital signal processing, long distance communication, in mining works, measurement of load and weight in industries etc.

Optical fibre sensors can also used for measurement of voltage and current of the transformers used in extra high voltage (EHV) applications because of breakdown of insulation. There are various fields of instrumentation systems such as biomedical, industrial, nuclear, rocket etc. In this book various transducers and measuring systems and process control systems are discussed.

EVALUATION COMPONENT

1. **Draw the block diagram of generalized measurement system. Explain the function of each block.**
2. **Define :**
   (a) Accuracy (b) Precision (c) Sensitivity (d) Linearity
3. **Compare accuracy and precision.**
4. **Write a short note on :**
   (a) Gross errors (b) Systematic errors (c) Residual errors
5. **Define :**
   (a) Transfer function (b) Sinusoidal transfer function
6. **What are the standard input signals used in time response analysis ? Draw the input signals and specify their laplace transforms.**
7. **Write a short note on :**
   (a) Time response analysis (b) Frequency response analysis
8. **Define the terms :**
   (a) Arithmetic mean (b) Average deviation (c) Standard deviation
9. **Following values were taken from the measurements of resistors in Kilo-Ohms 2·27 kΩ, 2·29 kΩ, 2·26 kΩ, 2·28 kΩ, 2·22 kΩ, 2·24 kΩ, 2·23 kΩ, 2·30 kΩ. Determine :**
   (i) Arithmatic mean (ii) Average deviation
   (iii) Standard deviation
10. **Specify the precautions necessary to avoid the errors in measurement systems.**
11. **List the merits of proper instrumentation system.**

Chapter **2**

MEASURING INSTRUMENTS

Learning Objectives

- To study basic concepts of instruments.
- To study classification of instruments.
- To define types of instruments.
- To study forces developed in instruments.
- To classify types of meters.
- To study different analog meters.
- To discuss working principle of various instruments.
- To analyze torque equation of different instruments.
- To study measurement of voltage, current, resistance, power.
- To study ammeter, voltmeter, ohmmeter in different configurations.
- To study AC and DC meters.
- To study different type of wattmeters, their working principles and calculation of power.
- To discuss various digital meters.
- To study energy meters.

In general to discuss and explain construction, working, calculations, advantages, disadvantages and applications of different instruments.

2.1 Introduction

Instrument is a device or equipment used to determine the present value of the quantity under measurement. The unknown quantity which is to be analyzed is measured by means of the instruments.

Generally, the parameters or quantities measured by the instruments are voltage, current, resistance, power, energy, frequency etc. Measuring instruments are involved in every field of engineering applications, laboratory works, research laboratories, industries etc.

In this chapter both type of instruments *i.e.* analog and digital instruments are discussed. In analog instruments the classic approach of D' Arsonval movement is used in which the pointer moves over a standard calibrated scale proportional to the input unknown quantity. Digital instrument uses analog to digital converters and the measured quantity is indicated on display devices in numerical form.

Under the heading of analog instruments, AC and DC types of ammeters and voltmeters, ohmmeters with several ranges is discussed.

Types of energymeters for electrical energy measurement and wattmeters for measurement of power in AC and DC circuits is discussed briefly.

Analog and digital multimeters are also discussed. These instruments are used for measurement of AC and DC voltage and current, resistances of several ranges.

Galvanometer is an instrument used for measuring small AC voltages and currents has also been discussed.

2.2 Classification of Instruments

Instruments may be divided into two types :

(A) Absolute Instruments

In such type of instruments no comparison with another instruments is followed. For absolute instruments there is no necessity of calibration. Absolute instruments provide the measured quantity in terms of constants of the instruments proportional to their respective deflection. Mostly used in **standard** institutions and bureaus.

(B) Secondary Instruments

In secondary instruments calibration is necessary, which is done by comparing with absolute or primary instruments or already calibrated instruments. Calibration of the instrument is essential for proper deflection and determination of unknown quantity. These instruments are used in every field of measurements.

Secondary instruments are classified as follows :

(1) **Indicating instruments** : As the name implies these instruments indicate the instantaneous value of the electrical quantity which is to be measured at the time of measurement. In these instruments the pointer moves over a standard calibrated scale. *e.g.* ammeter, voltmeter.

(2) **Recording instruments** : These instruments provide the trace or record of the quantity which is to be measured. This is carried out by using a ink pen connected to moving system and slides or moves on a graph paper or chart. This system provides continuous recording of the variations in the deflection system.

(3) **Integrating instruments** : These instruments measure and register the total amount of electricity or electrical energy supplied to any circuit during given time.

e.g. ampere hour meter and watt hour meter.

Mainly the instruments are classified into two types;

(a) Analog instruments

(b) Digital instruments

(a) Analog Instruments

These instruments provide the output readings or values of unknown input parameters which are continuous function of time. Analog instrument displays or indicates the reading by deflection means *i.e.* read out by deflection or the pointer over a specified standard scale.

Classification of analog instruments :

(i) **On the basis of quantity** : In this category analog instruments are classified on the basis of quantity (unknown parameter) which is to be measured by the instrument.

e.g. : If voltage is to be measured by an instrument, then voltage is measured by **voltmeter**. Similarly, current is measured by **ammeter**.

Accordingly, wattmeters for energy measurement, frequency meter for frequency measurement etc.

(ii) **On the basis of types of quantity (AC or DC)** : Type of the quantity which is applied to the input of an instrument for measurement purpose as mentioned below:

(1) DC current/voltage

(2) AC current/voltage

(3) Both AC and DC current/voltage.

(iii) **On the basis of method used for comparing unknown quantity with the unit of measurement** :

(a) **Direct measuring instruments** : In these type of instruments the quantity under measurement provides energy to the mechanism of the instruments. Due to this energy the instrument gets actuated, resulting unknown quantity gets indicated by deflecting the pointer over standard calibrated scale provided.

(b) **Comparison type instruments** : As the name implies, the unknown quantity gets compared with the (known quantity as a reference) standard.

e.g. : AC and DC bridges.

In indicating instruments pointer moves over a standard and calibrated scale. The instantaneous value of the quantity (under measurement) is measured. These instruments consist of a pointer over a calibrated scale and it is attached to pointer moving system with jeweled bearings. For the sliding of pointer over calibrated scale some force is essential. These forces are :

Deflecting force : This force is essential to move the instrument pointer from its zero position *i.e.* the position when supply is not applied. The deflecting force is produced by deflecting system by converting the electrical (energy) current into mechanical force. The deflection force is produced by deflecting system.

Controlling force : This force is produced by controlling system. The function of this system is to produce a force equal and opposite to the deflection force at final steady position of pointer in order to deflect the pointer at definite value for the fixed magnitude of current. The controlling force is also essential to bring the pointer back to zero position when the current is removed. *i.e.* the pointer once deflected would not return to its zero position on removing the current unless the controlling force gets developed. This force is developed by the spring control system or by the gravity control system. Without this force the system pointer would swing to maximum deflected position irrespective of the magnitude of current which is to be measured.

Damping force : This force is necessary to bring the pointer at rest position quickly. If this force is not provided then the pointer (moving system) remains in oscillating position about its final deflected position for sometimes before to rest at steady position.

According to movement of pointer the damping systems are :

(1) If the moving system rises slowly from zero towards final (reading) deflection then the system is overdamped.

(2) If the moving system moves quickly without any oscillations then the instrument is critically damped or dead beat.

(3) If the moving system is oscillating about its final value and the amplitude of oscillations is decreasing then the instrument is underdamped.

The damping force should be adjusted at such value so that pointer should move or deflect without any overshoot. Type of systems which provide damping force are :

(a) Air friction damping

(b) Fluid friction damping

(c) Electromagnetic damping

(d) Eddy current damping

(b) Digital instruments

In these type of instruments the measured quantity is indicated by means of numerical values. Actually, the information *i.e.* measured quantity may be in the form of analog, which gets converted into digital form (decimal numbers) by analog-to-digital converters. In digital instruments the *A* to *D* converters (dual slope integration technique) are used.

Digital instruments differ from analog instruments as these devices uses analog-to-digital converters and display devices to indicate numerical values. As the output is digital in nature these instruments can be easily interfaced with microprocessors and computers or can be stored into digital memories.

Digital instruments have the following characteristics : High accuracy, high input impedance, better resolution, versatility etc.

Mainly digital instruments are classified as follows :

(i) Panel meters (ii) Bench-type meters (iii) System meters

(i) **Panel meters** : These meters are useful for fixed range one function. In these instruments only one function having a fixed range gets measured and placed at one particular location. Read-out range for these instruments is 3-digit to 4¾ digits.

(ii) **Bench-type meters** : This type of meter displays the read-outs of a.c. and d.c. quantities *i.e.* a.c. and d.c. amperes, a.c. and d.c. volts, resistance over several ranges.

The bench type meters are hand held units having 3 1/2 digits read-out with 5% accuracy to 5 1/2 digits with 1 μV resolution.

(iii) **System meters** : System meters are multimeters. These instruments can read a.c. and d.c. voltage, current and resistance over several ranges. The modes available are d.c. volt, a.c. volt and resistance.

System type voltmeters (DVM's) or Multimeters (DMM's) can be interfaced with microprocessors for providing useful mathematical functions in addition to meter operation.

These type of meter utilizes dual slope *A* to *D* converters with selectable integration times from 0·01 to 100 power lines cycle. The speed is 330 readings per second with accuracy of ± 0·1%.

In system meters following mathematical functions are available.

(a)	NULL	This function subtracts first reading from every successive reading and difference gets displayed.
(b)	STAT	This function stores the readings and calculates the (statistics) mean and variance.
(c)	$(X - Z)/Y$	This function provides offsetting and scaling where '*X*' is the reading. *Z* and *Y* are the constants.
(d)	PASS/FAIL	It indicates the reading falling in the outside limit entered by the user in upper and lower limits will display HIGH or LOW.
(e)	$100\,(X - Y)/Y$	This is percentage deviation.
(f)	$20 \log (X/Y)$	Displays '*X*' in decibels relative to stored value of '*Y*'.
(g)	dB m (R)	This function indicates that the user have to enter resistance '*R*' and the readings are displayed as a function of power dissipated in '*R*' in terms of decibels reference to 1 mW.

2.3 Analog Instruments

Different type of these instruments are mentioned below :

(1) Moving iron instruments.

(2) Moving coil instruments : These are again classified as :

(a) Permanent-magnet form.

(b) Dynamometer form.

(3) Thermal

(4) Rectifier

(5) Electrostatic

(6) Induction

(1) Moving Iron Instruments

These instruments are classified into two types :

(a) Attraction type (b) Repulsion type

Moving iron instruments are used to measure current and voltage. Laboratory type ammeters and voltmeters are the moving coil instruments.

Working Principle : When a soft iron vane or plate of soft iron (moving element) is placed in magnetic field produced by stationary coil, the stationary coil is excited by current or voltage under measurement. Due to this it becomes an electromagnet and the iron vane moves in such a way to increase the flux of electromagnet. Thus the force produced is always in such direction so as to increase the inductance of coil.

(a) Attraction Type

Fig. 2.1 shows attraction type moving iron instrument. In this type the moving iron is a flat disc which is eccentrically pivoted. When the current flows through the coil, magnetic field gets produced and the moving

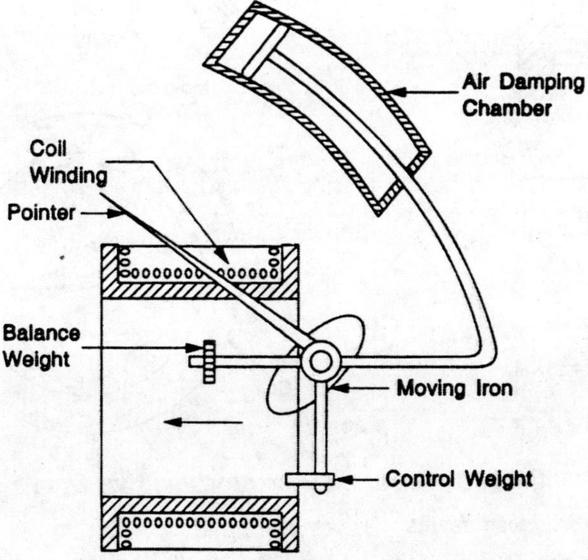

Fig. 2.1 Attraction type moving iron instrument

iron moves from weaker magnetic field outside the coil into stronger magnetic field inside it *i.e.* the coil gets attracted in. The controlling torque is produced by spring action, the damping is provided by air friction by using light aluminium piston.

(b) Repulsion Type

These type of instruments are most commonly used. In repulsion type meters the iron has two vanes (rods) or pieces inside the coil in which one is fixed and other is movable.

When the current flows through the coil both vanes get magnetized due to which a force of repulsion between two vanes gets developed, it results movement of vane. The repelling force is proportional to square of current but this performance of square law gets affected by hysteresis and frequency creating non-linear deflection.

Repulsion type instruments are classified into two types :

(i) **Radial vane type** : In radial vane type meter radial strips of iron are placed within coil. The fixed vane is connected to coil and movable is connected to the spindle of instrument as shown in Fig. 2.2 (a)

(ii) **Concentric iron vane type** : This type is also called as co-axial vane type because one vane is permanently attached to coil frame and other is rotating coaxially inside the fixed vane as shown in Fig. 2.2 (b)

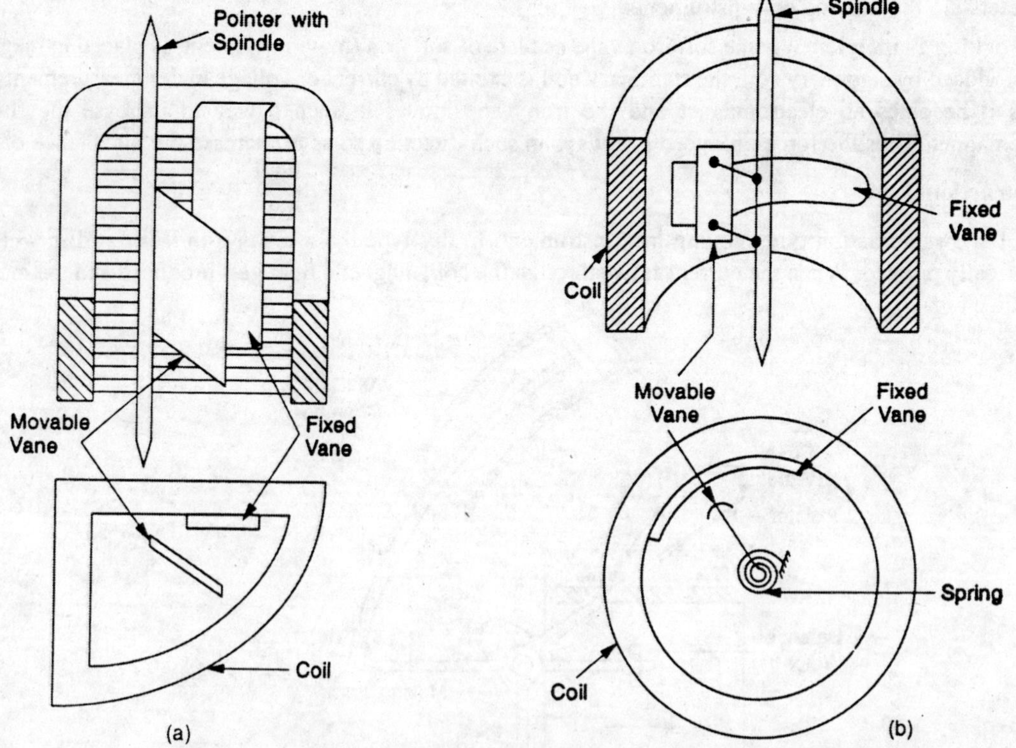

Fig. 2.2 (a) Radial Vane type (b) Concentric iron vane type.

Torque equation of moving iron instruments :

When current of small increment is applied to the instrument, torque gets developed in accordance with the energy. Due to this torque small amount of pointer deflection takes place with mechanical work.

Suppose 'I' is the initial current through the instrument, 'L' is instrument inductance, 'θ' is the deflection of pointer.

As the current increases by an increment dI it results increment in deflection by $d\theta$ and inductance by dL. The initial current is incremented by dI then applied voltage is given by :

$$e = \frac{d(LI)}{dt} = I\frac{dL}{dt} + L\frac{dI}{dt} \quad ...(2.1)$$

The electrical energy supplied is given by :

$$eI\,dt = I^2\,dL + LI\,dI \quad ...(2.2)$$

The energy stored in the inductance 'L' is given by :

$$e = \frac{1}{2}LI^2$$

$$\therefore \text{ Energy changes from } = \frac{1}{2}I^2L \text{ to } \frac{1}{2}(L + dL)\ (I + dI)^2 \quad ...(2.3)$$

Because of increment in the quantities, change in the stored energy is specified as below :

$$e = \frac{1}{2}(L + dL)\ (I^2 + 2I\,dI + dI^2) - \frac{1}{2}I^2L \quad ...(2.4)$$

$$e = LI\,dI + \frac{1}{2}L\,dI^2 + \frac{1}{2}I^2dL + dL\,I\,dI + \frac{1}{2}dI^2\cdot dL \quad ...(2.5)$$

By neglecting second and higher order terms in small quantities above equation is,

$$e = IL\,dI + \frac{1}{2}I^2\,dL \quad ...(2.6)$$

According to theory of conservation of energy,

Supplied electrical energy = Increase in stored energy + Mechanical work done

$$I^2\,dL + IL\,dI = IL\,dI + \frac{1}{2}I^2\,dL + T\,d\theta$$

$$T\,d\theta = \frac{1}{2}I^2\,dL$$

$$\therefore \quad \text{Deflecting Torque} \quad T = \frac{1}{2}I^2\frac{dL}{d\theta} \quad ...(2.7)$$

where ; T = Torque (N-m)

I = Current (ampere)

L = Inductance (Henry)

θ = Deflection of Pointer (Radians)

Advantages of moving iron instruments :

(1) They are applicable for a.c. and d.c. measurements.

(2) More accuracy because of high torque to weight ratio.

(3) Simple construction.

Errors in moving iron instruments : When the meter operates on a.c. supply frequency and eddy current errors are present.

When the meter operates on a.c. and d.c. supply, the errors are hysteresis error, temperature error and error due to stray magnetic fields.

(2) Moving Coil Instruments :

These can be classified into two types :

(a) Permanent magnet moving coil instruments (PMMC) : Widely used for direct current measurements.

(b) Dynamometer type moving coil instruments : Used for direct as well as alternating current measurements.

(a) Permanent magnet moving coil instruments

The working principle of these instruments is D''Arsonval movement.

Principle : Consider Fig 2.3 a rectangular coil is suspended in the air gap between the two poles of a permanent magnet and a soft iron cylinder. When current flows through this coil, deflecting torque gets developed due to magnetic field of permanent magnet and magnetic field of coil.

Let us consider, N = Number of turns of coil.

B = Flux density in air gap Wb/m^2.

F = Force (Newton).

I = Current in the moving coil (ampere).

L = Length of the coil. (meter)

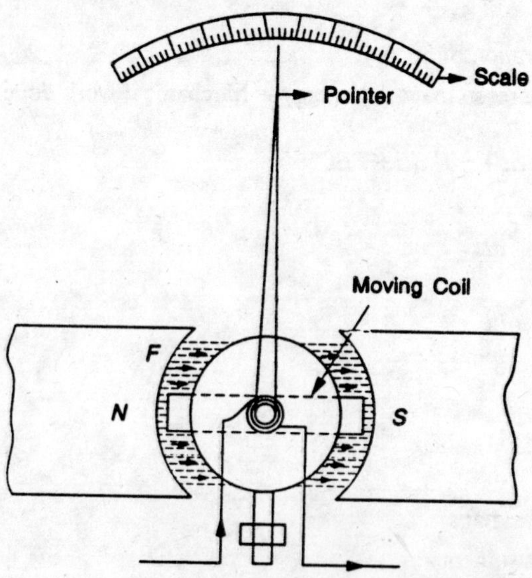

Fig. 2.3 : Principle of PMMC instrument.

The force 'F' is given by ;

$$F = N B I L \qquad ...(2.8)$$

The deflecting torque is constant because the magnetic flux density is uniform.

The torque in newton-meter is,

$$T = 2r N B I L \qquad ...(2.9)$$

where r is distance of wires from the sides of coil to the axis of rotation.

As the current flows through coil the deflection torque gets developed.

$$TD \propto I \qquad ...(2.10)$$

The controlling torque is proportional to deflection θ.

$$TC \propto \theta \qquad ...(2.11)$$

$$\therefore \quad TD = TC \qquad ...(2.12)$$

$$\therefore \quad \theta \propto I \qquad ...(2.13)$$

Thus the deflection is proportional to given current.

These types of instruments consists of a moving coil made up of copper wire wound on rectangular aluminium former. The coil is placed inside the permanent magnetic field. The permanent magnet is made up of U-shaped nickel-aluminium-cobalt-steel alloy, alcomax. The coil is supported with jewel bearings. Two phosphor bronze hair springs are attached for controlling the torque.

The damping torque is produced by movement of aluminium former in magnetic field of permanent magnet. The light weighted pointer having spindle arrangement slides over a standard scale on the dial of the meter.

Advantages of PMMC instruments :

(1) In these type of instruments high accuracy is possible because of high torque to weight ratio.

(2) These instruments are free from errors such as hysteresis and errors due to stray magnetic fields.

(3) Power consumption is low.

(4) Scale is uniformly divided.

(5) Measurement ranges for voltages and currents can be increased by using shunt and series resistances.

Errors : In PMMC instruments error may occur due to friction and temperature effects. The resistance of moving coil changes with temperature. Due to temperature the spring and permanent magnet gets weakened.

(b) Dynamometer Type Moving Coil Instruments (Electrodynamometer Type Instrument)

These instruments are used in laboratories as a.c. as well as d.c. voltmeter and ammeter in the range of power frequencies and low audio frequencies. Sometimes they are used as power factor meters and wattmeters The disadvantage of these instruments is high power consumption.

The electrodynamometer acts as transfer instrument by calibrating it firstly on d.c. then this calibration is transferred to a.c. instrument on a.c. measurements. That is the instrument is calibrated with d.c. source and the same is used for a.c. measurements. Fig. 2.4 shows electrodynamometer type instrument.

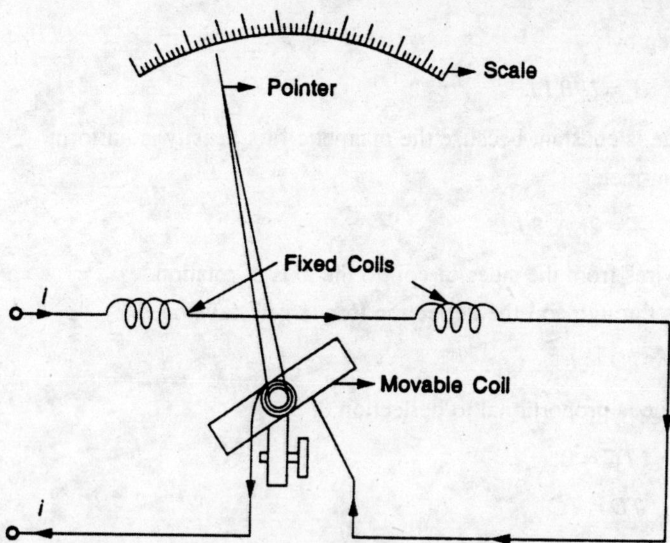

Fig. 2.4 : (a) Electrodynamometer type instrument

Operating Principle : In dynamometer type instruments instead of permanent magnet one or two fixed coils are used. These fixed coils are connected to the moving coil to produce fixed field. The current or voltage (in terms of current) is applied to fixed coils.

The current under measurement flowing through fixed coils produce the flux known as field flux necessary for deflection. The pointer is connected to moving coil which is placed inside the fixed coils. The deflection is controlled by spring as per D'Arsonval movement. Fig. 2.4(b) and (c) shows Electrodynamometer ammeter and voltmeter.

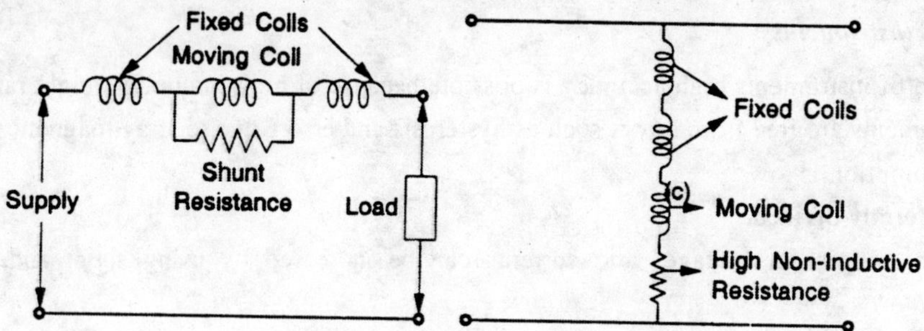

Fig. 2.4 (b) Electrodynamometer ammeter and
(c) Electrodynamometer voltmeter.

The deflection torque necessary for a.c. and d.c. measurement is as follows :

(1) For d.c. measurements the torque depends upon product of currents I_1, I_2 of fixed coils and the rate of change of mutual inductance.

(2) For a.c. measurements the torque depends upon the product of r.m.s. values of fixed coil current I_1, I_2 and cosine of angle between these currents.

But the instantaneous torque is proportional to squared current. In these type of instruments air friction damping is used.

The total assembly of fixed coils, moving coil, pointer mechanism, spring, damping mechanism is placed in a special alloy to protect the system from stray magnetic fields and external effects like temperature, shocks etc.

(3) Thermal Meters : (Electrothermic Instruments)

These instruments work on the principle that **if current which is to be measured flows through a fine wire, expansion of the coil wire takes place due to heating effect.**

The expansion of hot wire is used to produce the deflection of pointer.

Group of these meters is used to measure current at higher frequencies that is more frequency than moving iron and electrodynamometer type instruments. Thermal meters can be classified as follows

(a) Hot wire instruments

Fig. 2.5 shows hot wire type instrument. Current which is to be measured is passed through a fine wire which is made up of fine platinum irridium or phosper bronze. This wire is stretched between two terminals. Another wire is connected to fine wire through a thread. Through pulley the thread is connected to spring.

When the current which is to be measured flows through the fine wire, it gets heated resulting expansion of wire. This expansion causes deflection of pointer through thread and spring arrangement. This will indicate the value of current which is to be measured.

The expansion is proportional to square of the r.m.s. value of current. Thus hot wire instruments are used to measure r.m.s. values of current.

The drawbacks in hot wire instruments are overloading effect, need to compensate ambient temperature which may affect the system stability and error gets occurred, more power consumption during operation. Due to these drawbacks thermoelectric or thermocouple instruments are preferred instead of hot wire instruments.

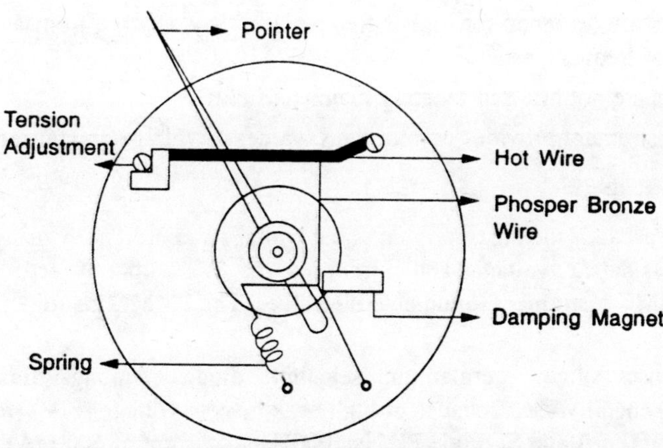

Fig 2.5 : Hot-wire type instrument.

(b) Thermocouple instruments : (Thermoelectric instruments)

As the name implies this instrument is assembled by using thermocouple section and permanent magnet moving coil action as shown in Fig. 2.6.

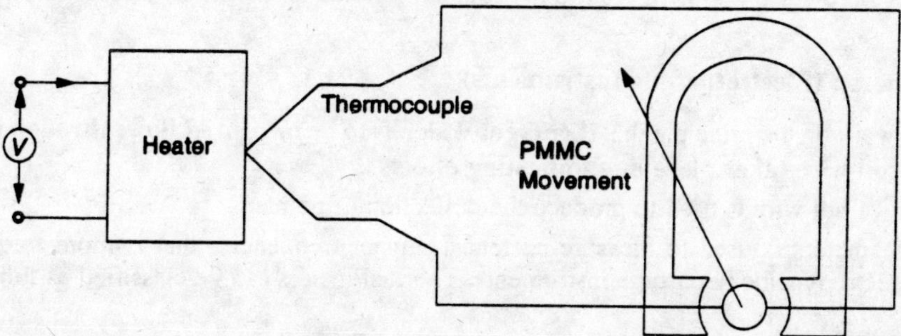

Fig. 2.6 : Thermocouple instrument.

Thermocouple instruments are used for the measurement of currents and voltages at high frequencies upto 50 MHz. So these instruments are also called RF instruments.

Working Principle : As this instrument contains thermocouple element it is necessary to study about it. Thermocouple works on the principle that **if two dissimilar metals are connected or welded together to form junction and different temperatures are applied to the ends (cold *i.e.* reference and hot) due to this difference in temperature, voltage gets developed.** The voltage is directly proportional to temperature difference. This developed voltage is applied to PMMC instrument for measurement.

The materials used for thermocouple are iron-constantan, cromel-alumel, platinum-rhodium etc.

The heat produced is directly proportional to current squared ($I^2 R$) flowing through the heater element and the temperature rise is proportional to square of r.m.s. current. Thus this instrument can read the r.m.s. value of current through the heater.

Advantages of thermoelectric instruments :

(1) These instruments are designed for higher frequencies, they are free from any frequency errors over wide range of frequencies.

(2) These instruments are not affected by stray magnetic fields.

(3) Thermoelectric instrument provides correct r.m.s. values of voltage and current.

(4) Rectifier Type Instruments :

This type of instruments are used for the measurement of a.c. voltage and current by using rectifier element. The rectifier element converts the a.c. voltages and current into d.c. The output of rectifier *i.e.* d.c. voltages and current is applied to PMMC instrument to indicate the values. The PMMC instrument uses D'Arsonval movement.

The rectifier element uses silicon, germanium, selenium, diodes. Amongst this silicon (Si) and germanium (Ge) diodes are generally used because of high peak inverse voltage (PIV) (for Ge diode PIV is 300 V and current rating of 100 mA and Si diode PIV is 1000V and current of 500 mA).

Depending upon the types of rectifiers following AC voltmeters are discussed.

(a) AC Voltmeter Using Half Wave Rectifier

Fig. 2.7 shows AC voltmeter using half wave rectifier, resistance R_s is multiplier, resistance R_m is meter resistance. In the circuit diode '*D*' is connected .

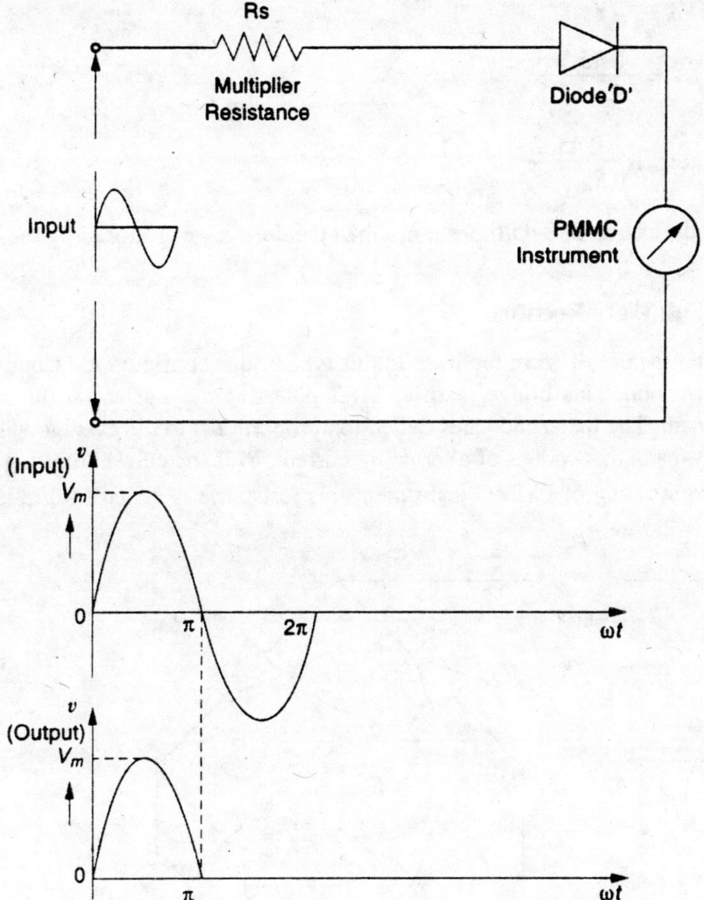

Fig. 2.7 : A.C. Voltmeter using half wave rectifier.

When $V_{dc} = V$ voltage is applied to rectifier, current through meter is

$$I = \frac{V}{(R_m + R_s)}$$; due to this current full scale deflection (f_{sd}) takes place.

When a.c. voltage $v = V_m \sin \omega t = \sqrt{2}\ V \sin \omega t$ is applied.

due to this, average current of meter is,

$$\therefore \quad I_{av} = I_{dc} = \frac{1}{2\pi}\int_0^{\pi} i\ d(\omega t) \qquad \text{...(2.14)}$$

$$= \frac{1}{2\pi}\int_0^{\pi} \frac{V_m \sin \omega t}{(R_m + R_s)}\ d(\omega t) \qquad \text{...(2.15)}$$

$$= \frac{I}{2\pi (R_m + R_s)}\int_0^{\pi} V_m \sin \omega t\ d(\omega t) \qquad \text{...(2.16)}$$

$$= \frac{V_m}{2\pi (R_m + R_s)}[-\cos \omega t]_0^{\pi}$$

but $$V_m = \sqrt{2}V$$

$$= \frac{0.45\ V}{(R_m + R_s)} \quad ...(2.17)$$

The current through meter is $\dfrac{0.45\ V}{(R_m + R_s)}$ ...(2.18)

This current produces deflection of 0·45 times more than the deflection produced by the meter with d.c. voltages.

(b) AC Voltmeter Using Full Wave Rectifier

Fig. 2.8 shows AC voltmeter using full wave rectifier. In this type bridge configuration (four diodes) is used to provide full wave rectification. This bridge rectifier gives pulsating d.c. current at the output which is supplied to PMMC instrument. The meter indicates deflection proportional to the average values of current. Meter is calibrated so as to read r.m.s. values of alternating current. R_s is the current limiting resistor which limits the current upto current rating of PMMC instrument this resistance is called Multiplier. R_m is meter resistance.

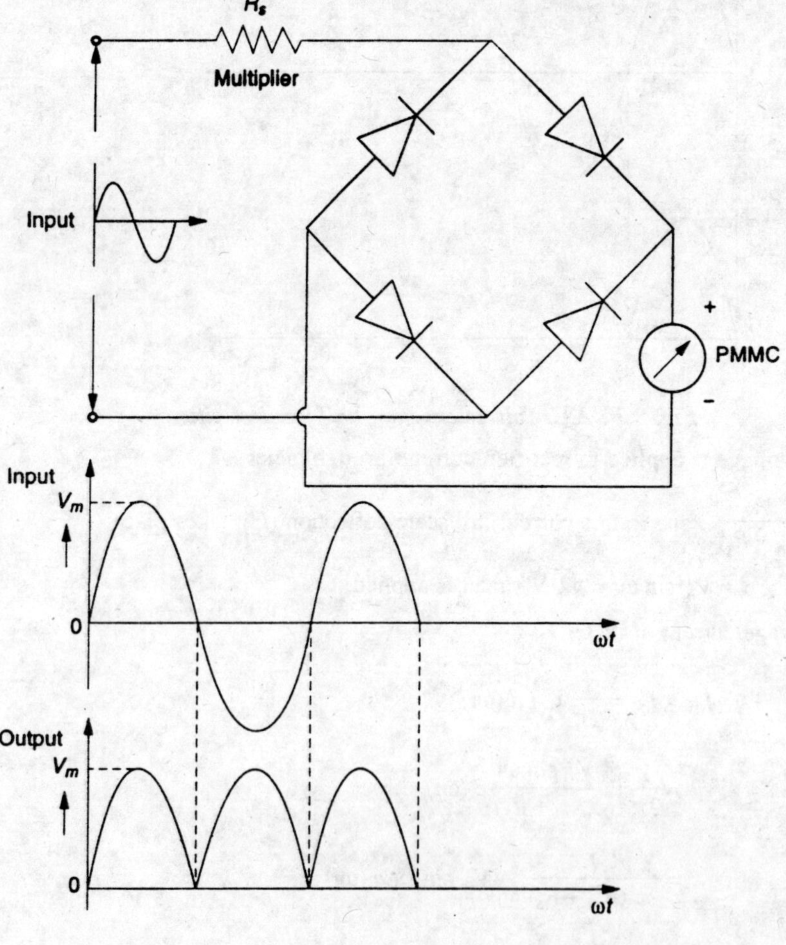

Fig. 2.8 : A.C. Voltmeter using full wave bridge rectifier.

When $V_{dc} = V$, voltage is applied to rectifier,

Current through meter $I = \frac{V}{(R_m + R_s)}$; due to this current full scale deflection (f_{sd}) takes place.

When a.c. voltage $v = V_m \sin \omega t = \sqrt{2}\ V \sin \omega t$ is applied due to this, average current of meter is,

$$I_{av} = I_{dc} = 1/\pi \int_0^{\pi} id\,(\omega t) \quad ...(2.19)$$

$$= 1/\pi \int_0^{\pi} \frac{V_m \sin \omega t}{(R_m + R_s)} d\,(\omega t) \quad ...(2.20)$$

$$= \frac{V_m}{\pi\,(R_m + R_s)} \int_0^{\pi} \sin \omega t\, d\,(\omega t) \quad ...(2.21)$$

$$= \frac{V_m}{\pi\,(R_m + R_s)} [-\cos \omega t]_0^{\pi} \quad ...(2.22)$$

$$= \frac{2\,V_m}{\pi\,(R_m + R_s)}$$

$$= \frac{0{\cdot}636\,V_m}{(R_m + R_s)} \quad ...(2.23)$$

but $\qquad V_m = \sqrt{2}\,V$

$$= \frac{0{\cdot}9\,V}{(R_m + R_s)} \quad ...(2.24)$$

Therefore, the current through meter is $\frac{0{\cdot}9\,V}{(R_m + R_s)}$.

This current produces deflection of 0·9 V times more than the deflection produced by the meter with d.c. voltages.

Sensitivity of rectifier type instrument :

Sensitivity of the instrument is defined in terms of current required for full scale deflection of meter and type of rectifier used in the instrument for a.c. measurement.

The d.c. sensitivity of an instrument is defined as ;

$$S_{dc} = 1/I_{fs}\ V/\Omega \quad ...(2.25)$$

Where I_{fs} is the current required for full scale deflection. The a.c. sensitivity is defined by considering the type of rectifier *i.e.* for half wave rectifier with sinusoidal input the sensitivity is,

$$S_{ac} = 0{\cdot}45\ S_{dc} \quad ...(2.26)$$

For full wave rectifier with sinusoidal input the sensitivity is,

$$S_{ac} = 0{\cdot}9\ S_{dc} \quad ...(2.27)$$

Example 2.1 : Calculate the value of R_s multiplier resistance for 10 V r.m.s. a.c. range on voltmeter as shown in Fig. 2.9.

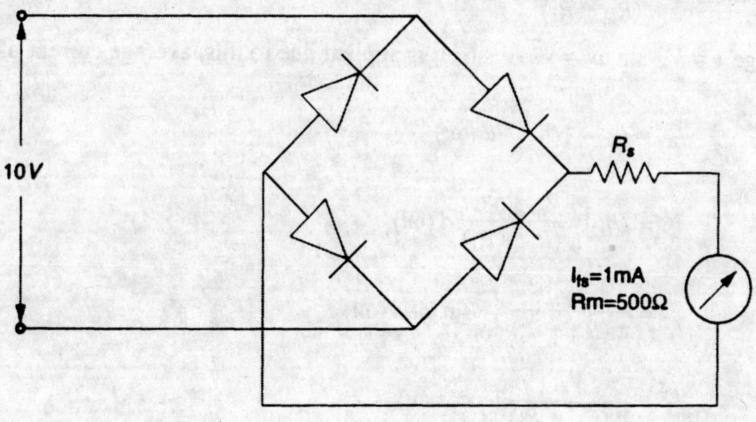

Fig. 2.9.

Solution : $S_{dc} = 1/I_{fs} = 1/1 \times 10^{-3} = 1 \text{ k}\Omega/\text{V}$

S_{ac} for bridge rectifier is

$$S_{ac} = 0{\cdot}9\ S_{dc}$$

$$= 0{\cdot}9 \times 1 \text{ k}\Omega/\text{V}$$

$$= 0{\cdot}9 \text{ k}\Omega/\text{V}$$

$$R_s = S_{ac} \times \text{Range} - R_m$$

$$= 0{\cdot}9 \text{ k}\Omega/\text{V} \times 10 - 500$$

$$R_s = 8{\cdot}5 \text{ k}\Omega$$

Advantages of rectifier type instruments :

(1) Useful for a.c. measurements in the power frequency range 60 and 400 Hz and in the audio frequency range.

(2) Practically uniform scale for most ranges.

(3) Accuracy of ± 5% under normal operating coinditions.

(5) Electrostatic Instruments

In these type of instruments no iron is used in the construction. As there is no iron, these instruments are free from the hysteresis, eddy current losses, no errors due to stray magnetic fields because these instrument works on electrostatic principle.

They are widely used in the measurement of high voltages at very high impedance. Also, they are used in measurement of current and power quantities.

Working Principle : Consider in Fig. 2.10 (a) two metal plates *P* and *Q*. Plate *P* is fixed and plate *Q* is movable which is connected to fixed point '*O*' through spring '*S*'. These two plates *P* and *Q* are oppositely charged.

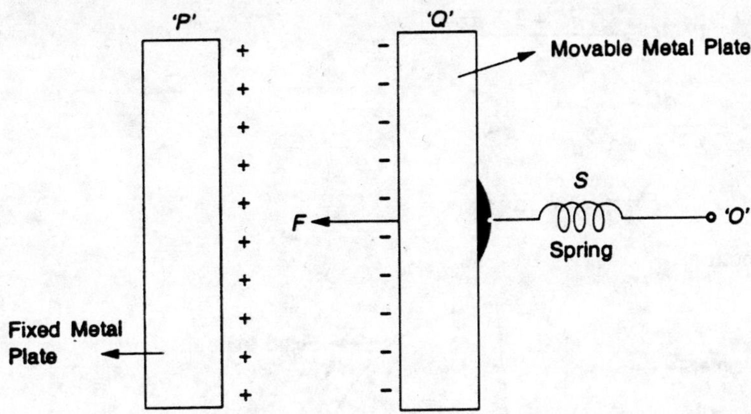

Fig. 2.10. (a) Working Principle of Electrostatic Instrument

The total input electrical energy is utilized in increasing the energy stored across capacitor and to perform mechanical work or task *i.e.* mechanical force which causes the displacement of spring.

Let us consider voltage 'V' is applied to the plates 'P' and 'Q' due to this, force of attraction 'F' gets developed between two plates. Due to this force the movable plate gets attracted towards fixed plate 'P'. This phenomenon constitutes a capacitor *i.e.* two metal plates seperated by some distance and potential difference is applied to it. The capacitance between two plates is 'C' Farad. Energy across capacitor is $\frac{1}{2}CV^2$ Joule.

If small voltage change 'dV' is applied to metal plates, plate 'Q' moves towads fixed plate 'P' by small distance dx. The current across capacitor is ;

$$i = \frac{dq}{dt} = \frac{d(CV)}{dt} \quad \text{(The charge on capacitor } q = CV) \qquad ...(2.28)$$

$$= C\frac{dV}{dt} + V\frac{dC}{dt}.$$

The input energy is $Vidt$

$$\therefore \qquad Vi = V\left[C\frac{dV}{dt} + V\frac{dC}{dt}\right] \qquad ...(2.29)$$

$$Vidt = [V^2\,dC + VCdV] \qquad ...(2.30)$$

The change in stored energy is ;

$$\frac{1}{2}(C + dC)(V + dV)^2 - \frac{1}{2}CV^2$$

$$= \frac{1}{2}(C + dC)(V^2 + 2\,VdV + dV^2) - 1/2\,CV^2$$

$$= \frac{1}{2}V^2dC + CVdV \qquad ...(2.30)$$

$$V^2\,dC + VC\,dV = \frac{1}{2}V^2\,dC + CV\,dV + F\,dx \qquad ...(2.31)$$

$$V^2\,dV = \frac{1}{2}V^2\,dC + F\,dx \qquad ...(2.32)$$

$$V^2\,dC = \frac{V^2\,dC + 2\,F\,dx}{2} \qquad ...(2.33)$$

$$V^2\,dC = 2\,F\,dx$$

$$2\,F\,dx = V^2\,dC$$

$$F = \frac{1}{2}\,V^2\,dC/dx$$

This is for linear motion.

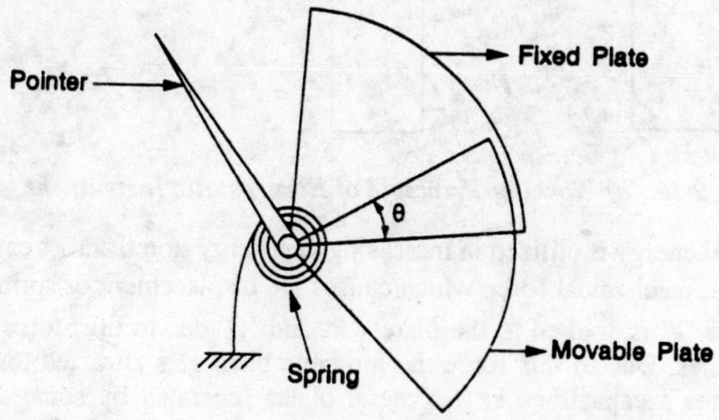

Fig. 2.10 (b) Rotary motion

For rotary motion : Consider Fig. 2.10(b). As the motion is rotary, angular displacement (θ) is to be considered. The deflection torque is,

$$Td = \frac{1}{2}\,V^2\,dC/d\theta \qquad ...(2.34)$$

But $Tc = K\,\theta$ (controlling torque)

$$K\,\theta = \frac{1}{2}\,V^2\,dC/d\theta$$

$$\theta = \frac{1}{2}\,\frac{V^2}{K}\,\frac{dC}{d\theta} \qquad ...(2.35)$$

Advantages of electrostatic instruments :

(1) These instruments are useful for a.c. and d.c. measurements.

(2) Less power consumption

(3) Useful at higher voltages.

(4) As these instruments are electrostatic instruments there is no possibility of stray magnetic fields.

Disadvantages :

(1) Scale is not uniform.

(2) These instruments are expensive and large in size.

(3) At higher voltages it is necessary to connect a series resistance with the instrument to avoid short circuit between two plates.

2.4 Analog Ammeters, Voltmeters and Ohmmeters

(1) Analog Ammeters

(A) **DC ammeter** : Fig. 2.11 shows basic ammeter circuit which is a PMMC D'Arsonval galvanometer. Here R_{sh} is shunt resistance, its function is to bypass the major part of current when large currents are to be measured.

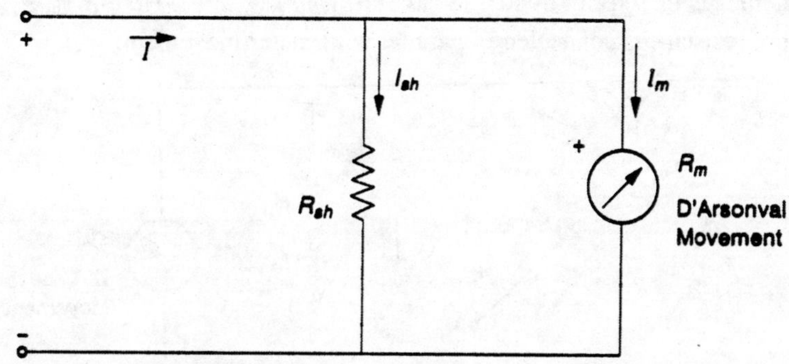

Fig. 2.11. D. C. Ammeter

Here ; R_{sh} = Shunt resistance in (Ω)

R_m = Internal resistance of coil (Ω).

$I_m = I_{fsd}$ = Full scale deflection current (A)

I_{sh} = Shunt current (A)

I = Current to be measured (A).

Voltage drop across shunt resistance = $I_{sh} \cdot R_{sh}$

Voltage drop across meter resistance = $I_m \cdot R_m$

As the (R_{sh}) shunt resistance is parallel to meter following condition gets satisfied.

$$\therefore \quad I_{sh} \cdot R_{sh} = I_m R_m$$

$$\therefore \quad R_{sh} = \frac{I_m \cdot R_m}{I_{sh}}$$

But $$I_{sh} = I - I_m$$

$$\therefore \quad R_{sh} = \frac{I_m \cdot R_m}{(I - I_m)}$$

$$\therefore \quad \frac{I}{I_m} = 1 + \frac{R_m}{R_{sh}}$$

The ratio I/I_m is called multiplying power of shunt 'm'.

$$\therefore \quad m = \frac{I}{I_m}$$

$$\therefore \quad R_{sh} = \frac{R_m}{(m-1)}$$

The shunt resistance is used in the system may be a resistance wire. The temperature coefficient of resistance wire and instrument should be low. The shunt should carry the current without increasing the temperature. Constantan and manganin material is used for shunt resistances. The value of resistance should not be changed with time.

(B) **Multirange ammeters** : If number of shunt resistance (R_{sh}) are increased the range of instrument gets increased. In multirange ammeter circuit, number of shunt resistance (R_{sh}) are connected and are selected by means of selector switch 'S' as shown in Fig. 2.12. Here R_{sha}, R_{shb}, R_{shc}, R_{shd} and R_{she} are the shunt resistances connected in parallel with meter movement.

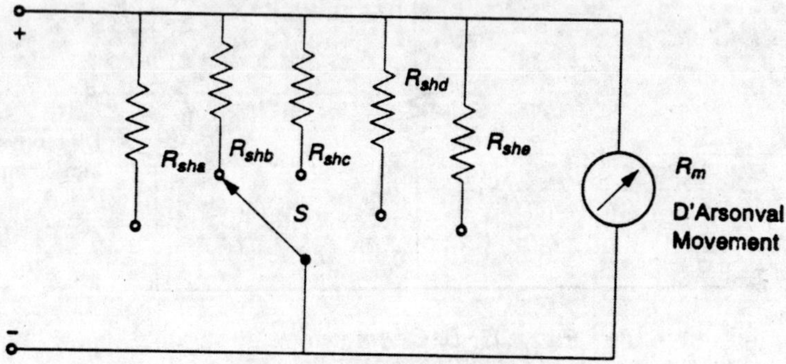

Fig. 2.12. Multirange Ammeter

They are high precision resistances.

The current through resistance is I_a, I_b, I_c, I_d and I_e that is current ranges.

$R_{sha} = R_m/(m_a - 1)$ $R_{shb} = R_m/(m_b - 1)$

$R_{shc} = R_m/(m_c - 1)$ $R_{shd} = R_m/(m_d - 1)$

$R_{she} = R_m/(m_e - 1)$

Here switch 'S' is the multiposition switch having low contact resistance and high current carrying capabilities. The current range is 1 to 50 ampere.

(2) Analog Voltmeter

(A) **DC voltmeter** : Fig. 2.13 shows basic DC voltmeter. Here the basic D'Arsonval movement is converted into DC voltmeter by adding a series resistance R_s that is also known as **Multiplier** resistance. The function of this resistance is to limit the current through movement and it should not exceed the full scale deflection current. The D'Arsonval movement and multiplier is connected across the voltage which is to be measured.

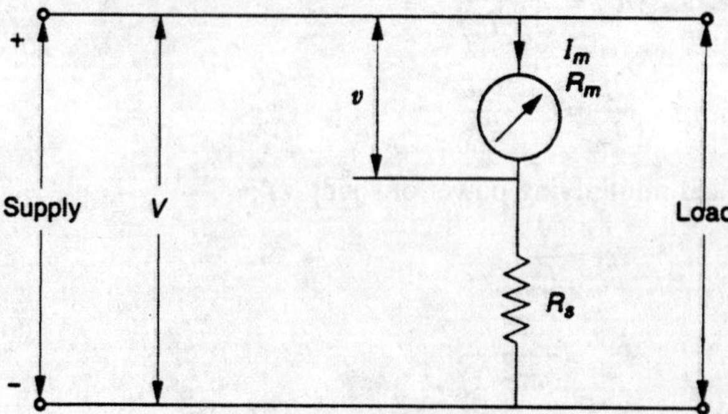

Fig. 2.13 : D.C. voltmeter.

Let ; R_s = Multiplier resistance (Ω).

R_m = Internal resistance of movement (Ω).

$I_m = I_{fsd}$ = Meter full scale deflection current (ampere).

v = Voltage across the meter movement for current I_m(V).

V = Instrument full range voltage.

$$v = I_m R_m$$

$$V = I_m (R_m + R_s)$$

$$R_s = \frac{V - (I_m R_m)}{I_m} \quad ...(2.36)$$

$$R_s = \frac{V}{I_m} - R_m \quad ...(2.37)$$

The multiplying factor for multiplier is,

$$m = V/v = \frac{(R_m + R_s)}{R_m}$$

$$m = (1 + R_s/R_m)$$

Multiplying resistance $R_s = (m - 1) R_m$ ...(2.38)

The materials used for multiplier resistance are constantan and manganin. The resistance should not be affected by change in temperature and change in time.

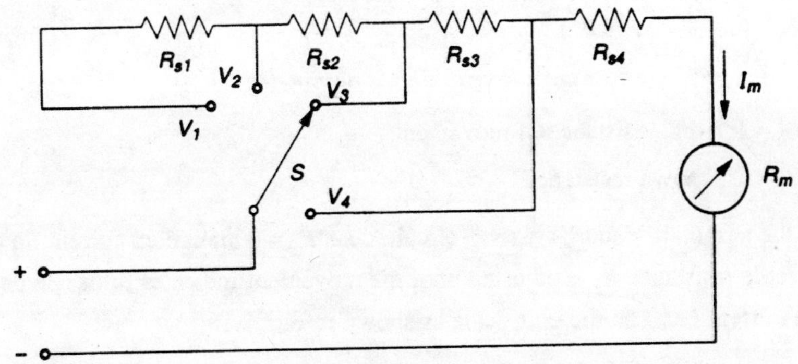

Fig. 2.14 : Multirange voltmeter.

(B) **Multirange voltmeter** : Fig. 2.14 shows multirange voltmeter. It consists of the multiplier resistances R_{s1}, R_{s2}, R_{s3} and R_{s4} connected in series. The range selector switch '*S*' selects the value of multiplier resistance connected in series with the movement.

V_1, V_2, V_3 and V_4 are the desired voltage ranges.

$$R_{s1} = (m_1 - 1) R_m \; ; \quad R_{s2} = (m_2 - 1) R_m$$

$$R_{s3} = (m_3 - 1) R_m \; ; \quad R_{s4} = (m_4 - 1) R_m$$

where ; $m_1 = V_1/v$, $m_2 = V_2/v$, $m_3 = V_3/v$, $m_4 = V_4/v$

(3) The Ohmmeter

It is possible to measure the values of unknown resistance by the basic D'Arsonval movement with voltage source and resistive network. This instrument is widely used for the measurement of resistances like heater, electronic circuits, coils etc. They are also applicable in continuity checking, testing of semiconductor diodes, laboratory work etc.

(A) **Series type ohmmeter** : Fig. 2.15(a) shows the series type ohmmeter. It consists of a basic D'Arsonval movement connected in parallel with resistor R_2. Series resistance R_1 and unknown resistance R_x is connected with battery. The unknown resistance R_x is connected between terminals X and Y.

From the Figure,

R_1 = Current limiting resistance

R_2 = Zero adjust resistance

E = Internal battery

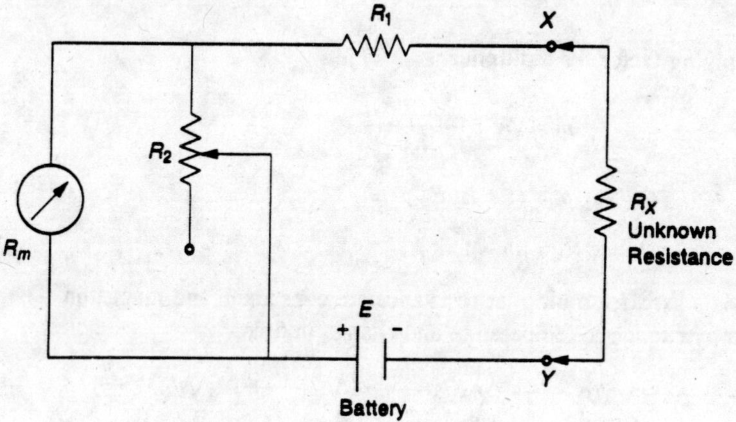

Fig. 2.15 (a) : Series ohmmeter.

R_m = Internal resistance of movement

R_x = Unknown resistance

Case 1 : When the terminals X and Y are short circuited *i.e.* $R_x = 0$ maximum current flows in the circuit. In this case, the variable resistance R_2 is adjusted until the movement indicates full scale deflection current (I_{fsd}). The full scale current *i.e.* 0 on the ohm scale as shown in Fig. 2.15(b).

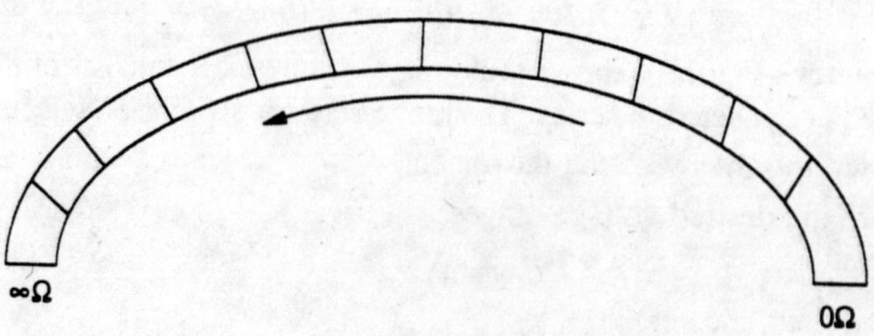

Fig. 2.15(b)

Case 2 : When the terminals X and Y are open circuited *i.e.* $R_x = \infty$ the current decreases to zero. The current indicated is zero current *i.e.* ∞ on ohm scale as shown in Fig. 2.15(b). The position of pointer over the scale is marked as "∞" ohms.

(B) **Shunt type ohmmeter** : The Fig. 2.16(a) shows shunt type ohmmeter. The unknown resistance R_x is connected between the terminals X and Y, it is parallel to the movement (meter). The battery 'E' is in series with adjustable resistance R_1 and meter. Switch S is used to make and brake the circuit connections with battery 'E'.

Case 1 : When the terminals X and Y are short circuited *i.e.* $R_x = 0$ the meter reading is zero. The pointer position is marked as "0 " ohms.

Case 2 : When the terminals X and Y are open circuited *i.e.* $R_x = \infty$ then current flows through the meter. The position of pointer is marked as "∞" ohms.

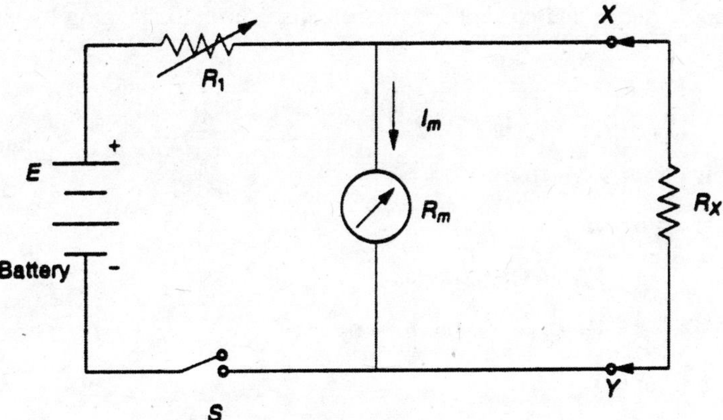

Fig. 2.16(a) : Shunt type ohmmeter.

The approximate value of R_1 is selected so as to get full scale current.

Thus this type of ohmmeter has **zero** at the left hand side which indicates no current and **infinite** at right side which indicates full scale current deflection as shown in Fig. 2.16(b).

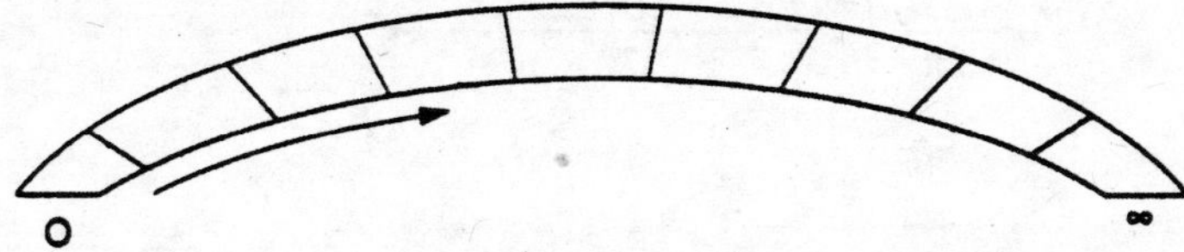

Fig. 2.16(b)

2.5 Wattmeters

Wattmeter is used for the measurement of electrical power.

Power in DC Circuits

Power in d.c. circuits is given by the product of voltage (V) and current (I) in watts.

$$P = V \, . \, I \text{ watt} \qquad ...(2.39)$$

In d.c. circuits power taken by the load is measured by connecting a voltmeter and a ammeter by applying d.c. source to the circuit. The d.c. power is product of the readings of voltmeter and ammeter. Instead of using voltmeter and ammeter for power measurement, wattmeter is used. Because of wattmeter, accurate and precise reading is possible, it also avoids the instrument handling and labour.

Power in AC Circuits

The instantaneous value of a.c. power varies in accordance with the current and voltages during the cycle. In a.c. circuits the power at any instant is given by ;

$$p = vi \quad \text{...(2.40)}$$

where ; p = Instantaneous power

v = Instantaneous voltage

i = Instantaneous current

In a.c. circuits both voltage and current are sinusoidal so current is lagging by phase angle ϕ.

$$v = V_m \sin \omega t \quad \text{...(2.41)}$$

$$i = I_m \sin (\omega t - \phi) \quad \text{...(2.42)}$$

The instantaneous power is given by ;

$$p = vi \quad \text{...(2.43)}$$

$$p = V_m \sin \omega t \, I_m \sin (\omega t - \phi)$$

$$p = V_m I_m \sin \omega t \sin (\omega t - \phi) \quad \text{...(2.44)}$$

but $\omega t = \theta$

$$p = V_m I_m \sin \theta \sin (\theta - \phi) \quad \text{...(2.45)}$$

The mean power is given by ;

$$p = 1/2\pi \int_0^{2\pi} V_m I_m \sin \theta \sin (\theta - \phi) \, d\theta \quad \text{...(2.46)}$$

$$p = \frac{V_m I_m}{2\pi} \int_0^{2\pi} \frac{\cos \phi - \cos (2\theta - \phi)}{2} \, d\theta$$

$$p = \frac{V_m I_m}{2} \cos \phi \quad \text{...(2.47)}$$

$$p = VI \cos \phi \quad \text{...(2.48)}$$

where $\cos \phi$ is power factor ; V and I are r.m.s. values of voltage and current.

For the measurement of power **Wattmeter** is used because of power factor is present in a.c. circuits. In d.c. circuit power may be measured by using ammeter and voltmeter but in a.c. circuits power factor ($\cos \phi$) is present so it is essential to use wattmeter for power measurement. Various types of watt meter are discussed below :

(1) Electrodynamometer Type Wattmeter

The basic principle of these instruments is similar to electrodynamometer type moving coil instruments as explained in section 2.3.2 (b).

It consist of two fixed coils called as field coils. These two fixed coils are connected in series with load and carries the current, these coils are called current coils. Another coil *i.e.* moving coil is connected across the voltage so these coils are named as **voltage coil** or **pressure coil**. The moving coil carries current proportional to the voltage so a high non-inductive resistance is connected in series with this coil.

Fig. 2.17 shows the construction of electrodynamometer type wattmeter. The fixed coils are made up of laminated heavy wires to avoid eddy current losses in conductors. These are designed to handle the current of 100 A. The moving coil is placed on a pivoted spindle and covered by current coil. The two coils are air cored. In such instruments spring control action and air friction damping is used. A knife edge pointer with mirrored scale is also used.

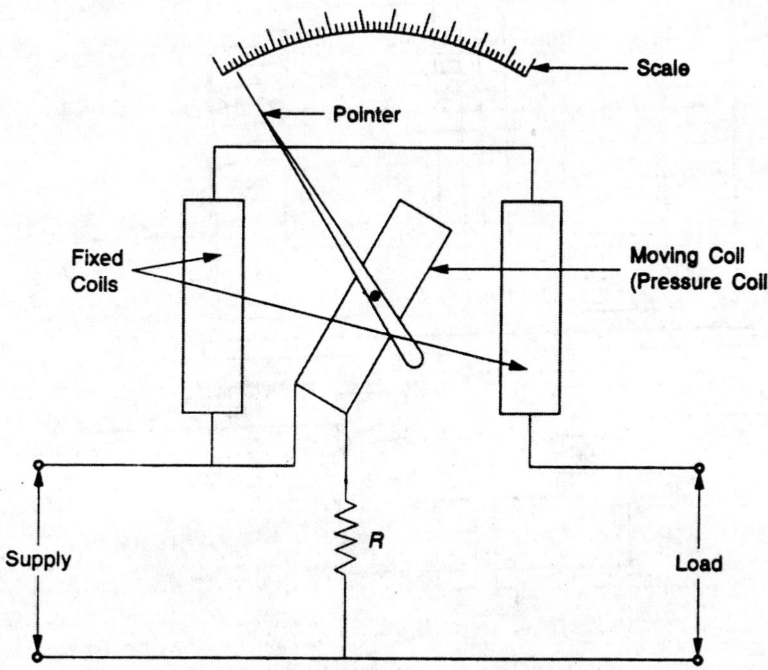

Fig. 2.17 : Electrodynamometer wattmeter

(2) Induction Wattmeters

Induction wattmeters are applicable in those applications where the frequency and supply voltage are fixed.

Fig. 2.18 shows induction type wattmeter. The wattmeter consists of two laminated electromagnets which provides two electromagnetic circuits. One is supplied by load current called series magnet. The other is excited by current proportional to the voltage of circuit in which power is to be measured, this magnet is called as shunt magnet.

In the air gap a thin aluminium disc is placed which cuts the flux developed by these two magnets and hence two eddy currents are produced in the disc. The flux and eddy current produces deflection torque. Two copper rings are fitted on the central limb of shunt magnet and adjusted so as to develop a resultant flux in the magnet lags by 90 degrees behind to applied voltage.

The two (voltage) pressure coils are connected in series so as to transmit the flux to central limb. Two coils are wound on the series magnet to magnetize the appropriate core in the same direction. Both the coils are placed in such a manner that their flux gets cut by the moving disc. A spring is connected to the spindle of moving system which carries the pointer.

Current rating of these instruments is 100 A. Current transformer is used with the wattmeter for higher current ratings.

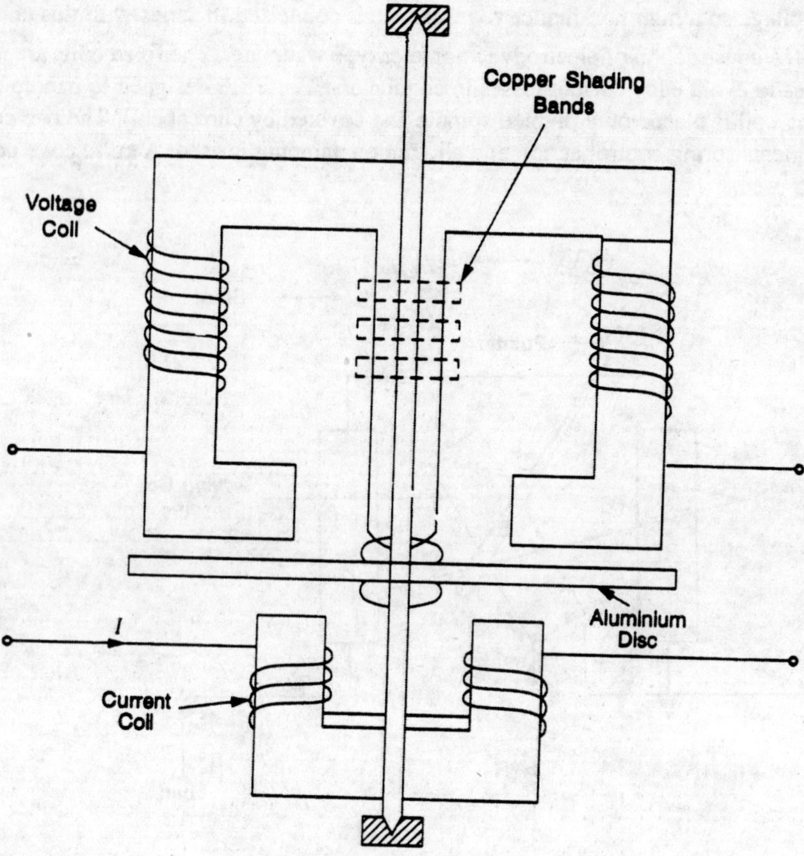

Fig. 2.18 : Induction wattmeter.

2.6 Galvanometer

Galvanometer is an instrument used for measuring the magnitudes of small currents and voltages. These instruments are mainly used as a null detector (to indicate zero current) in bridge and potentiometer circuits for resistance measurements.

D'Arsonval Galvanometer

Fig. 2.19 shows the construction of D'Arsonval galvanometer. It consists of a current carrying moving coil having number of turns with circular or rectangular shape. This coil is placed inside the magnetic field between air gap of permanent magnet and core. For circular coil the iron core is in spherical shape and for rectangular coil the iron core is cylindrical. The iron core produces strong magnetic field by providing low reactance path. The damping is provided by coil wound on metal former.

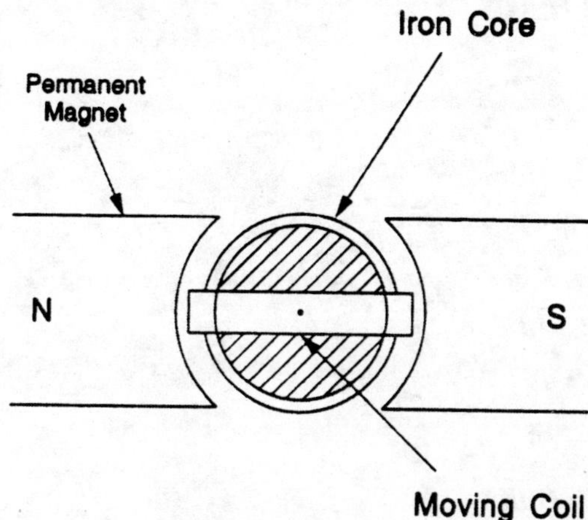

Fig. 2.19 : D'Arsonval Galvanometer.

When current flows through the coil, electromagnetic torque gets developed due to this torque the coil rotates. The electromagnetic torque is balanced by the mechanical torque which is produced due to spring action. Damping can be obtained by connecting a low resistance across galvanometer terminals. For critical damping the value of damping torque is adjusted by selecting proper value of resistance.

The coil is supported by flat ribbon suspension, first is lower suspension, it carries current to the coil. The upper suspension is gold or copper wire in the form of ribbon. In some galvanometers taut suspensions are used instead of ribbon suspensions. This type of galvanometer provides high sensitivity and are less affected by shocks, vibrations overloads and temperature changes. In taut suspension type the moving coil is suspended by two torsion ribbon which are placed at sufficient tension by tension springs.

On the suspension a small mirror is placed on which a beam of light is thrown or bombarded. This light is reflected on a scale over which the deflection is to be measured.

Torque equation for galvanometer :

Let us consider,

$l \cdot d$ = Length of coil in (m).

N = Number of turns of coil

B = Flux density in air gap (Wb/m^2)

i = Current through moving coil (ampere)

K = Constant of suspension (Nm/rad.)

θf = Final deflection of moving coil (rad.)

α = Angle between direction of magnetic field and conductor.

$$\theta f = N B i l \sin \alpha \qquad ...(2.49)$$

but $\sin \alpha = 90°$ because field is radial.

$$= N B i l \qquad ...(2.50)$$

Deflecting torque T_d = Force × Distance

$$T_d = N\,B\,i\,l\,d$$

$$T_d = N\,B\,A\,i \qquad ...(2.51)$$

As $A = ld$; Area of coil (m^2)

Deflecting torque $T_d = Gi$

where ; $G = N\,B\,A = N\,B\,l\,d$

G is displacement constant of galvanometer.

$$\text{Controlling torque } T_c = K\,\theta f \qquad ...(2.52)$$

For final steady state deflection $T_c = T_d$; ...(2.53)

$$\therefore \quad T_c = T_d$$

$$\therefore \quad \theta f\,K = Gi$$

$$\therefore \quad \theta f = Gi/K \qquad ...(2.54)$$

This is the torque equation for D'Arsonval galvanometer.

2.7 Energymeters

The term energy is defined as power delivered or consumed over a time interval. It is given by ,

energy = power × time.

By electrical means the energy may be mathematically expressed as

$$W = \int_0^t vi\,dt \qquad ...(2.54a)$$

v →Voltage in Volts.

i →Current in Amperes.

It is the energy developed as a work or heat dissipated over time interval 't'.

Units of Energy—The unit of energy is Joule or Watt second *i.e.* Energy is 1 watt over time interval of one second. In terms of hour energy can be expressed as watt hours *i.e.* 1 watt over time interval of one hour. Energy may be expressed as kilowatt hour *i.e.* Energy consumed when power is delivered at the rate of 1000 watt for one hour.

For the measurement of energy in a.c. circuits induction type energy meters are used. Generally these type of meters are used in domestic and industrial applications for measurement of energy. There are two types of induction, wattmeters

(a) Single phase induction type energy meter.

(b) Two element energy meter —It is used for 3-phase, 3-wire circuits or systems.

These meters are discussed in the next sections.

(A) Single Phase Induction type energy meter

This meter is used for the measurement of energy in single phase circuits or systems.

Construction—Fig. 2.20 shows constructional details of single phase induction type energymeter.

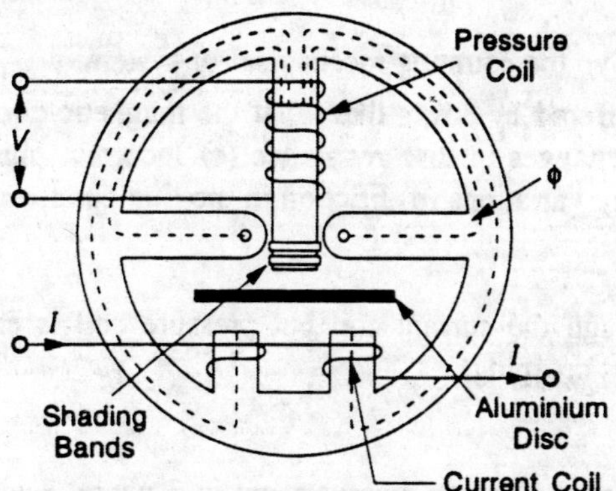

Fig. 2.20. Single Phase induction type energy meter.

(i) Driving System—It consists of two electro-magnets having core which is made up of silicon steel laminations. Coil mounted on one electromagnet is excited by load current is known as current coil. Whereas the second coil is placed on second electromagnet which is connected across the supply and carries a current proportional to supply voltage. This second coil is called pressure coil.

(ii) Shading Bands—At the central limb of the core copper shading bands are provided so as to adjust the flux produced by shunt magnet exactly in quadrature with applied voltage. In this arrangement of meter these two electromagnets are called as series and shunt magnets.

(iii) Moving System—It is consists of an aluminium disc mounted on a light alloy shaft. This disc is placed in the air gap of series and shunt magnets. Rotor is provided which runs on a steel pivot screwed to the foot of shaft. Separate pinion is provided to the shaft with counting mechanism.

(iv) Braking System—Braking systems is provided for controlling the speed of moving system. In this instrument the braking is achieved by placing permanent magnet near to the aluminium disc. Due to this arrangement aluminium disc moves in magnetic field and braking torque gets developed. Braking torque can be adjusted by adjusting the position of permanent magnet.

(v) Counting system—Function of this system is to count or record continuously a number proportional to revolutions made by moving system. By using proper arrangment and pinion the rotor shaft drives five to six pointers. This arrangement rotates round dials marked with ten equal divisions.

Working

Step 1—When the supply voltage is applied across the pressure coil, current I_p flows through it and produces flux ϕ_{pt}. The current I_p is proportional to supply voltage and lags it by 90° approximately because pressure coil is highly inductive and has small resistance.

The flux ϕ_{pt} has two components ϕ_p and ϕ_g. Flux ϕ_g flows across side gaps because reluctance of this path is small. The reluctance to flux ϕ_p is large so magnitude is small. This flux ϕ_p is transferred to aluminium disc and driving torque gets developed. So flux ϕ_p is proportional to voltage '*V*' and lags by 90° approximately.

Step 2— The load current '*I*' flows through current coil and produces flux ϕ_s which is proportional to load current in the same phase. The eddy current I_{es} interacts with flux ϕ_p to produce a torque and eddy current I_{es} interacts with ϕ_p to produce another torque.

Step 3—These two torques are opposite in direction due to which the resulting torque will be the difference of two torques.

This torque is responsible to drive the counting mechanism or system.

Single phase energymeter is suffered by errors like (a) If the magnetic circuits are not symmetrical it results into creeping. (b) Random changes in disc resistance (c) Incorrect phase angle due to change in resistance with temperture, frequency variations (d) Friction in moving parts.

"Creeping" in energymeters

If there is no current flowing through the current coil but pressure coil is excited, the meter indicates continuous rotations this is known as creeping.

(B) Two Element Energy meter

This type of energy meter is used for the energy measurement in 3-phase 3-wire circuits or systems. Fig. 2.21 shows constructional details of two element energy meter.

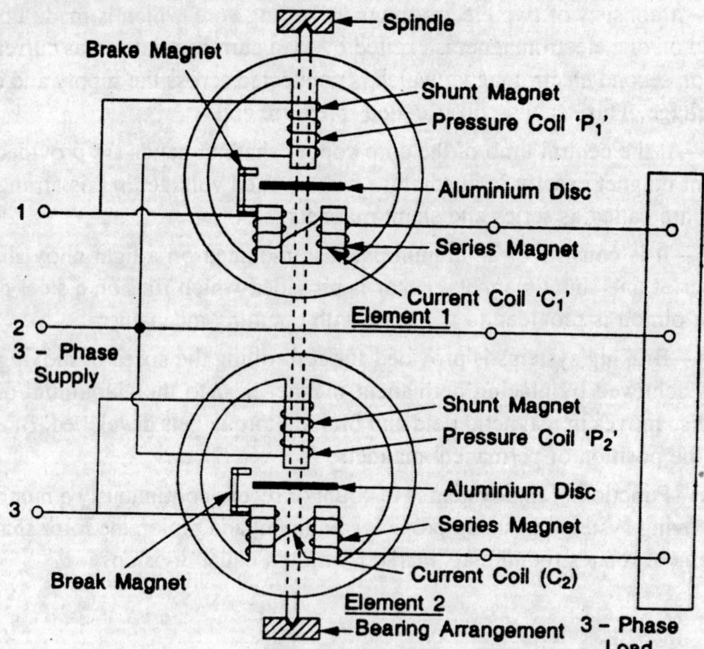

Fig. 2.21. Two element energymeter

It consists of two meter elements *i.e.* element 1 and element 2 having two seperate individual aluminium discs. Similarly two pressure coils 'P_1' and 'P_2' are connected in parallel whereas the current coils C_1 and C_2 are connected in series fashion due to which the torque produced by these two elements are opposite to each other. The important aspect is that driving torque of two elements should be equal for equal amount of power passing through them. For balancing the torque of both elements an adjustable magnetic shunt is provided. This magnetic shunt is adjusted to a position such that if two torques are equal and opposite then there is no disc rotation.

2.8 DIGITAL INSTRUMENTS

Now a days digital instruments are widely used in measuring systems. Analog instruments are replaced by digital instruments. In digital instruments the unknown quantity (quantity under measurement) is displayed by digits or in the form of digits instead of deflection of pointer over calibrated scale. Important designing elements of these type of instruments are Analog-to-Digital and Digital-to-analog converters, Display devices such as LED, LCD etc.

Important advantages of digital instruments are high accuracy, small size and weight, high input impedance. Their output is visual read out and can also be interfaced with external devices like microprocessor, printer, computer etc. By using coloured display devices like red, green, amber etc. good resolution can be achieved in dark surroundings. Basic types of digital instruments are discussed in the section 2.2 (b) of this chapter.

(A) Digital Voltmeter (DVM)

Digital voltmeters are used to display the ac and dc voltages in numerical readout form. If proper signal conditioning stage gets interfaced at the input stage of DVM, it can be used for ac voltage, dc voltage and current, and ohms. Important features of DVM are ;

(a) Input range is +1V to +1000V.

(b) Automatic range selection and overload indication.

(c) Accuracy ± 0·005%

(d) Small size and weight

(e) Input resistance is 10 MΩ and input capacitance 40 pF.

(f) Additional facility to measure Current and Ohms.

The Digital voltmeters (DVM) techniques are classified into two main types ;

(a) Non integrating Type (b) Integrating Type.

This classification is on the basis of technique or method used for converting the analog dc signal into digital form.

(a) Non Integrating Type—These types of DVM are classified as—

(1) Potentiometric Type—(i) Servo (ii) Successive approximation (iii) Null Balance.

(2) Ramp Type—(i) Linear, (ii) Stair case.

(b) Integrating Type—(1) Voltage-to-frequency converter.

(2) Potentiometric to integrating.

(3) Dual slope.

In the next section Linear Ramp type and Dual slope integrating types of DVM are discussed.

(B) Ramp Type (DVM)

Basically the linear-ramp type technique is a voltage-to-time conversion. Here linear ramp is used to convert an analog dc signal into digital form for presentation. Basic elements used in this technique are ramp generator, comparator unit and counter with display assembly. Fig. 2.22 shows block diagram of ramp type DVM.

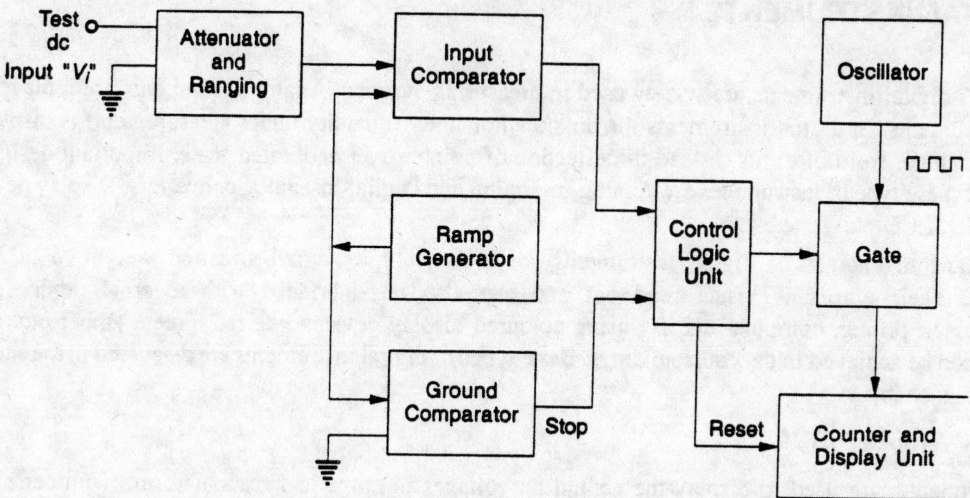

Fig. 2.22 Block diagram of Ramp type DVM.

At the start of measurement, ramp voltage is generated by ramp generator. This ramp voltage is continuously compared with the voltage which is to be measured. Generally, the ramp voltage ranges from + 12V to − 12V. The output of ramp generator is connected to both comparators *i.e.* input signal. At the stage when both ramp and input voltage level become equal the output state of input comparator gets changed.

The second comparator *i.e.* ground comparator compares the ramp voltage with ground. If both ramp and ground potential become equal the output of this comparator changes state *i.e.* it generates stop pulse which closes the gate.

The output of both comparators are applied to control logic unit which controls the gate between oscillator and counter units. Initially the counter should be reset to zero before measurement and set the logic to specified state. This logic is such that after reset first signal from any one of the comparator will open the gate between oscillator and counter. At the next signal gate will close. During the period in which gate is open the counter accumulates the count according to rate set by the oscillator. The magnitude of count indicates the magnitude of input voltage which is displayed by the display unit. Thus the voltage is converted into time and count of time represents the magnitude of voltage.

For example, if 4·927 V is applied at the input of DVM (Fig. 2.22) When the ramp signal passes through voltage equal to input voltage, logic "opens" the gate. If ramp passes through zero volt the logic causes to close the gate. At the end of ramp, a signal is transferred to counter and display assembly. Suppose oscillator frequency is 200 kHz and slope is 1·0 V/ 5m sec then ;

$$\text{Gate period} = 4{\cdot}927 \times 0{\cdot}005$$
$$= 0{\cdot}024635 \text{ sec.}$$

$$\therefore \text{ Number of pulses} = 0{\cdot}024635 \times 200 \times 1000$$
$$= 4927 \text{ pulses.}$$

This type of voltmeter is easy to design and are cost effective. Output can be transmitted over long distance. Accuracy depends upon offset and drift of the two comparators. Linearity depends on slope of ramp and frequency stability.

(C) Dual Slope Integrating Type DVM

Basically it is a voltage-to-time conveter. It uses the dual slope technique. It produces positive and negative ramps using integration method due to which the noise problem gets eliminated. Working principle of this method is discussed as follows.

The input voltage 'V_i' is integrated with the slope of the integrator output proportional to test input voltage. After a fixed time duration t_1 the input voltage is disconnected. Now the input of integrator is connected to negative voltage $- V_r$. In this case the output of integrator has negative slope proportional to magnitude of input voltage. Means in this technique the discharging time is measured using reference voltage level. This principle is shown in Fig. 2.23.

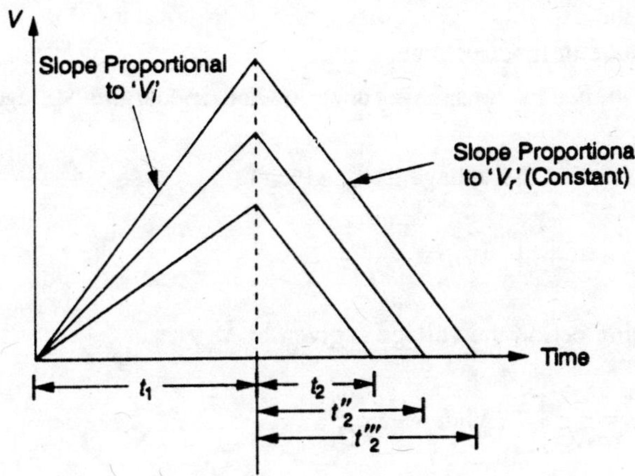

Fig. 2.23 Working principle of Dual Slope integrating type DVM.

Fig. 2.24 shows the block diagram of dual slope integrating type DVM. It consists of an OP-AMP

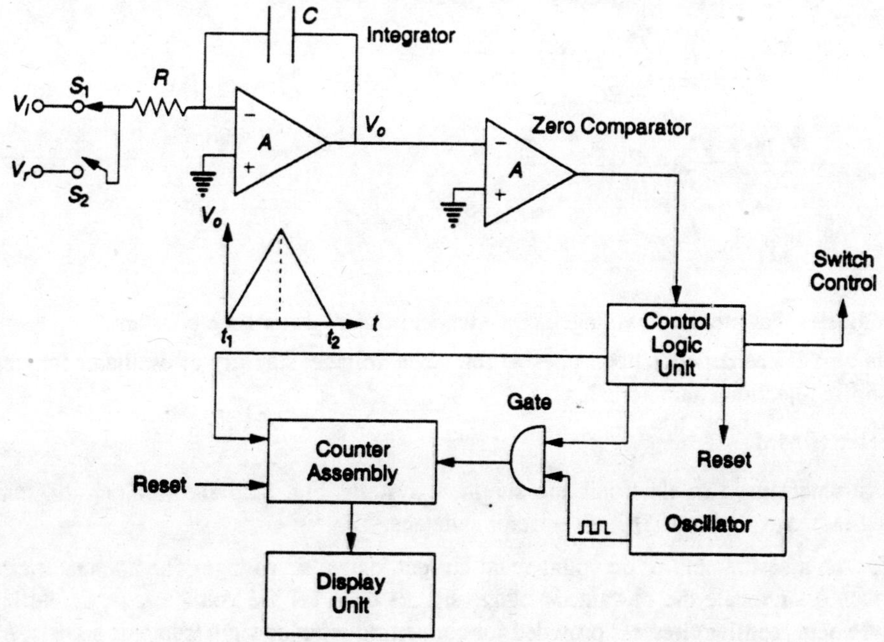

Fig. 2.24 Block diagram of dual slope integrating type DVM

integrator with RC combination. At the starting a pulse resets the counter to logic level 'O'. When switch 'S_1' is closed and 'S_2' is open the capacitor 'C' starts to charge. As the integrator output exceeds zero, the output voltage of comparator changes the state and enables (starts) the gate through control logic.

Due to which the oscillator pulses are fed to counter assembly. When the counter reaches to maximum count it gets activated at logic '1'. At this stage the voltage 'V_i' gets disconnected and 'V_r' is connected to integrator. The integrator output has negative slope and this output decreases linearity to 0 volt. This is the discharge time 't_2' which is proportional to input voltage. At this stage the comparator output changes state and closes (locks) the gate (the capacitor 'C' is discharged with constant slope). The pulses counted by counter represents the magnitude of input voltage.

The working can also be studied by considering charging and discharging voltage equations of capacitor 'C'.

Step 1—During charging period the voltage is V_{oc} given by ;

$$V_{oc} = -\frac{1}{RC}\int_0^{t_1} V_i\,dt = -\frac{V_i t_1}{RC} \quad ...(2.55)$$

Step 2—During discharging period the voltage is given by V_{od}

$$V_{od} = \frac{1}{RC}\int_0^{t_2} -V_r\,dt = -\frac{V_r t_2}{RC} \quad ...(2.56)$$

Step 3—Subtract both voltage levels during changing and discharging period.

$$V_{od} - V_{oc} = \frac{-V_r t_2}{RC} - \left(\frac{-V_i t_1}{RC}\right) \quad ...(2.57)$$

$$0 = \frac{-V_r t_2}{RC} - \left(\frac{-V_i t_1}{RC}\right) \quad ...(2.58)$$

$$\therefore \quad \frac{V_r t_2}{RC} = \frac{V_i t_1}{RC} \quad ...(2.59)$$

$$\therefore \quad V_i = V_r \frac{t_2}{t_1} \quad ...(2.60)$$

This equation indicates that measured voltage is independent of integrator time constant.

Accuracy of this type DVM depends upon internal reference voltage, stability of oscillator frequency. But it has excellent noise rejection characteristics.

(D) Digital Multimeter (DMM)

DMM is versatile instrument used in electronic measurement systems Fig. 2.25 shows block diagram of digital multimeter (DMM). S_1A, S_1B and S_1C are selector switches.

DMM is used for the measurement of dc voltage and current, ohms, ac voltage. The attenuator circuit is provided to eliminate or attenuate the unwanted voltage signals or to set the voltage at particular level. For ac voltage measurement rectifier circuit is provided for converting input ac signals into dc signals. After rectification input ac voltage signals are measured.

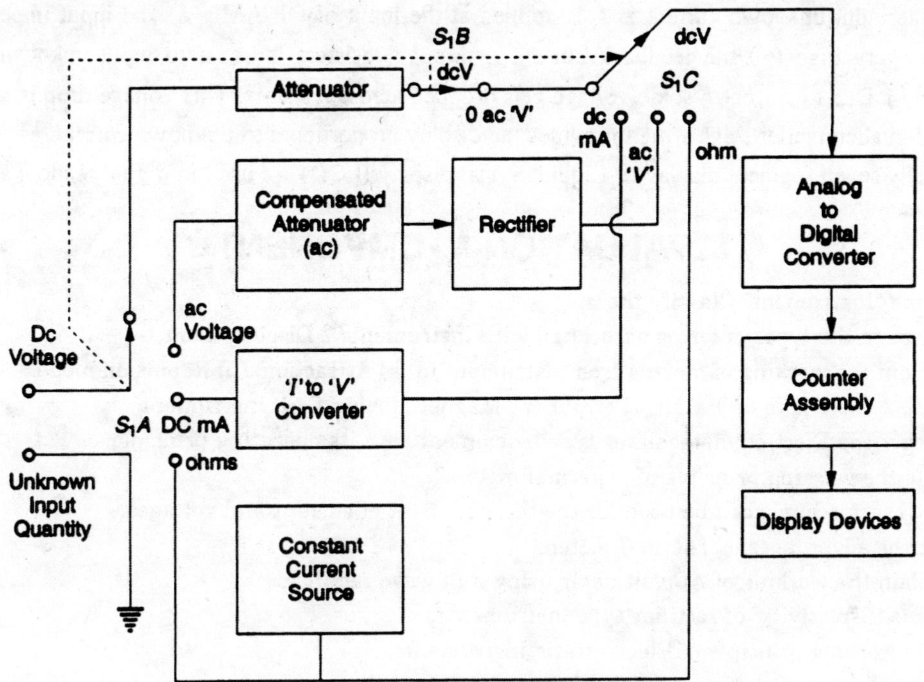

Fig. 2.25 Block diagram of Digital Multimeter (DMM).

The seperate constant current source is used in measurement of unknown resistance. Resistance is measured by passing a known current through unknown resistance. The known value of current is adjusted from constant current source.

Due to flow of Current the voltage gets dropped across a resistor. This voltage is applied to Analog-to-Digital converter for indicating (displaying) the values of unknown resistance.

In this block diagram current (I) to-Voltage (V) converter stage is used for converting the unknown current quantity into voltage *i.e.* the unknown current is measured by converting into voltage forjn. Generally, OP-AMP current to voltage converter is used which is shown in Fig. 2.26.

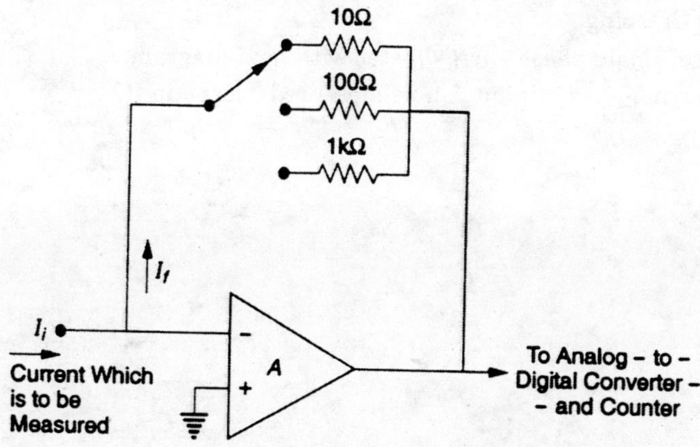

Fig. 2.26 I to V converter

Here the input unknown current is I_i is applied at the input of OP-AMP. As the input impedance of OP-AMP is very high (∞) the feedback current I_f through resistors is equal to input unknown current (I_i) *i.e.* ($I_f \cong I_i$). Due to current I_f some voltage gets dropped across resistors. This voltage drop is applied to analog-to-digital converter stage which provides the display proportional to unknown current.

Generally seven segment display or Liquid crystal display (LCD) are used in digital readout stage.

EVALUATION COMPONENT

1. Define "Instrument" Classify them.
2. What are the types of forces associated with instruments ? Discuss them.
3. Explain the working of moving iron instrument in (a) Attraction, (b) Repulsion mode.
4. Write a short note on PMMC (Permanent Magnet Moving Coil) instrument.
5. Discuss the Electrodynamometer type instrument with its operating principle.
6. State the working principle of "Thermal meters".
7. How thermocouple can be used for the measurement of current and voltages.
8. List the advantages of Thermal meters.
9. Explain the working of AC voltmeter using Full wave rectifier.
10. Define "Sensitivity" of rectifier type instruments.
11. State working principle of Electrostatic Instruments.
12. How multirange ammeters can be designed ? Discuss it.
13. What is the function of multiplier resistance in DC voltmeter ?
14. Why Ohmmeter is necessary ? Compare Series and Shunt type Ohmmeters.
15. How Electrodynamometer type wattmeter is used for measurement of Power ?
16. How D'Arsonval Galvanometer works ? Explain it in detailed manner.
17. List the advantages of Digital instruments.
18. Classify Digital voltmeters (DVM).
19. With neat block diagram explain the working of Ramp type DVM.
20. What do you mean by dual slope integration method ? How it can be implemented for the measurement of unknown voltage ?
21. State the applications of DMM. With neat block diagram explain working of DMM.
22. How energy can be measured ?
23. Define (i) Energy (ii) Creeping.
24. Explain the working of single phase energy meter with heat diagram.
25. How the energy through 3-phase circuit can be measured ? Explain it.

Chapter **3**

TRANSDUCERS

Learning Objectives

- To define the term Transducer.
- To compare transducer and sensor.
- To classify and elaborate types of transducers.
- To distinguish transducer and inverse transducer.
- To discuss selection parameters of transducer for particular application.
- To define working principle of resistive transducers.
- To explain construction and working of linear and rotary POT's.
- To calculate linearity and sensitivity of POT's.
- To explain working principle of inductive transducer.
- To discuss linear displacement measurement using LVDT and Rotary displacement measurement using RVDT.
- To compare LVDT and RVDT.
- To study synchros.
- To elaborate working principle of photoelectric transducers.
- To define photoconductive effect.
- To discuss working of photomultiplier, photodiode, phototransistor with their V-I characteristics.
- To elaborate piezoelectric effect in piezoelectric transducers.
- To analyze equation of emf across piezoelectric crystal when force is applied.
- To define strain, piezoresistive effect, gauge factor.
- To derive expression for gauge factor.
- To classify and study different type of strain gauges.
- To define temperature and temperature scales.
- To define "Seebeck Effect".
- To explain thermocouple, temperature probes.
- To study temperature-resistance relationship in thermistors.
- To explain construction and working of thermistors.
- To study RTD.

3.1 Introduction

In most of the industrial applications and automation systems it is desirable to convert the physical quantities or non-electrical quantities like temperature, pressure, velocity, liquid level etc. into electrical signals for measurement, monitoring and controlling purpose. This conversion is performed by the transducers.

Transducer is a device which accepts the energizing inputs from one or more transmitting signals and converts them into another form (rated signal) of energizing signals to various transmitting systems. Briefly, transducer is defined as a device which converts one form of energy into another form. This energy may be mechanical, electrical, acoustical, optical, etc.

Depending upon method of conversion, nature of output signal, applications etc. the transducers are classified into different types which are discussed in this chapter. In instrumentation systems according to the application and requirements, selection of appropriate and accurate transducer is essential which is also discussed.

Construction, working principle, features of the transducers like resistive, capacitive, piezoelectric, photo-electric etc. are discussed in detailed manner.

3.2 Significance of Transducers in Measurement and Instrumentation System

In electronic instrumentation systems different type of instruments and components are used for measurement, recording and controlling the signals generated in different processes. The three main section of instrumentation systems are :

(1) Sensing device (input device).

(2) Signal conditioning or signal processing stage.

(3) Output stage.

First stage (Sensing Device)

If the quantity under measurement *i.e.* the quantity involved in the process is non-electrical, under such circumstances the particular quantity should be converted into electrical signal by the device **transducer** for measurement and controlling. Thus the term transducer is defined as a device which is used for measurement of physical quantities by electrical means.

Here, it is essential to discuss the term **Sensor**. The sensor is a part of transducer which senses or responds to a physical quantity or measurand. The response of sensor is converted to electrical signals by the transduction stage (transducers).

Second stage (Signal Conditioning Stage)

Function of this stage is amplification, filteration, conversion etc. of the output of transducer so that it can be compatible with the output device or stage.

Third stage (Output Stage)

The output of signal conditioning stage is displayed or monitored on display devices, recorders. These devices are CRO, magnetic tape, printer, computer etc.

3.3 Classification of Transducers

The transducers are classified as follows.

(a) **On the Basis of Method Used for Transduction** : In this type the transducers are classified on the basis of principle of transduction *i.e.* resistive, capacitive, inductive etc.

(b) **Active and Passive Transducers** : Active transducers are defined as those transducers which does not require any external or auxiliary power source to generate the equivalent output *i.e.* voltage or current. They are self generating transducers.

e.g. : Conversion of pressure into e.m.f. by using piezoelectric crystal transducer.

Passive transducers are defined as those transducers that require external or auxiliary power supply for their operation of conversion of physical quantity into electrical signal. They are externally powered transducers.

e.g. : Measurement of displacement using potentiometer.

(c) **Analog and Digital Transducers** : Analog transducer converts the input physical quantity into analog output which is a continuous function of time.

e.g. : Thermocouple.

Digital transducer converts the input physical quantity into digital form *i.e.* in the form of pulses having logic 0 and logic 1 levels. *e.g.* Rotary Encoder.

(d) **Transducers and Inverse Transducers** : Transducer is a device which converts the physical quantities into electrical signals. (By electrical means)

Inverse transducer is a device which converts the electrical quantities (signals) into non-electrical quantities.

e.g. : When electrical charge or voltage is applied to the surface of piezoelectric crystals, the dimensions of crystal gets changed due to deformation. This deformation results into mechanical displacement.

In measurement systems, at the output stage measuring devices such as Ammeter or Voltmeter are used for the measurement or current of voltage. These devices receives the input quantity current or voltage (electrical quantity) and provides mechanical displacement (deflection) in accordance with the magnitude of input quantity.

(e) **Primary and Secondary Transducers** : Consider following example to study the working of primary and secondary transducer.

Fig. 3.1 shows method for pressure measurement using L.V.D.T. Here L.V.D.T. does not accept pressure signal for converting it into electrical voltage. Hence initially, pressure is converted into linear displacement using Bourdon tube. This displacement is converted into electrical voltage using L.V.D.T. In this method C-type Bourdon tube is used.

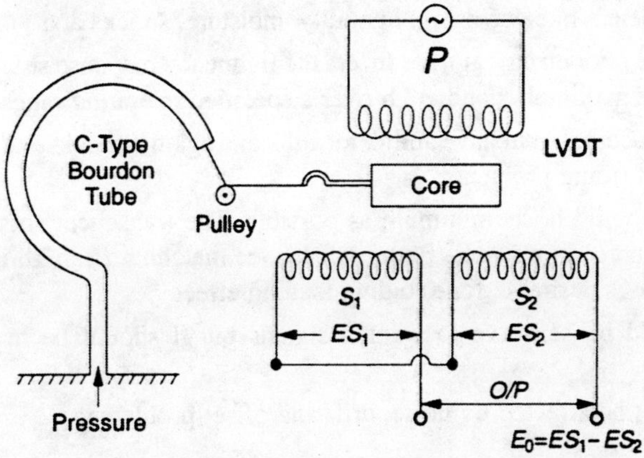

Fig. 3.1 Pressure measurement using Bourdon Tube and L.V.D.T.

Step 1. First pressure is converted into linear displacement by using Bourdon tube.

Step 2. The linear displacement is converted into voltage using L.V.D.T.

Hence, the Bourdon Tube acts as a primary transducer because it senses the pressure and converts it into displacement which is suitable for L.V.D.T. and it acts as secondary transducer.

If a particular physical quantity is not suitable for the main transducer for obtaining electrical signal then, it is sensed by a first stage of transducer called primary transducer. The secondary transducer converts the output of primary transducer into electrical signal.

3.4 Selection Parameters for a Transducer

While selecting a transducer for particular application it is important to study complete electrical, mechanical characteristics and physical characteristics of a transducer.

In instrumentation systems the transducers are used to provide usable output *i.e.* electrical signal associated with particular application. Following parameters or requirements should be considered for selecting a transducer.

(1) Physical quantity which is to be measured *i.e.* first consider the input quantity which is to be applied at the input of transducer.

(2) Determine type and range of the measurand. The range of input quantity and range of transducer should be compatible to provide good resolution.

(3) **Linearity** : The transducer should provide linear characeristics between input (physical parameter) and output (resulting electrical signal).

(4) **Stability** : Under operating or working conditions, the transducer should provide high degree of stability.

(5) **Repeatability** : The transducer should have ability to produce the exact output signal when the same measurand is applied repeatedly under same environmental conditions of the system.

(6) **Sensitivity** : Transducer should convert the physical quantity into electrical signals with maximum sensitivity *i.e.* it should be highly sensitive to produce electrical output.

(7) **Accuracy** : The transducer should work with high degree of accuracy under operating conditions. Calibration and repeatability are the important parameters for good accuracy.

(8) **Environmental Conditions** : For selection and installation of particular transducer, consider the environmental parameters like corrosion, humidity, moisture, shocks and vibrations, friction etc.

(9) **Frequency Response** : For electrical transducers the frequency response should be flat and stable, it should also provide maximum bandwidth over a specified operating range.

(10) **Ruggedness** : Transducers should have ability to withstand against shocks and vibrations. It should also tolerate the overloading problems.

(11) **Loading Effect** : It should be as minimum as possible. The transducer should not produce any loading effect on the preceding stages. Better impedance matching *i.e.* high input impedance and low output impedance is desirable for avoiding loading effect.

(12) The transducer should be sensitive to wanted signals but it should be minimum sensitive to unwanted signals.

(13) Transducer should not be affected by noise, drift and offset problems.

3.5 Resistive Transducers

The electrical resistance of a metal wire is given by :

$$R = \frac{\rho l}{A} \qquad \text{...(3.1)}$$

where ; R = Resistance (Ω)

ρ = Resistivity of the conductor material (Ω – m)

l = Length of the conductor (m)

A = Uniform cross-sectional area of conductor (m^2)

The principle used in resistive transducers is that the resistance of sensing element changes in accordance with the physical quantity (variable) which is to be measured. The examples of resistive transducers are (1) Potentiometers (translational and rotational) (2) Strain gauges (discussed in section 3.10)

In potentiometers, the translational and rotational displacement is measured by change of resistance due to change in the length of conductor.

Strain gauges use the principle resistance of the conductor or semiconductor changes in accordance with that applied strain.

In this section all type of resistive transducers are discussed.

(a) **Potentiometers** : They are the passive transducers and also known as **POT**. A resistive potentiometer consists of resistance element, over this element a sliding (moving) contact called as wiper is provided. Linearity of potentiometer depends upon uniformity of the resistance winding.

The contact motion of the wiper may be translational or rotational or combination of both *i.e.* translational and rotational. The combinational type of potentiometers are called **Helipots** because the shape of resistive element is like Helix. Helipots are used for the measurements of rotary and translatory displacements.

The resistive element of the potentiometer is excited either by a.c. or d.c. voltage. Thus, **Potentiometer** is a passive transducer because it requires external power supply for its operation. The output voltage is a linear function of the input displacement provided to the wiper of the potentiometer.

Constructional Details of Resistive Element

The resistive elements are :

(1) Wire wound resistive element

(2) Nonwire wound resistive element

(1) **Wire wound resistive element** : This type of potentiometer carries high amount of current at higher temperatures. Materials used for wire wound resistive elements are Copper, Nickel, Platinum etc.

The resolution of potentiometer depends upon resistive material used. This resistive material is placed in small area to provide high value of resistance in that particular area. Generally wire wound resistance element is used.

The resistance wire is wound on a mandrel or card having circular or helical shape as shown in Fig. 3.2(b) for rotational POTS. For translatory POTS the resistance wire is wound on a mandrel or card. In this type of potentiometers the resistance changes in small steps according to the displacement applied and contact of the wiper with resistive element or turns. Different types are shown in Fig. 3.2

(a), (b) and (c). For translatory POTS the mandrel is linear (straight) whereas for rotary POTS the mandrel is circular shaped over which the resistance wire is wound. These diagrams 3.2 (a), (b) and (c) show the constructional details of potentiometers.

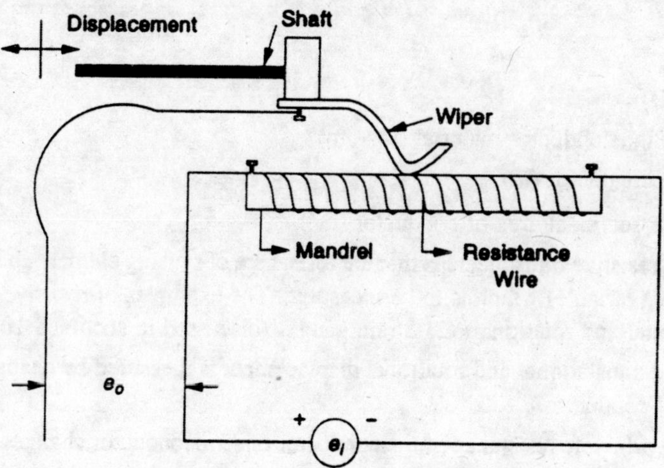

Fig. 3.2(a) : Linear (translational) potentiometer.

(2) **Nonwire wound resistive element (continuous potentiometers) :** As the name implies, in these type of potentiometer the resistive element is not in wirewound manner but it is in the form of flat strip or film having smooth surface. Over this smooth surface the wiper slides or moves. Due to this construction the resolution of nonwire wound resistive element is improved.

These resistive elements are highly sensitive to temperature changes and operate on moderate range of currents only. The wiper contact resistance is very high.

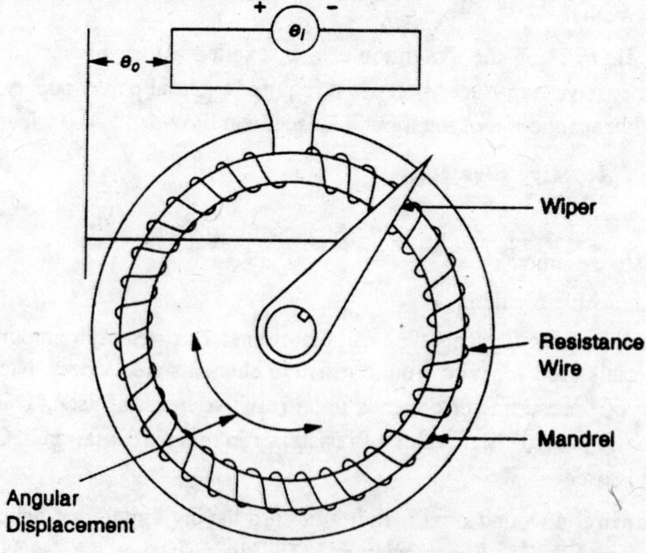

Fig. 3.2(b) : Rotary potentiometer

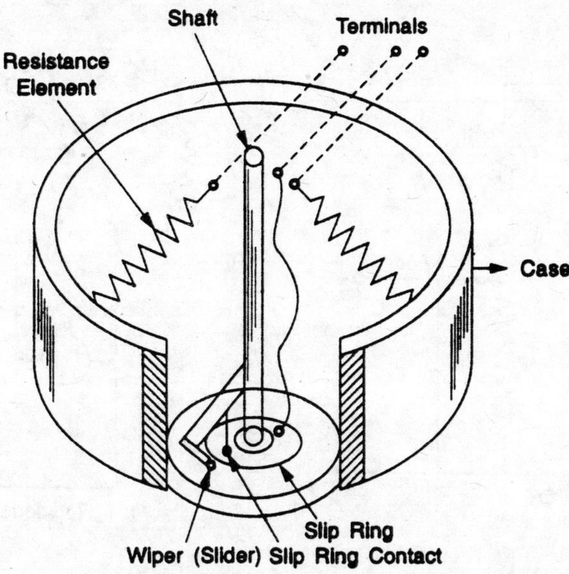

Fig. 3.2(c) : Potentiometer displacement transducer.

Materials used for these resistive elements are :

Cermet : (Ceramic + Metallic particles). They are made up of ceramic and some metallic particles. These resistive elements are generally used for a.c. supply and work at higher temperatures.

Conducting plastic : It is made up of plastic binder and conducting powder. This type of potentiometers are also applicable for handling a.c. supply.

Translatory POTS have strokes range of 2 mm to 50 cm. The range for rotary POTS is about 10 to 60 full turns.

Fig. 3.3(a), (b) and (c) show the diagrams of translational, rotational and helipot potentiometers.

Consider, the translational POT as shown in Fig. 3.3(a)

Analysis of potentiometers Translatory POTS are also called linear potentiometers.

Let, e_i = Input voltage applied to potentiometer (V)

e_0 = Output voltage of potentiometer (V)

x_i = Wiper displacement from its (rest) zero position.

x_t = Total length of translational POT (m)

R_p = Total resistance of the potentiometer (Ω).

$$\underset{(e_0)}{\text{Output voltage}} = \frac{\text{Resistance at output terminals}}{\text{Resistance at input terminals}} \times \text{Input voltage}$$

$$e_0 = \left[\frac{R_p\,(x_i/x_t)}{R_p}\right] . \, e_i \qquad ...(3.2)$$

$$e_0 = (x_i/x_t) \times e_i \qquad ...(3.3)$$

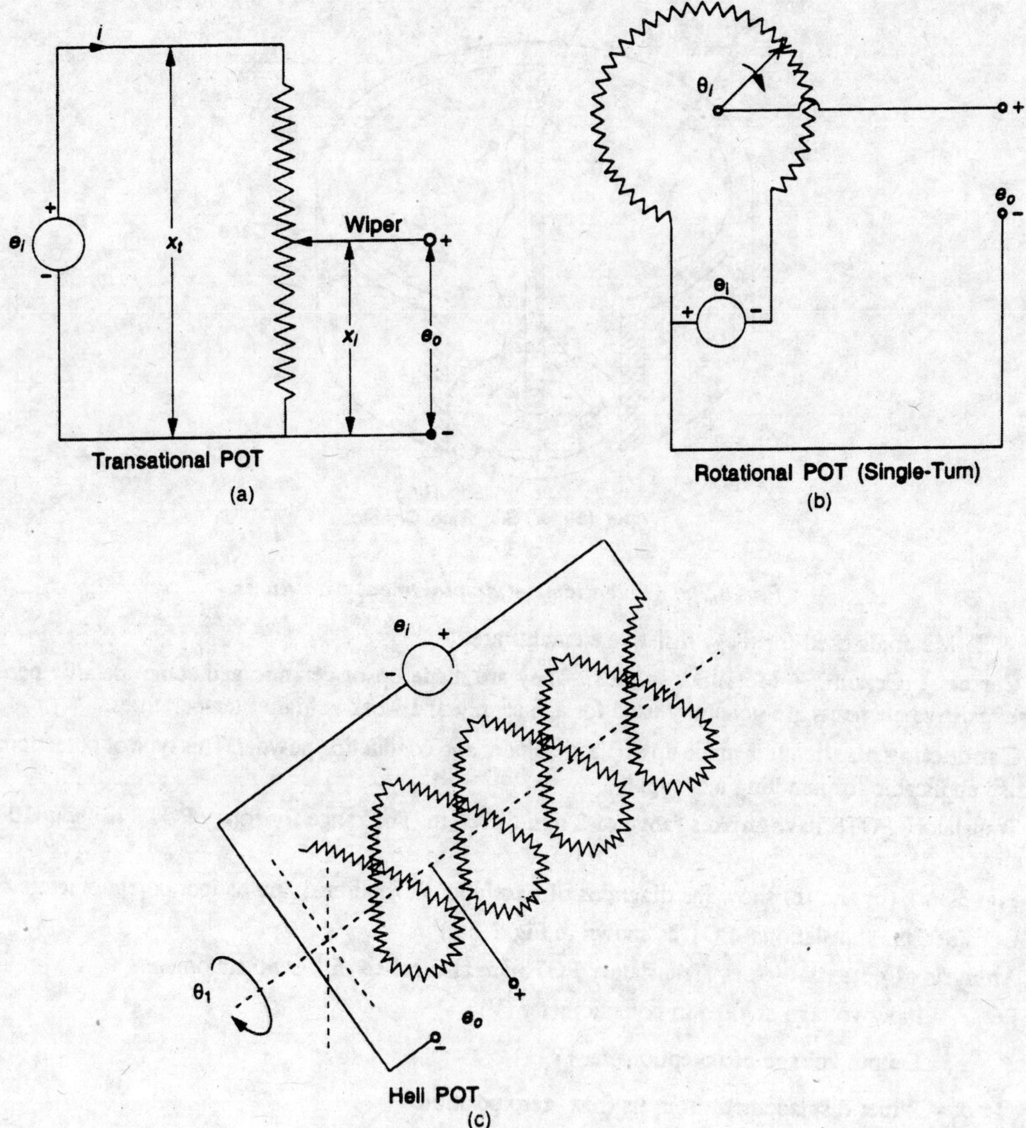

Fig. 3.3 (a), (b) and (c).

$$\text{Sensitivity } (S) \rightarrow S = \frac{\text{Output}}{\text{Input}}$$

For translatory POT's $\rightarrow$ $S = \dfrac{e_0}{x_i} = \dfrac{e_i}{x_i}$ ...(3.4)

Under ideal condition sensitivity is constant and POT has linear relationship between Input and Output. For rotational potentiometers, consider Fig. 3.3(b)

Let ; e_i = Input voltage applied to potentiometer (V)

e_0 = Output voltage of potentiometer (V)

θ_i = Input angular (rotational) displacement in degrees

θ_t = Total displacement of wiper in degrees

R_p = Total resistance of the potentiometer (Ω)

$$e_0 = \left(\frac{\theta_i}{\theta_t}\right) \times e_i \quad ...(3.5)$$

Power Ratings : For potentiometers the power ratings and heat dissipation capacity parameters are co-related.

Let ; P = Power (Watt)

$e_{i\,max}$ = Maximum input voltage applied to potentiometer (V)

R_p = Total resistance of potentiometer (Ω)

Power is given by :

$$P = e_{i\,max}^2 / R_p \quad ...(3.6)$$

$$e_{i\,max} = \sqrt{R_p\, P} \text{ Volt} \quad ...(3.7)$$

This equation indicates that the maximum input voltage which is to be applied across the potentiometer.

The single turn wire wound potentiometers are 5 cm in diameter with power ratings of 5 W. The total resistance range is 100 Ω to 100 kΩ.

Linearity and sensitivity of potentiometers : For obtaining good linearity the resistance of potentiometer 'R_p' should be as low as possible when using a meter for reading the output voltage. This meter has a fixed value of input resistance 'R_m' as shown in Fig. 3.4.

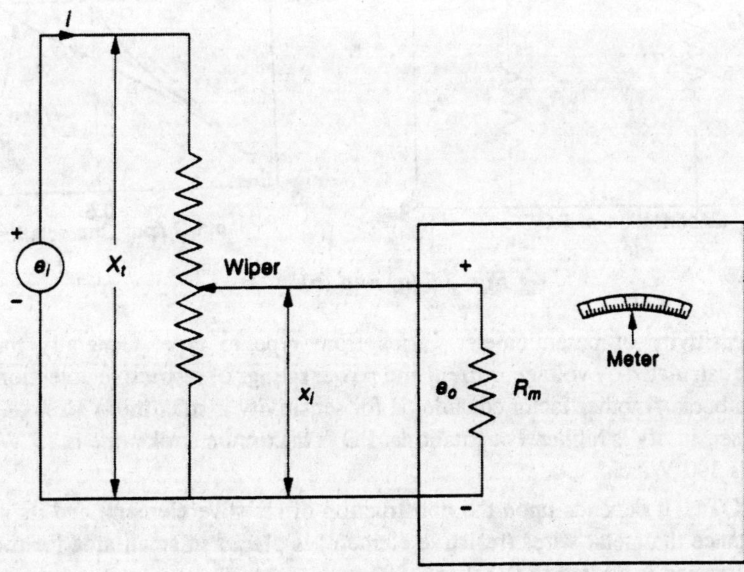

Fig. 3.4.

But for higher sensitivity output voltage 'e_0' should be high, it is possible when high input voltage 'e_i' is applied. Because there is limitation of power dissipation, as the input voltage is limited by resistance of potentiometer. For low power dissipation the input voltage should be small and resistance of potentiometer should be high.

(1) For higher sensitivity input voltage should be high which demands high value of resistance R_p.

(2) For good linearity resistance of potentiometer should be low but if it is made low the power dissipation increases.

So it is necessary to adjust the linearity and sensitivity by selecting proper value of potentiometer resistance 'R_p' because (*i*) If R_p is small linearity gets improved (*ii*) If R_p is small it handles low input voltages e_i (*iii*) If R_p is small, linearity improves but power dissipation increases too also sensitvity gets decreased because as e_i is small e_0 will be smaller.

Linearity can be improved by connecting a resistance $R_{L_{in}}$ with R_m. Generally, the value of $R_{L_{in}}$ is made equal to R_m without using a meter of high resistance. This arrangement is shown in Fig. 3.5(a).

Fig. 3.5(b) shows the performance characteristics (input-output characteristics) when $R_{L_{in}}$ is connected and when it is not connected with meter. It indicates that the linearity gets improved by connecting $R_{L_{in}}$ but not by increasing the value of R_m.

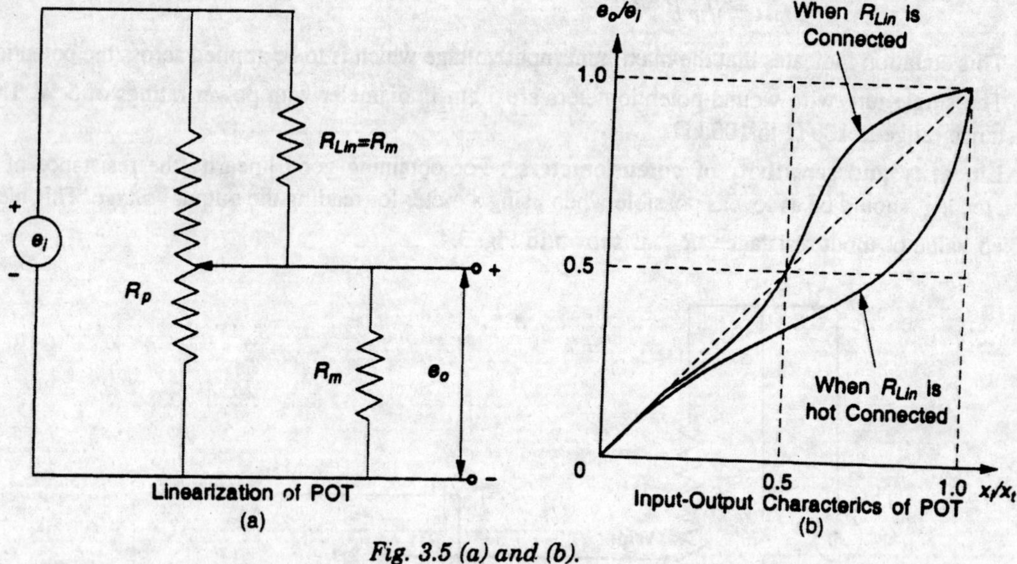

Fig. 3.5 (a) and (b).

Although the sensitivity of potentiometer varies from type to type. Generally the sensitivity of potentiometers can be calculated by voltage, current and power ratings of respective potentiometer prescribed in manufacturers data book. Another factor considered for sensitivity is **maximum strokes.** If the device is of shorter strokes its sensitivity is higher. For rotational POTs maximum stroke rate is 15 V/degrees and for transational POTs it is 300 V/inch.

Resolution of POTs : It depends upon the construction of resistive element, and its size. For getting higher value of resistance the metal wires (resistive element) is placed in small area for rotary POT's. For transitional POT's resolution is limited to 0·001 to 0·002 in.

The best angular resolution is given by :

$$\text{Angular resolution} = \frac{0{\cdot}12 \text{ to } 0{\cdot}24}{D} \text{ degrees ; } D\text{–in Inches}$$

The non wire wound POTs has improved resolution and life. Because their construction is in the form of strips or films and provides smooth surface and better performance. For high resolution high resistance materials like carbon, conductive-plastic resistance element are used. For translational devices resolution is 25-50 μm.

Selecting potentiometer for particular application the parameters to be discussed are resolution, operating temperature *i.e.* higher or lower temperature conditions during operation, humidity, shocks, vibrations etc.

The device potentiometer is affected by the unwanted signal called **Noise**. This noise signal affects the resolution and linearity of the potentiometer. Noise is produced due to (1) the output voltage variations or fluctuations occur when the arm slides over resistive element. (2) sometimes the wiper bounces on resistive element at higher speeds (r.p.m.). (3) dust or dirt particles collected between wiper surface and resistive element (winding) also produces noise.

The advantages of potentiometer are simple operation, applicable for measurement of higher order displacements, low cost, higher electrical efficiency due to which potentiometer has better controlling facility.

3.6 Capacitive Transducers

The term **Capacitor** is defined as two metal plates separated by some distance '*d*' and dielectric media is placed between the two metal plates. When a voltage or potential difference is applied to the metal plates, equal and opposite charge gets developed on the metal plates.

These transducers are used for the measurements of rotational or translatory motions in accordance with the change in capacitance variable, this change in the capacitance gets converted into an equivalent electrical signal. Capacitive transducers can be used for measurements of displacement, vibrations, pressure, sound, level, etc.

Capacitive transducer works on the principle of capacitance of parallel plate capacitor which is given by :

$$\text{Capacitance } C = \frac{\varepsilon_0 \, \varepsilon_r A}{d} \text{ Farad} \qquad \text{...(3.8)}$$

where ; A = Area of plates (m^2).

d = Distance (separation) between two plates (m)

ε_0 = Permittivity of free space or vaccum ($8{\cdot}854 \times 10^{-12}$ F/m).

ε_r = Relative permittivity.

From above equation it is observed that capacitance can be changed by changing the area, changing the distance between two plates or by changing the dielectric constant. These changes can be obtained by displacement, vibration, pressure, sounds, level measurements etc. as discussed earlier.

Capacitive transducer works on principle that capacitance gets changed due to any one of following parameter ;

(a) Change in area '*A*'

(b) Change in distance '*d*'

(c) Change in dielectric constant

If physical parameters like displacement, force, pressure etc., in any process is applied to capacitve transducer its capacitance gets changed because any one of the parameters (as stated above *i.e.* (a) (b) and (c)) gets varied. This change in capacitance can be measured by bridge arrangment or by any other suitable method with proper impedance matching network because its output impedance is very high. Output impedance of capacitive transducer is capacitive reactance which is given by $X_c = \frac{1}{2\pi fC}$;

where f = Frequency of input signal (Hz)

C = Capacitance (Farad)

(A) Capacitve Transducer Using Change in Area of Plates

It is found that the capacitance is directly proportional to area of plates '*A*'. Fig. 3.6 shows the capacitive transducer used for displacement measurement using the concept of change in area of plates.

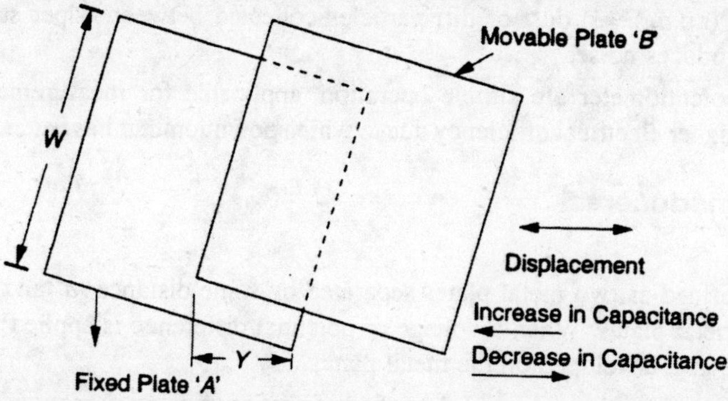

Fig. 3.6 Capacitive Transducer (Change in capacitance due to change in area)

This type of transducer consists of two metal plates *i.e.* plate '*A*' which is fixed whereas plate '*B*' is movable. The linear displacement which is to be measured gets changed resulting to change the capacitance.

The capacitance is given by $C = \frac{\varepsilon A}{d} = \frac{\varepsilon WY}{d}$ Farad ...(3.9)

where, Y = Length of overlapping part of plates (m)

W = Width of overlapping part of plates (m)

$$\text{Sensitivity 'S'} = \frac{\partial C}{\partial Y} = \frac{\varepsilon W}{d} \text{ Farad/m.} \quad ...(3.10)$$

There is linear relationship between capacitance and displacement because sensitivity is constant which is shown in Fig. 3.7

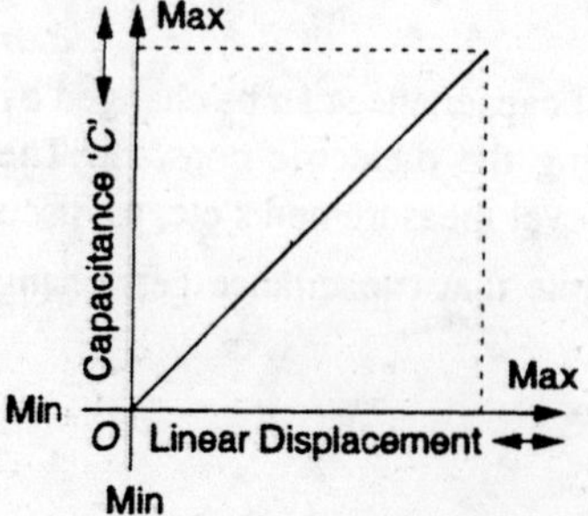

Fig. 3.7 Curve between capacitance and linear displacement.

This method is useful for linear displacement measurement of range 1 mm to several cm.

(B) Capacitve Transducer used for Angular Displacement Measurement

This method uses the concept of change in capacitance by bringing a change in area due to angular displacement. Fig. 3.8 shows capactive transducer used for angular displacement measurement. It consists of two plates in which one plate is fixed and other is movable. A shaft is provided for applying the angular displacement. θ is the angular displacment which is to be measured. Fig. 3.8 shows capacitive transducer used for angular displacement measurement.

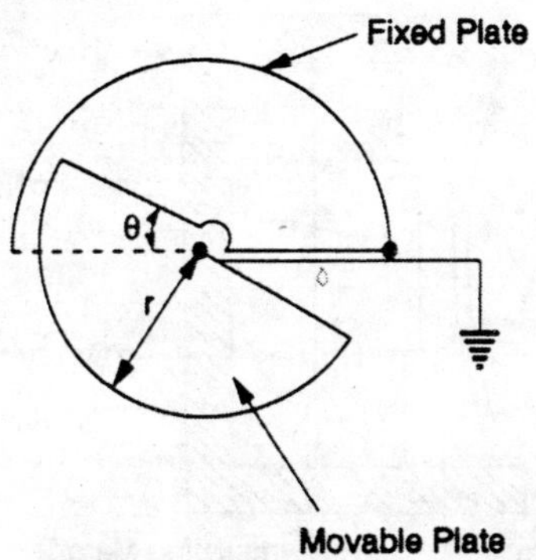

Fig. 3.8 Capactive Transducer

When the angular displacement which is to be measured is applied to the movable plate through shaft the effective area between two plates get changed which results to change in the capacitance. The maximum value of capacitance is given by ;

$$C_{max} = \frac{\varepsilon A}{d} = \frac{\varepsilon \pi r^2}{2d} \text{ Farad} \qquad ...(3.11)$$

But we have to determine the capacitance with respect to angular displacement θ.

$$\therefore \quad C_{max} = \frac{\varepsilon \theta r^2}{2d} \text{ Farad} \qquad ...(3.12)$$

where θ = Angular displacement

r = radius

When θ = 180° *i.e.* two plates overlap each other the capacitance is maximum.

Fig. 3.9 shows the capacitance displacement curve between angular displacement (θ) and capacitance (*C*) It indicates the linear relationship between capacitance variation and angular displacement.

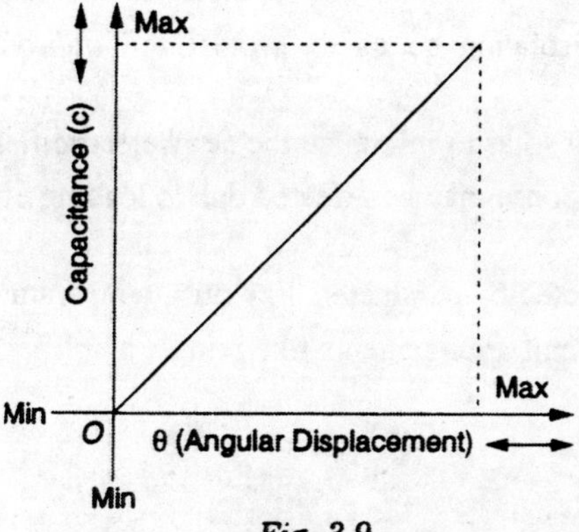

Fig. 3.9

$$\text{Sensitivity } 'S' = \frac{\partial C}{\partial \theta} = \frac{\varepsilon r^2}{2d} \text{ F/radian} \qquad ...(3.13)$$

(C) Capacitive Pressure Transducer

Fig. 3.10 shows construction of capacitance pressure transducer. It uses the principle of change in capacitance due to change in distance between two plates when pressure is applied to diaphragm.

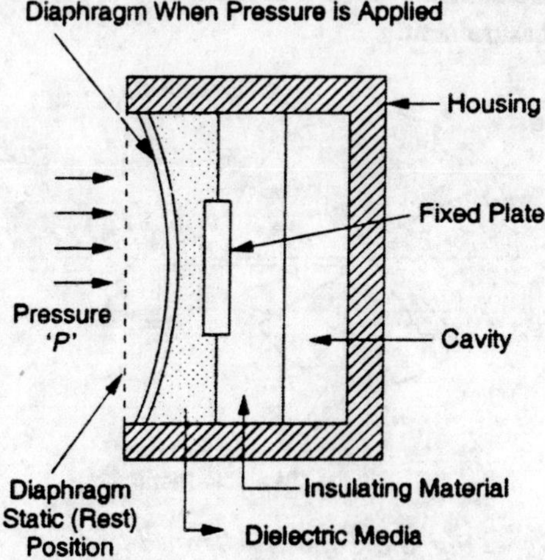

Fig. 3.10 Capacitive Pressure Transducer.

It consists of an airtight housing in which a metallic diaphragm is placed. At the inner side there is a fixed plate. In between the diaphragm and fixed plate dielectric media is placed. When the pressure is applied to diaphragm it moves towards the fixed plate resulting to change the capacitance. This capacitance is calibrated into voltage proportional to applied pressure.

Advantages

The capacitive transducers has following advantages

(1) As these transducers have high input impedance so loading effect is minimum.

(2) They require small force for operation so these are useful for small displacement, pressure measurements and require small power.

(3) Good frequency response

(4) Less affected by stray magnetic fields.

Disadvantages

(1) Requires good quality isolation so as to protect the transducer (metal plates) from stray capacitances.

(2) Guard rings are necessary so as to minimize the stray electric fields.

(3) Sometimes frequency response may be affected due to loading effects from connecting links and cables.

(4) Performance may be affected by parameters like dust, temperature, moisture, vibrations etc.

(5) They require complex circuit arrangements like bridge, amplifier etc. for measurement purpose.

3.7 Inductive Transducers

These are passive transducers. An inductive transducer converts the displacement or positional change (translational or rotational) into change in inductance of inductor(s) or coil. The variable inductance transducer works on the following principle (*i.e.* how the change in inductance occurs).

(1) Change in self inductance.

(2) Change in mutual inductance.

By using Inductive transducers, displacement can be measured by varying (a) Number of turns, (b) Geometric configuration, (c) Permeability of magnetic circuit.

These principles are discussed below :

(1) Change in Self Inductance

Self inductance of the coil relates with current flowing through the coil and induces flux in the same coil. Consider Fig. 3.11 (a) and (b).

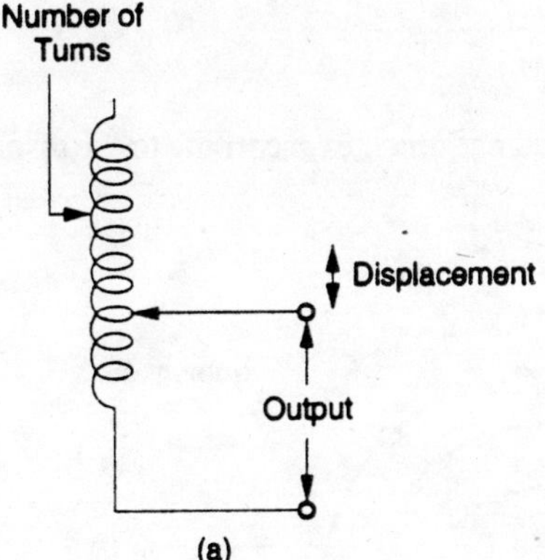

Fig. 3.11 (a) Linear inductive transducer.

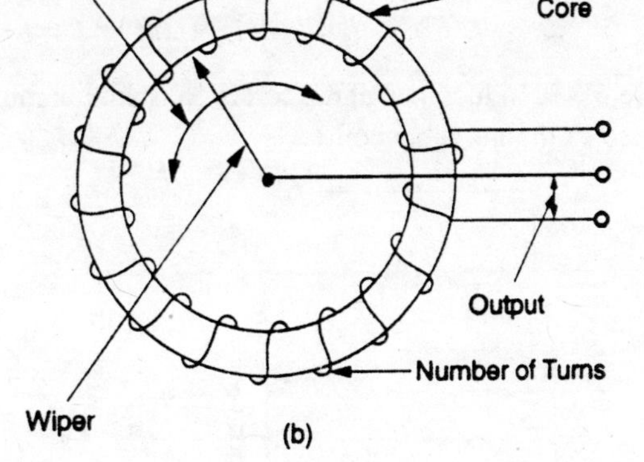

Fig. 3.11 (b) Rotary inductive transducer

The self inductance of coil is given by :

$$L = \frac{N^2}{S} \ H \qquad ...(3.14)$$

where ; N = Number of turns of the coil.

S = Reluctance of magnetic circuit.

$$S = \frac{l}{\mu_0 \mu_r A} \qquad ...(3.15)$$

l = Length of coil (m).

A = Cross-sectional area of coil (m^2).

μ_0 = Free space permeability.

μ_r = Relative permeability.

This equation indicates that the change in inductance may be due to :

(1) change in number of turns (*N*).

(2) change in permeability (μ).

(3) change in geometric configuration.

(2) Change in Mutual Inductance

Transducer works on the principle of change in mutual inductance utilizing multiple coil arrangements as shown in Fig. 3.12 (a) and (b).

The mutual inductance (*M*) between two coils is given by :

$$M = K \cdot \sqrt{L_1 \cdot L_2} \quad \text{...(3.16)}$$

where ; L_1 and L_2 are self inductance of coils. *K* is coefficient of coupling.

It is possible to convert the mutual inductance (*M*) into self inductance (*L*) by connecting the two coils in series. In this condition the self inductance varies between $L_1 + L_2 + 2M$ to $L_1 + L_2 - 2M$ by keeping one coil movable and other fixed :

$$L = L_1 + L_2 + 2M \;\; ; \;\; L = L_1 + L_2 - 2M \quad \text{...(3.17)}$$

Here, self inductance of the coil is constant but mutual inductance changes according to the displacement provided to the movable coil.

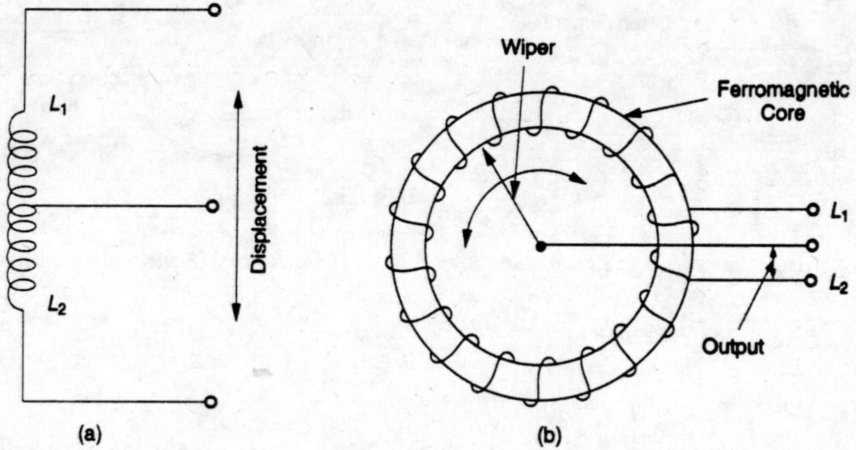

Fig. 3.12 (a) and (b) shows the arrangement of translational and rotatory displacements.

In next sections different inductive transducers are discussed.

(a) Linear Variable Differential Transformer (LVDT)

LVDT is variable inductance transducer used for measurement of linear displacements. It converts the linear motion into electrical signals.

Fig. 3.13(a) shows the construction of LVDT. It consists of a primary winding '*P*' and two secondary windings 'S_1' and 'S_2' These windings are wound on a cylindrical former. The former should be mechanically rigid to absorb shocks, vibrations and temperature. The secondary windings are wound on the opposite sides of primary winding. A movable core is placed inside the hollow former for sliding. Generally, the core is made-up of nickel-iron alloy for reducing the eddy current losses.

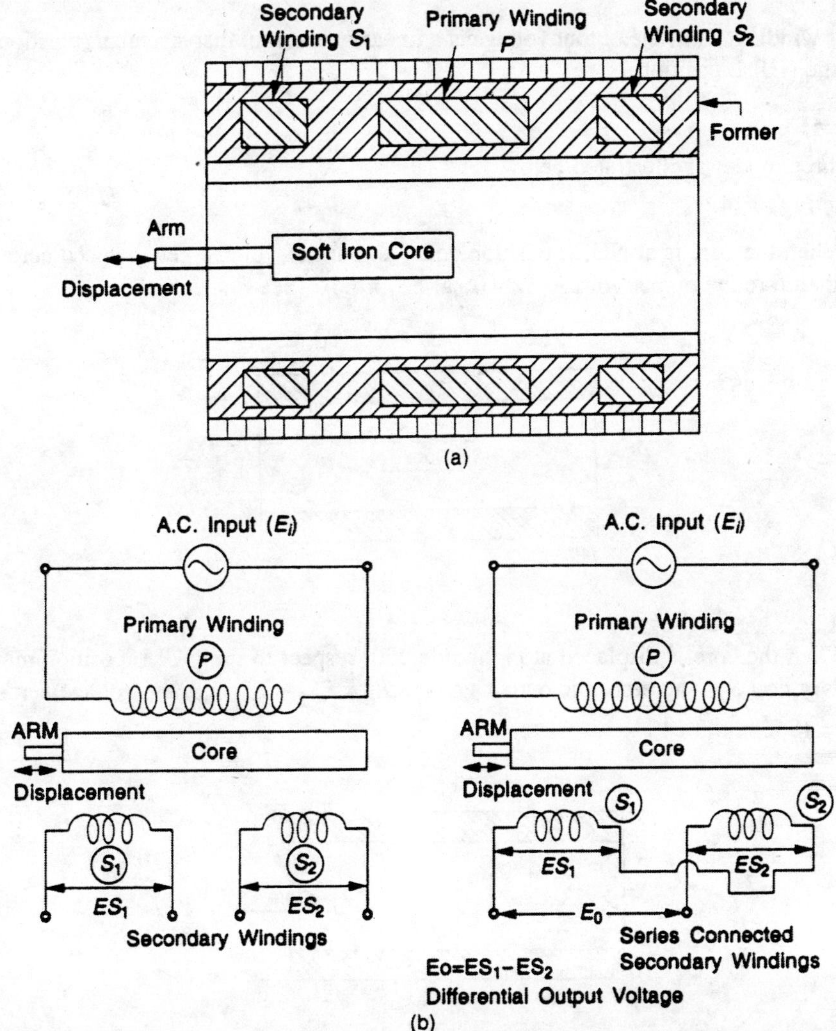

Fig. 3.13 : (a) Construction of linear variable differential transformer (b) Connections of LVDT windings.

A non-magnetic rod is connected to the core for applying the displacement or linear motion which is to be measured. The sinusoidal voltage having range of 5 to 25 Volts and 60 Hz to 20 kHz frequency is applied to the primary winding.

This complete assembly is placed in a stainless steel housing to get electrostatic and electromagnetic shielding. Due to this stray electric and magnetic effect get minimized.

As the primary winding is excited by sinusoidal voltage, an alternating magnetic field get induced. This magnetic field induces alternating voltages in two secondary windings. Voltage induced depends upon the position of magnetic core with respect to NULL condition. The NULL position at which output voltage E_0 is zero.

The two secondary coils are connected in series opposite mode as shown in Fig. 3.13(b). Let the voltage induced at secondary coil S_1 is ES_1 and to that S_2 is ES_2. Output voltage is E_0.

When the core moves from NULL position then the mutual inductance is more for one winding and smaller for other winding. Thus the output voltage is a linear function of displacement (position) of the core with respect to the NULL position.

Working of LVDT

Is divided into three modes as illustrated below.

Consider the Fig. 3.14.

Mode 0 : When the core is at NULL position equal amount of voltage gets induced across secondary coils S_1 and S_2 therefore the output voltage $E_0 = 0$ *i.e.* $ES_1 = ES_2$. (See Fig. 3.14(a))

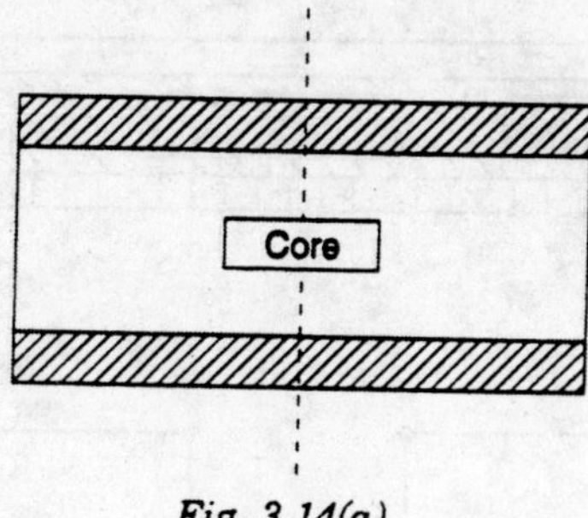

Fig. 3.14(a)

Mode 1 : When the core is displaced at **right** side with respect to the NULL position more flux gets linked in secondary coil S_2. Therefore the output voltage $E_0 = ES_2 - ES_1$, *i.e.* the voltage across coil S_2 is more than coil S_1. (See Fig. 3.14(b))

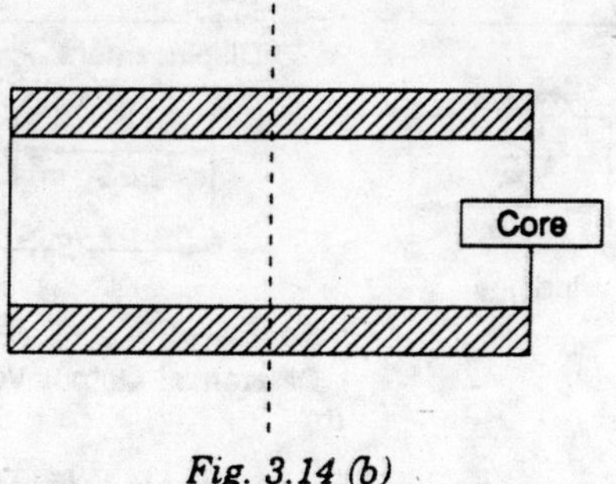

Fig. 3.14 (b)

Mode 2 : When the core is moved to **left** side with respect to the "NULL" position more flux gets linked in secondary coil S_1. Therefore the output voltage $E_0 = ES_1 - ES_2$, *i.e.* the voltage across coil S_1 is more than coil S_2. (See Fig. 3.14 (c))

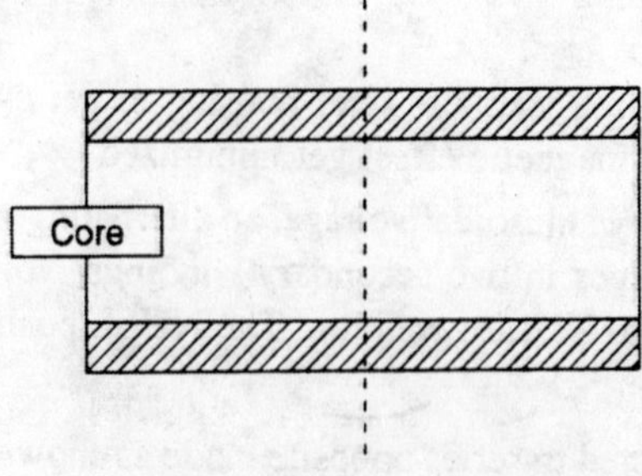

Fig. 3.14 (c)

This indicates that the output *i.e.* differential voltage is out of phase of by 180° when the core passes through NULL position. The output voltage E_0 across secondary windings is proportional to the linear displacement applied to the movable core. The output voltage may be applied to recorder or controller for analysis.

The displacement can be measured by comparing the amplitude and phase of differential output voltage with the amplitude and phase of the excitation or source voltage and direction of movement.

Fig. 3.15 shows the transfer characteristics (output voltage V_S displacement) of LVDT. It indicates that the output voltage E_0 is linear with the displacement of core but there is phase shift of 180° when core moves through NULL position.

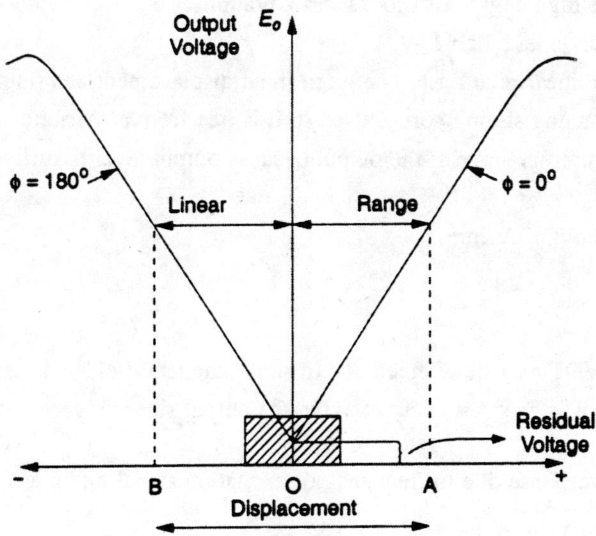

Fig. 3.15 : Output voltage Vs displacement characteristics.

This characteristics is linear from $O - A$ and $O - B$. Then it becomes non-linear

(i) At point 'O' $ES_1 = ES_2$ that is $E_0 = 0$ *i.e.* NULL Position.

(ii) When core is moved to left, ES_1 is more than ES_2 *i.e.* positive E_0 so phase angle ϕ is 0°. (Position *A*)

(iii) If core is shifted to right, ES_2 is greater than ES_1 so E_0 is negative. The output voltage is phase shifted by 180°. (Position *B*)

At NULL position the output voltage is $E_0 = 0$ but there is small voltage at NULL position because of residual voltage. This residual voltage is due to following reasons :

(1) stray electric and magnetic fields between windings.

(2) harmonics present in the excitation voltage (E_i).

(3) temperature effects.

(4) improper isolation between windings.

These problems of residual voltage can be minimized by using low pass filters, proper grounding circuits, shunt capacitive and resistive circuits, isolation techniques etc.

Applications of LVDT :

(1) It is used as a secondary transducer for the measurement of force, pressure, load.

e.g. : Pressure measurement using Bourdon tube and LVDT. Here Bourdon tube acts as a primary transducer which converts the pressure into displacement. This displacement is converted into voltage by means of LVDT as a secondary transducer.

(2) Mainly LVDT is used for displacement measurement with the range of 1·25 mm to 250 mm. It converts directly displacement into electrical voltage signal.

(3) Used in thickness measurement, level indicators, numerical machines etc.

Major advantages of LVDT :

(1) LVDT can tolerate high degree of shocks and vibrations.

(2) Power consumption is less than 1 W.

(3) Good linearity *i.e.* linear relationship between input displacement and output voltage.

(4) Separate non-contacting sliding core is used so it is free from any frictional losses.

(5) No necessity of amplifier stage at the output because output level is sufficiently high.

(6) Good resolution.

(7) High sensitivity about 40 V/mm.

(8) Low hysteresis.

Disadvantages of LVDT :

(i) Performance of LVDT may be affected due to stray magnetic fields and temperature.

(ii) Demodulator circuit is necessary for receiving the output signal if receiving instrument operates on d.c.

(iii) Limited dynamic response due to frequency of excitation signal and mass of the sliding core.

Sensitivity of LVDT (S)

The sensitivity of LVDT depends upon the constructional parameters, excitation supply frequency, primary winding current and the linearity, low voltage at NULL position.

Sensitivity is defined as :

$$S = \frac{e_0 \, e_i}{d} \qquad \text{...(3.18)}$$

where ; d = displacement (cm)

e_0 = output voltage across secondary winding (Volt)

e_i = excitation voltage at primary winding (Volt)

Generally, S = 1 to 5 V/V/cm.

(b) Rotary Variable Differential Transformer (RVDT)

This type of transducer is used for the measurements of angular displacements. It converts the angular displacements into electrical signals. The operating principle of RVDT is similar to LVDT. Construction of RVDT is same as LVDT *i.e.* one primary winding P and two secondary windings S_1 and S_2. Only the core is cam shaped and rotated inside the windings through a shaft. Fig. 3.16 shows the constructional details of RVDT.

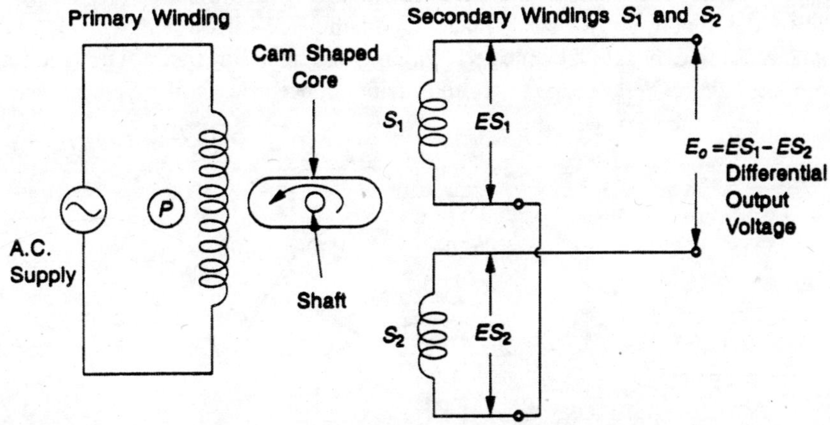

Fig 3.16 : Constructional details of RVDT.

The core is rotated in clockwise and anticlockwise directions with respect to NULL position. When the core is at null position the output voltage is zero because the voltages induced in secondary windings S_1 and S_2 are equal and in opposite direction. If any angular displacement which is to be measured with respect to null position is applied to the shaft differential voltage gets developed across the secondary windings at output. When the core is rotated in clockwise direction the voltage gets increased, if the core is rotated in anticlockwise direction voltage increases but there is phase opposition in the voltage developed in the two directions of the core.

In both directions output voltage is increasing but there is phase opposition w.r.t. each other according to position of shaft (clockwise or anticlockwise).

Comparison of LVDT and RVDT

	LVDT		RVDT
1.	LVDT converts linear (motion) displacements into electrical signals.	1.	RVDT converts angular (rotary) displacements into electrical signals.
2.	Its core is placed in a horizontal former. (in linear fashion)	2.	Its core is cam shaped.
3.	Used for converting linear motions into electrical voltage signals. (Left and Right side displacements).	3.	Used for converting angular (clockwise and anticlockwise) motions into electrical voltage signals.

(c) Synchros

Basically synchro is an electromagnetic device. It converts the angular position or displacements into electrical signals.

It is used in angular measurements, telemetering systems, angular data transmission, servomechanisms, control component, etc.

The basic structure of synchro is called as **synchro transmitter**. Fig. 3.17(a) shows the constructional details of synchro transmitter. It consists of a stator made up of laminated silicon steel with slots. These slots are provided for balanced three phase *Y*-connected winding mounted in stator with 120° apart. The construction of rotor is dumb-bell shaped wound with concentric coil. Let ar a.c. voltage $V_r(t) = V_r \sin w_c t$ is applied to the rotor winding.

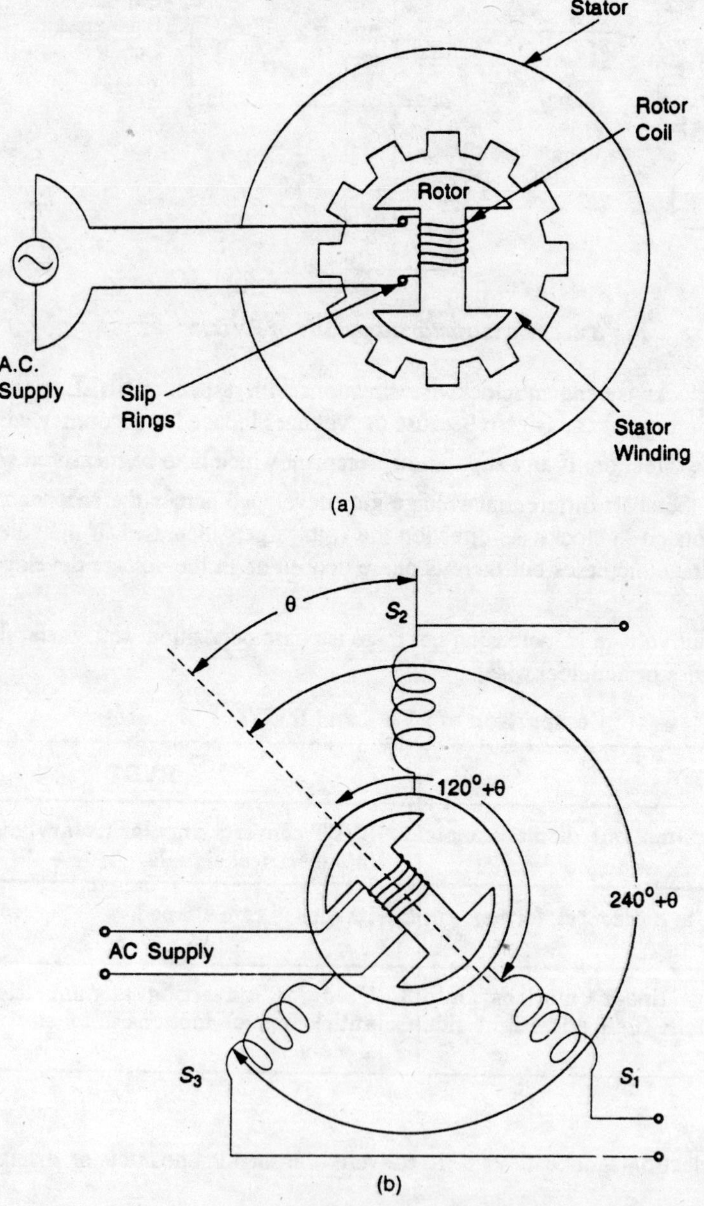

Fig. 3.17 : (a) Constructional features of synchro transmitter.
(b) Schematic diagram of synchro transmitter.

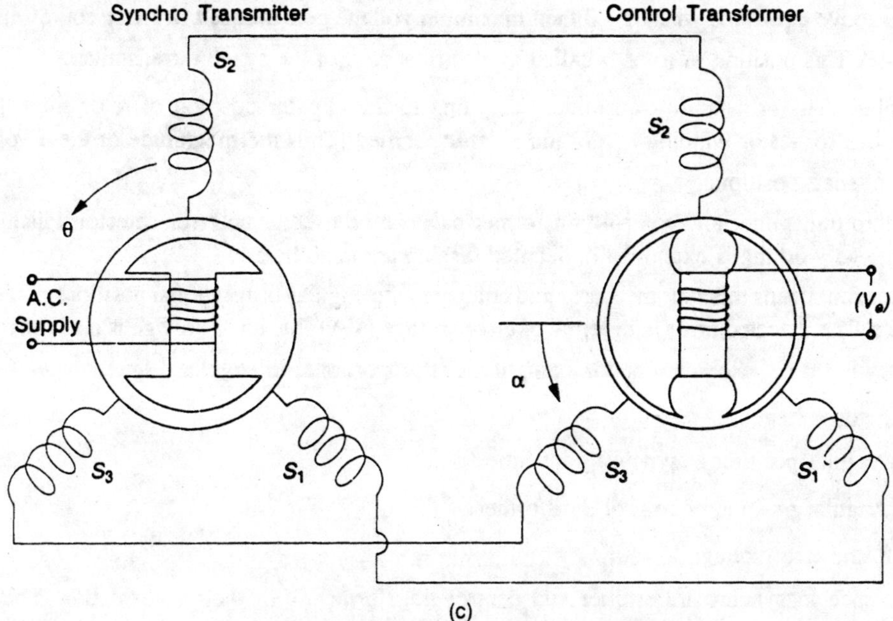

Fig. 3.17 : (c) Synchro error detector.

The rotor winding is supplied with a.c. voltage. Due to this voltage the magnetizing current flows and time varying flux gets produced. The same flux gets distributed in the air gap with stator. According to transformer action the voltage gets induced in the stator coils. The synchro transmitter acts as single phase transformer having rotor as a primary winding and stator as a secondary winding.

The flux linking in stator coil is proportional to the cosine of angle between rotor and stator coil axes. So the voltage gets induced in stator coil, as stated in earlier discussion.

Consider Fig. 3.17(b).

Let, S_1, S_2 and S_3 are the stator coils.

v_{s1n}, v_{s2n} and v_{s3n} are the voltages induced in stator coils S_1, S_2 and S_3 respectively, with respect to neutral.

θ = angle between rotor axis and stator coil S_2.

$$v_{s1n} = KV_r \sin w_c t \cos(\theta + 120°) \quad ...(3.19)$$

$$v_{s2n} = KV_r \sin w_c t \cos\theta \quad ...(3.20)$$

$$v_{s3n} = KV_r \sin w_c t \cos(\theta + 240°) \quad ...(3.21)$$

The stator terminal voltages are :

$$v_{s1s2} = v_{s1n} - v_{s2n} = \sqrt{3}\ KV_r \sin(\theta + 240°) \sin w_c t \quad ...(3.22)$$

$$v_{s2s3} = v_{s2n} - v_{s3n} = \sqrt{3}\ KV_r \sin(\theta + 120°) \sin w_c t \quad ...(3.23)$$

$$v_{s3s1} = v_{s3n} - v_{s1n} = \sqrt{3}\ KV_r \sin w_{ct} \sin\theta \quad ...(3.24)$$

From the above equations when $\theta = 0$ then maximum voltage gets induced in stator coil S_2 at the same instant $v_{s3s1} = 0$. This position of rotor is called as electrical zero of the synchro transmitter.

This implies that for synchro transmitter according to the angular position of rotor shaft the output voltage equations for stator winding S_1, S_2 and S_3 are specified. Thus the magnitude of these voltages are the functions of shaft position.

The synchro transmitter and control transformer pair can be used as an error detector which is shown in Fig. 3.17(c) and working is explained in detailed fashion as stated below.

Error detection means this pair measures and compares the angular or rotational positions of a load with some reference. The detected error is in terms of error voltage (V_e). This error voltage is a.c. voltage having same frequency to that of excitation and the amplitude is proportional to angular difference *i.e.* $(\theta_r - \theta_c)$.

where ; V_e = Error voltage.

θ_r = Angular position of synchro transmitter.

θ_c = Angular position of control transformer.

when $\theta_r = \theta_c$ the error voltage $V_e = 0$. ...(3.25)

The difference in synchro transmitter and control transformer is in their construction. The rotor of synchro transmitter is dumb-bell shaped whereas the rotor shape of control transformer is cylindrical for uniform air gap.

3.8 Photoelectric Transducers

The generation of electrons due to influence of light is called as photoelectric effect. The photoelectric effect includes :

(A) Generation of electrons from metallic surface.

(B) Generation of electron-hole pairs in semiconductor materials when subjected to radiations.

The photoelectric devices are :

(i) Photo emissive in which as the radiations falls on the of cathode, it causes the emission of electrons form the surface of cathode.

(ii) Photoconductive in which the resistance of a material changes in accordance with the illumination.

(A) Photomultiplier Tube

In many industrial applications, nuclear radiation detection, television pick up devices and astronomy it is necessary to measure very weak light intensities. In such applications very sensitive photomultiplier tube is used. This tube uses the criterion of amplifying the current from photoelectric surface by using secondary emission.

The Fig. 3.18(a) shows a photomultiplier tube. It consists of a photo cathode, an anode and number of electrodes called as dynodes. This complete assembly is placed in evacuated glass enclosure.

When light impinges upon a photosensitive cathode, it emits the photoelectrons, these electrons are attracted to the first dynode D_1. Electrons which strikes the dynode causes the secondary emission of electrons. Then emitted electrons gets attracted towards next dynode D_2, again the secondary emission takes place. This process is repeated at every subsequent dynodes $D_2, D_3, ..., D_n$. The high velocity is provided to

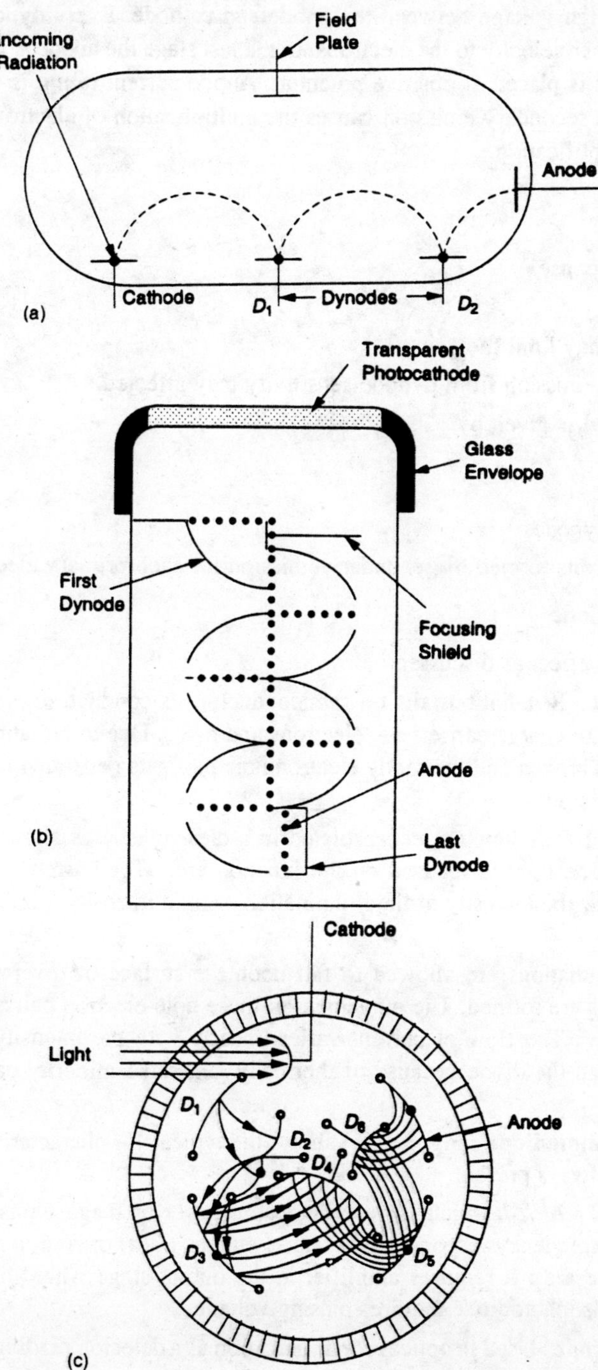

Fig. 3.18(a) Working principle of photomultiplier.
(b) Secondary emission of electrons in linear mode. (Linear Photomultiplier)
(c) Arrangement of photomultiplier tube in circular fashion.

the electrons by applying high voltage between the anode and cathode. Every dynode is placed at higher voltage so as to provide higher velocity to the electrons. In the last stage the anode A is placed which collects all the electrons. The anode is placed at positive potential. Anode current rating is 100 μA to 1 mA. This discussion indicates that the secondary emission causes the multiplication of electrons. The dynode section provides the increase in amplification.

Advantages :

(1) Stable performance.

(2) Good frequency response.

Limitations :

(1) Imperfect vacuum may limit the gain.

(2) Due to thermoionic emission from dynode sensitivity may affected.

The total amplification A is given by,

$$A = g^n \qquad \text{...(3.26)}$$

where ; n = are number of dynodes.

g = number of electrons formed by secondary emission for each primary electron.

(B) Semiconductor Photodiode

It uses photoconductive effect as discussed below.

Photoconductive effect : If radiation falls on semiconductor, its conductivity increases. Conductivity depends upon concentration of charge carriers *i.e.* electrons and holes. Due to radiation energy the covalent bonds in semiconductor gets broken and thermally electron hole pair gets generated *i.e.* current. The current increases and resistance decreases.

Photodiode consists of a P–N junction encapsulated in a clear plastic as shown in Fig. 3.19(a). This junction is reverse biased. The light is focused on the junction area. The current across a reverse biased p–n junction is determined by the intensity of the illumination or light intensity. Fig. 3.19 (b) shows symbol of photodiode.

When light shines or radiations are allowed to fall upon the surface of reverse biased p–n junction additional electron-hole pairs are formed. The movement of these hole-electron pairs in properly connected circuit results in current flow. The flow of current varies linearly with the intensity of light. In the dark leakage current flows through the diode because of thermally generated minority carriers. This current is called as Dark current.

***V-I* characteristics of photodiode** : Fig. 3.19(c) shows the typical V-I characteristics of photodiode. In dark the leakage current is always present.

The output resistance $R = \Delta V/\Delta I$ which is very high upto order of mega ohms. The photodiode has following advantages-better frequency response, lower noise effects, linear operation and fast response time. Disadvantages of photodiode are : it requires amplifier at the output stage when the intensity of light or illumination is at lower levels, photodiode requires biasing voltage.

Applications of photodiode : Used in optical communication as a detector, reading of film sound track's, object counters, in switching circuits etc.

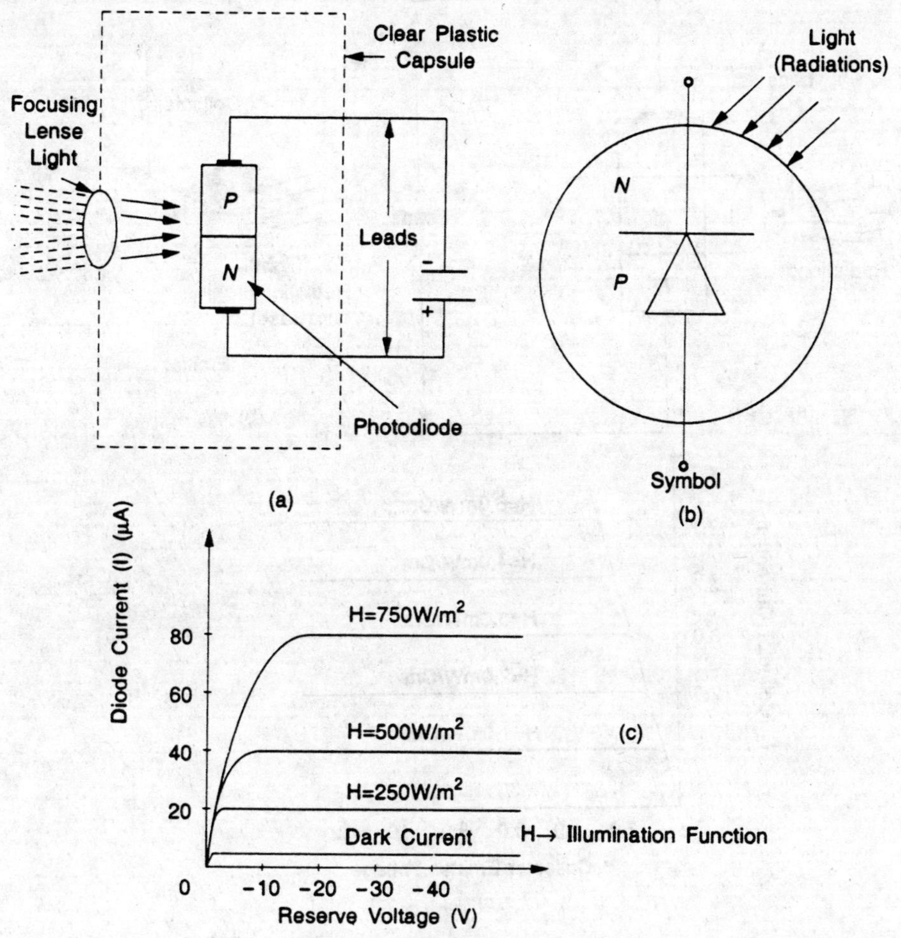

Fig 3.19 : (a) Construction of photodiode, (b) Symbol (c) V-I characteristics

(C) Phototransistor

Phototransistor is a light detector which is a combination of photodiode and transistor amplifier.

The Fig. 3.20(a) shows NPN phototransistor. Here a lens is used to concentrate the light on P-type wafer. This P-type is sandwiched between an N-type emitter and collector. Phototransistor has two terminals–collector and emitter, the base current is generated by means of illumination falling on base – emitter junction. This generated current is amplified by transistor action. Due to this the sensitivity of the device is increased.

***V-I* characteristics of phototransistor** : Fig. 3.20(c) shows a graph of collector current I_c as a function of collector–emitter voltage and as a function of illumination H. Totally, the current depends upon illumination level, voltage applied to the transistor thus it acts as a constant current source.

The advantages of phototransistor are high switching action, higher gain, low power consumption.

Applications : It is used in digital applications, object counters, industrial circuits, alarm units etc.

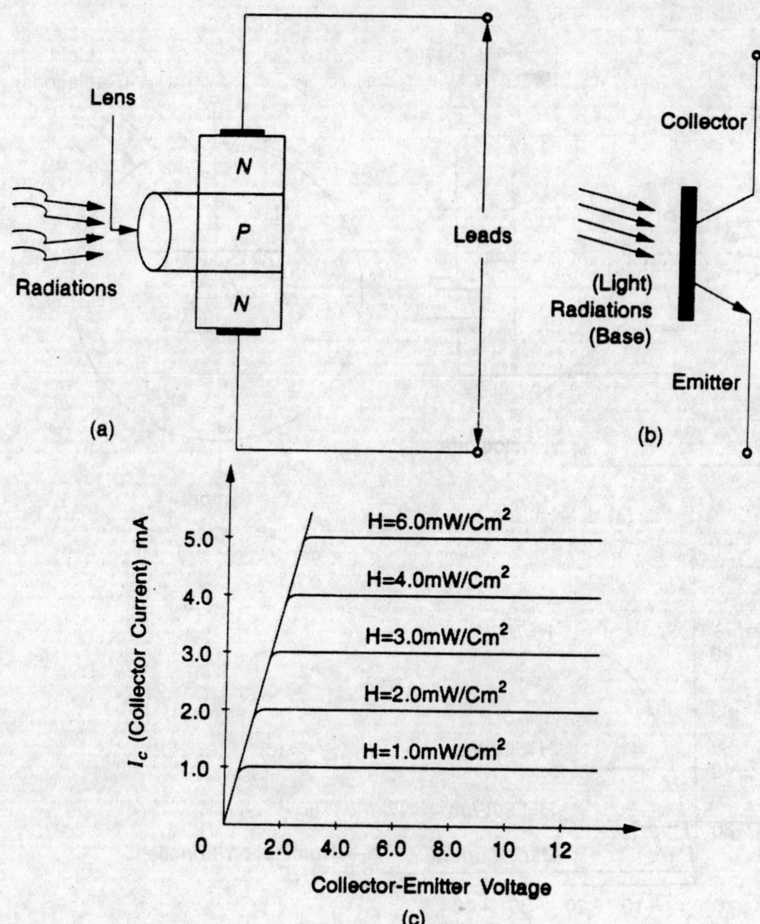

Fig. 3.20 : (a) Construction, (b) Symbol of phototransistor, (c) V-I characteristics.

3.9 Piezoelectric Transducer

Fig. 3.21(a) shows piezoelectric transducer. These transducers works on the principle of **Piezoelectric effect** which is explained as follows.

When a mechanical force is applied to piezoelectric material (crystal) the dimensions of these materials gets changed resulting the generation of electric charge or electric potential across the surface of crystal.

In the reverse manner, if electric charge or electric potential is applied across the surface crystal it results the change in dimensions of crystal due to deformation. This principle is used in piezoelectric transducers for the measurement of pressure, force and acceleration *i.e.* mechanical input and electrical output.

The piezoelectric materials used are classified as follows :

(1) **Natural** : Quartz crystal, Rochelle salt.

(2) **Synthetic group** : Lithium sulphate, barium titanate, dipotassium tartarate, ethylene diamine tartrate etc.

Polymer films can be used as piezoelectric materials.

For piezoelectric effect the crystal should have natural asymmetrical charge distribution. Because of this

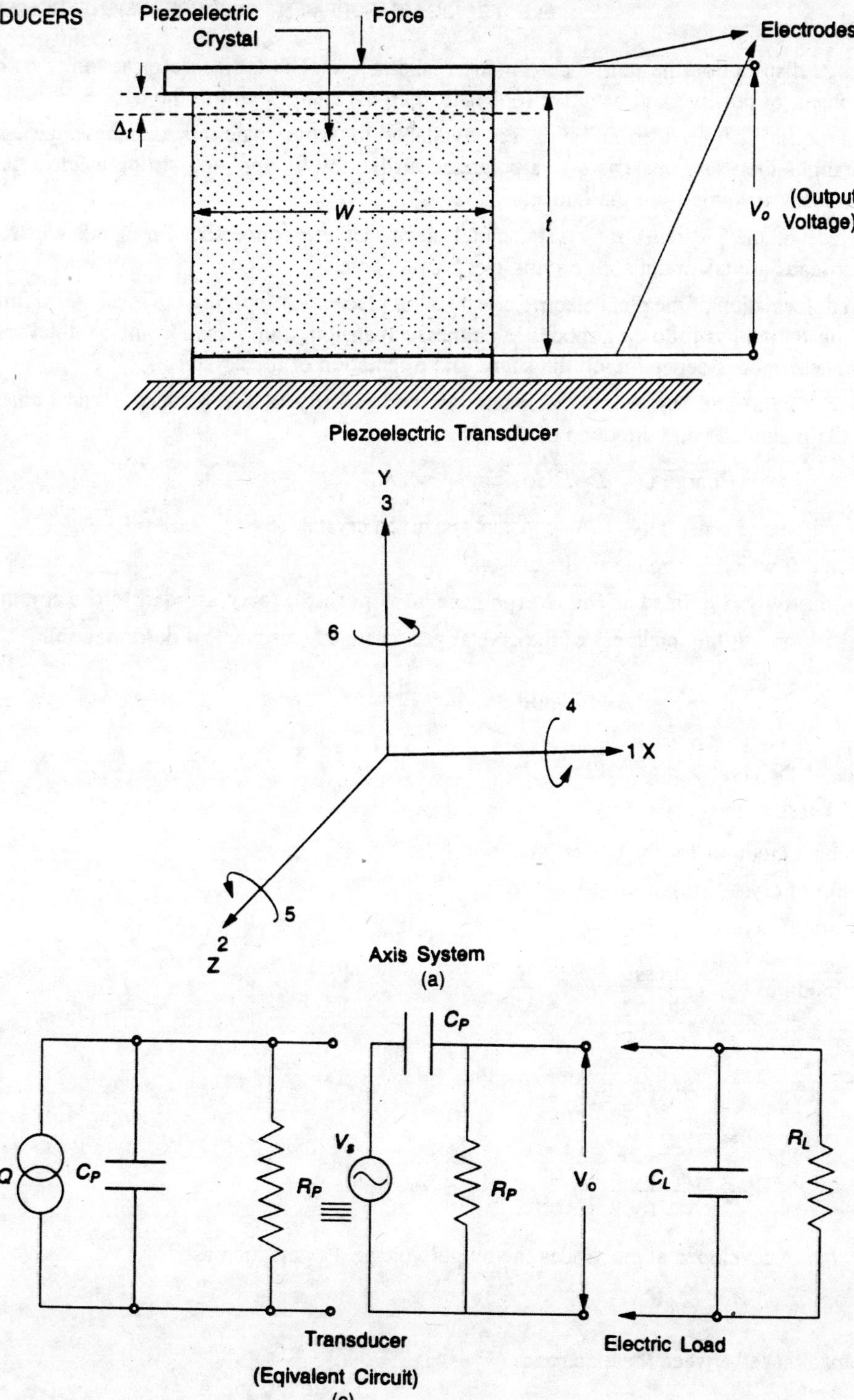

Fig. 3.21 : (a) Piezoelectric transducer, (b) Axis system, (c) Equivalent circuit.

asymmetric charge distribution the lattice deformation takes place. The lattice deformation is nothing but relative displacement of positive and negative (opposite charges) charges within the lattice.

If the asymmetric charge distribution is not available in some materials (synthetic group) *e.g.* : ferroelectric ceramics then such materials gets polarized artificially by applying strong electric field. This electric field produces asymmetry in the lattice.

On the surface of the piezoelectric crystal metal electrodes are connected. From this electrodes the metallic wires or leads are taken out for deriving electrical charge.

Mechanical deformation of the piezoelectric crystal takes place after applying the mechanical force. This deformation of the crystal has following modes : thickness expansion, transverse expansion, thickness shear and face shear. These modes depend upon the shape and orientation of the crystal.

Consider Fig 3.21(a) and (b). Here, the magnitude and polarity of the induced surface charges are proportional to the magnitude and direction of the applied force F.

$$\text{Charge } Q = d \times F \text{ Coulomb} \qquad ...(3.27)$$

where ; d = Crystal charge sensitivity (C/N) constant for given crystal.

F = Force (N) which is applied to the crystal.

The charge senstivity is defined as **the charge generated per unit force applied to the crystal.**

Due to applied force F the thickness of the crystal gets changed (mechanical deformation).

$$F = \frac{A \,.\, Y}{t} \Delta t \text{ Newton} \qquad ...(3.28)$$

where ; A = Area of the crystal (m^2) = WL

t = Thickness of the crystal (m)

Y = Young's modulus for the crystal material (N/m^2)

W = Width of crystal (m)

L = Length of crystal (m)

$$\text{Young's modulus } (Y) = \frac{\text{stress}}{\text{strain}} = (F/A)\, \frac{1}{\Delta t/t}$$

$$Y = \frac{F \,.\, t}{\Delta t \,.\, A} \text{ N/m}^2$$

$$Q = d \times F$$

$$Q = \frac{d \,.\, A \,.\, Y}{t} \cdot \Delta t \qquad ...(3.29)$$

Due to the charge developed at electrodes the output voltage V_0 gets increased.

$$V_0 = Q/C_p \qquad ...(3.30)$$

where ; C_p = Capacitance between the electrodes (F)

$$C_p = \frac{\varepsilon_r \, \varepsilon_0 \, A}{t} \qquad ...(3.31)$$

$$V_o = \frac{Q}{C_p} = \frac{d.F}{\varepsilon_0 \varepsilon_r A/t} = \frac{d.t.F}{\varepsilon_0 \varepsilon_r A} \quad ...(3.32)$$

But ; F/A = Pressure (N/m^2) (P)

$$V_0 = \frac{d.t}{\varepsilon_0 \varepsilon_r} \cdot P$$

$$V_0 = g.t.P \quad ...(3.33)$$

where ; $\frac{d}{\varepsilon_0 \varepsilon_r} = g$: which is voltage sensitivity of the crystal in V_m/N (constant for given crystal).

The voltage sensitivity (g) is defined as **the field produced per unit stress.**

$$V_0 = g.t.P \quad ...(3.34)$$

$$g = \frac{V_0}{t.P} = \frac{V_0/t}{P} \quad ...(3.35)$$

V_0/t = Electric field strength V/m

P = Pressure applied to the crystal.

Suppose ; $\varepsilon = \varepsilon_0/t$ = electric field

$$g = \varepsilon/P = \frac{\text{Electric field}}{\text{Stress}} \quad ...(3.36)$$

The charge sensitivity ;

$$d = \varepsilon_r \varepsilon_0 g \ \ C/N \quad ...(3.37)$$

For barium titanate the value of voltage sensitivity (g) is ; $g = 12 \times 10^{-3}$ V_m/N and charge sensitivity is 150 pc/N.

Equivalent circuit of piezoelectric transducer : Fig. 3.21(c) shows the equivalent circuit of piezoelectric transducer. The voltage V_0 is developed due to generation of charges, which is given by ;

$$Q = d \times F$$

This charge is generated across the capacitance C_p and leakage resistance R_p of the crystal. The circuit detail is voltage supply V_s in series with C_p and R_p.

Let the piezoelectric transducer be loaded by C_L and load resistance R_L. (C_L is the capacitance of load, capacitance of cable and stray capacitance)

$$\frac{V_0}{F} = d\left[\frac{j\omega R_L}{1 + j\omega (C_P + C_L) R_L}\right] \quad ...(3.38)$$

where ; $R_L < R_p$.

The impedance of the transducer is high due to this reason high impedance amplifier is used for buffering instead of the high voltage gain amplifier. In such type of transducer the charge amplifiers are used. Otherwise the transducer are neatly connected to the virtual earth point of the amplifier having feedback capacitor.

The features of piezoelectric transducer are :

(1) more stability.

(2) not affected by temperature variations and humidity.

(3) maximum output.

(4) useful for dynamic parameter measurements.

(5) quartz crystal has good stability so it is used as frequency generator or oscillator.

Applications : In vibration pickups, accelerometer, pressure sensors, velocity measurement, sound pressure measurement.

3.10 Strain Gauges

Strain is defined as elongation or compression per unit area. Strain gauges works on the principle of **Piezoresistivity**.

If a metal wire or conductor is stretched or compressed its resistance changes because of change in length, change in resistivity and change in cross-sectional area. This effect is called as piezoresistive effect.

Suppose ; A = Uniform cross-sectional area of wire (m^2)

L = Length of wire (m)

ρ = Resistivity (Ω – m)

R = Resistance (Ω)

are the dimensions of wire.

Therefore, the resistance of such wire is

$$R = \frac{\rho L}{A} \qquad ...(3.39)$$

If the conductor is stretched or compressed (strained) the above mentioned parameters get changed as follows :

ΔA = Change in cross-sectional area

ΔL = Change in length of wire

$\Delta \rho$ = Change in resistivity

ΔR = Change in resistance

According to the principle of strain gauges find out ∂R by differentiating equation (3.39) with respect to applied stress 'S'.

$$\therefore \quad \frac{\partial R}{\partial S} = \frac{\rho}{A}\frac{\partial L}{\partial S} - \frac{\rho L}{A^2}\frac{\partial A}{\partial S} + \frac{L}{A}\frac{\partial \rho}{\partial S} \qquad ...(3.40)$$

Divide equation (3.40) throughout by $R = \frac{\rho L}{A}$

$$\frac{1}{R}\frac{\partial R}{\partial S} = \frac{1}{L}\frac{\partial L}{\partial S} - \frac{1}{A}\frac{\partial A}{\partial S} + \frac{1}{\rho}\frac{\partial \rho}{\partial S} \qquad ...(3.41)$$

From equation (3.41) it is concluded that the change in resistance is due to –(1) per unit change in length $\Delta L/L$, (2) per unit change in area $\Delta A/A$, (3) per unit change in resistivity $\Delta \rho/\rho$.

Area of wire is given by $A = \frac{\pi}{4} D^2$.

Where D = Diameter of the wire.

$$\therefore \quad \frac{\partial A}{\partial S} = \frac{\pi}{4} . 2D . \frac{\partial D}{\partial S} \qquad ...(3.42)$$

Substitute equation (3.42) in equation (3.41).

$$\therefore \quad \frac{1}{R}\frac{\partial R}{\partial S} = \frac{1}{L}\frac{\partial L}{\partial S} - \frac{2}{D}\frac{\partial D}{\partial S} + \frac{1}{\rho}\frac{\partial \rho}{\partial S} \qquad ...(3.43)$$

The Poisson's ratio is given by

$$\gamma = -\frac{\partial D/D}{\partial L/L} = \frac{\text{Lateral strain}}{\text{Longitudinal strain}}$$

$$\therefore \quad \frac{\partial D}{D} = -\gamma \times \frac{\partial L}{L} \qquad ...(3.44)$$

$$\therefore \quad \frac{1}{R}\frac{\partial R}{\partial S} = \frac{1}{L}\frac{\partial L}{\partial S} + \frac{2\gamma}{L}\frac{\partial L}{\partial S} + \frac{1}{\rho}\frac{\partial \rho}{\partial S}$$

For small variations,

$$\frac{\Delta R}{R} = \frac{\Delta L}{L} + 2\gamma\frac{\Delta L}{L} + \frac{\Delta \rho}{\rho} \qquad ...(3.45)$$

The term gauge factor is defined as ratio of per unit change resistance to per unit change in length.

$$\therefore \quad \text{Gauge factor } G_f = \frac{\Delta R/R}{\Delta L/L} \qquad ...(3.46)$$

$$\therefore \quad \frac{\Delta R}{R} = G_f . \frac{\Delta L}{L} = G_f \times \varepsilon$$

$\varepsilon \Rightarrow$ Strain = $\Delta L/L$

∴ The gauge factor is,

$$= 1 + 2\gamma + \frac{\Delta\rho/\rho}{\varepsilon} \qquad ...(3.47)$$

$$\therefore \quad G_f = \frac{\frac{\Delta R}{R}}{\frac{\Delta L}{L}} = 1 + 2\gamma + \frac{\Delta\rho/\rho}{\Delta L/L} \qquad ..(3.48)$$

Here ; 1 = Change in resistance due to change of length.

2γ = Change in resistance due to change in area.

$\frac{\Delta\rho/\rho}{\varepsilon}$ = Change in resistance due to piezoresistive effect.

If change in resistivity of material due to strain is neglected the gauge factor is,

$$G_f = 1 + 2\gamma \qquad ...(3.49)$$

For, Nickel gauge factor is − 12·1

Nichrome gauge factor is + 2·0

Soft iron gauge factor is + 4·2

Carbon gauge factor is + 20

Platinum gauge factor is + 4·8

Example 3.1 : A resistance wire strain gauge is constructed using nichrome material with gauge factor $G_f = 2{\cdot}0$ calculate its Poisson's ratio. (Neglect piezoresistive effects).

Solution :

The gauge factor is specified as $G_f = 1 + 2\gamma + \frac{\Delta\rho/\rho}{\varepsilon}$.

As mentioned in problem piezoresistive effect is neglected.

$$\therefore \quad G_f = 1 + 2\gamma$$

$$\therefore \quad \gamma = \frac{G_f - 1}{2}$$

Given data $G_f = 2{\cdot}0$

$$\therefore \quad \gamma = \frac{2-1}{2}$$

$$\therefore \quad \gamma = \frac{1}{2} = 0{\cdot}5 \qquad (\therefore \text{ Poisson's ratio} = 0{\cdot}5)$$

Metals used for strain gauges are copper-nickel, nickel-iron, nichrome-constantan alloys about 0·03 mm in diameter with gauge factor of 2 to 4 and the gauge current rating is 5 to 40 mA. Strain gauges are mainly used in (1) the measurement and analysis of force, pressure, torque acceleration, displacement (2) in stress measurement and processing of machine and structures.

(a) Classification of Strain Gauges

(A) **Bonded strain gauges** : They are useful for stress analysis and transducer construction.

(i) **Bonded wire strain gauge** : A gird or fine wire of resistance is cemented to a base (sheet) of paper or sheet or teflon or bakelite. Particular resistance wire is covered on top to protect it from mechanical damages and faithfully handle the compression and elongation (stress). The specifications of strain gauges varies with application. Strain gauges should have high gauge factor G_f, low resistance temperature coefficient, no hysteresis and resistance should have linear function of strain.

Different types of wire strain gauges are (1) Linear (2) Rossette (3) Torque (4) Helical as shown in Fig 3.22 (a), (b), (c) and (d).

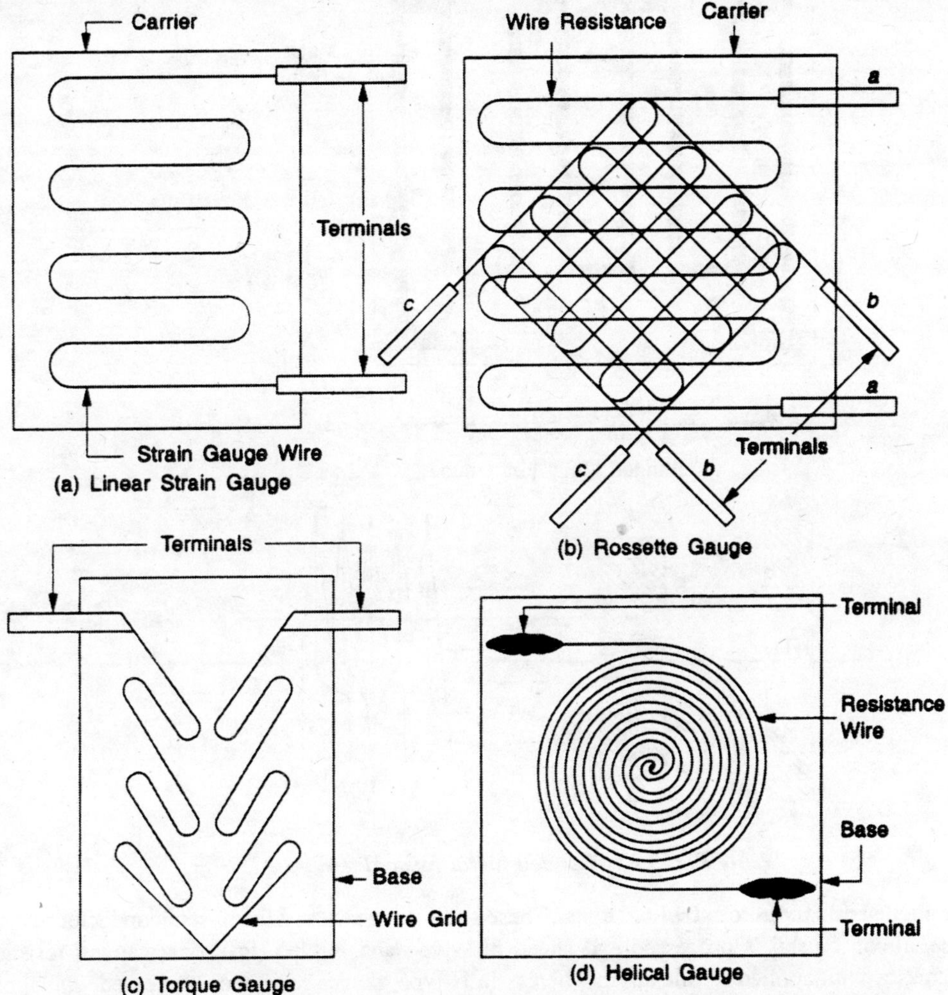

Fig. 3.22 : Types of resistance wire strain gauges.

(ii) **Bonded metal foil strain gauges** : They are made up of the similar material used for bonded wire strain gauges. The gauge elements are formed by photoetching process having dimensions less than 0·0002 inch. thick. Fig. 3.23 shows foil type strain gauges.

In foil type strain gauges surface area is increased for the same volume which facilitates much greater heat dissipation capacity and better bonding. Foil type gauges are mounted on a flexible insulating carrier film about 0·025 mm thick made up of polymide, glass and phenolic etc.

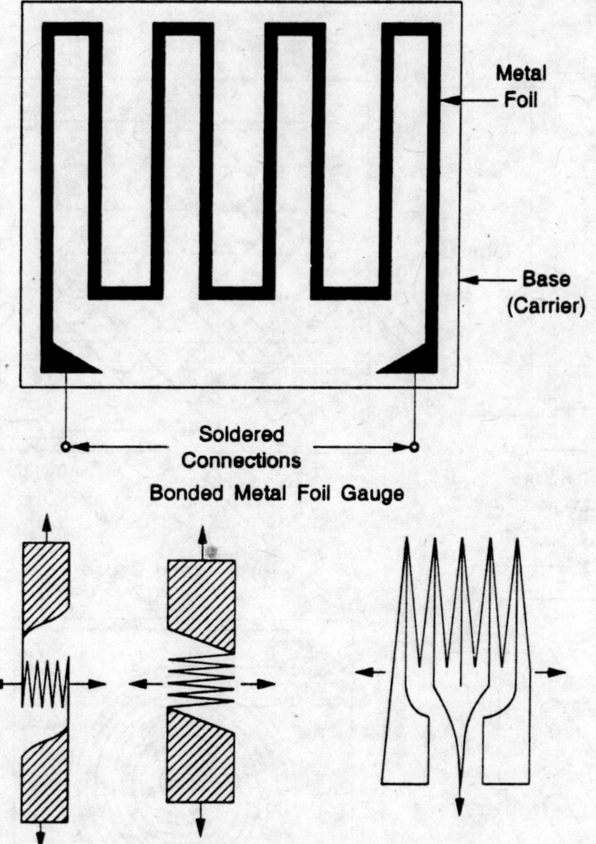

Fig. 3.23 : Bonded metal foil gauges.

(iii) **Bonded semiconductor strain gauges** : These strain gauges uses **Silicon** semiconductor material. The silicon crystal is processed and sliced in P-type and N-type. In P-type gauges, resistance increases with applied tensile strain whereas in N-type gauges resistance decreased with applied strain. Fig. 3.24. shows bonded semiconductor strain gauges.

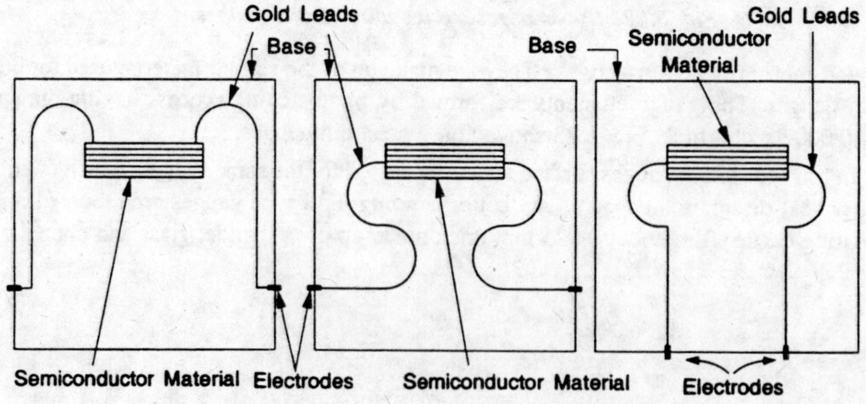

Fig. 3.24 : Semi-conductor strain gauge..

Major advantage of these gauges is high gauge factor of + 130. High gauge factor is affected by high temperature sensitivity, non-linearity and mounting difficulties. Small strain rating in terms of microstrain can be measured by using proper temperature compensation.

In this type of strain gauges the semiconductor material is sandwiched between protective matrix of teflon and leads are taken out.

Advantages of semiconductor strain gauges are :

(1) high gauge factor about ± 130 (2) possible to measure very small strain of the order of 0·001 microstrain (3) excellent hysteresis characteristics.

Disadvantages are –very sensitive to change in temperature, poor linearity, more expensive.

Rossettes : The combination of gauges is called as rossette. They are applicable for specific stress analysis or transducer applications.

Characteristics of strain gauges :

(1) The gauge factor G_f of strain gauge should be high so as to achieve higher sensitivity.

(2) Resistance of strain gauge should be as high as possible, due to which unwanted variations in the resistance in measurement gets minimized.

(3) For avoiding the errors and achieving more accuracy the strain gauges should have low resistance temperature co-efficient.

(4) Strain gauge should have linear relationship between strain and change in resistance due to applied strain.

(5) Strain gaugess should not have hysteresis effects.

(6) They should have linear frequency response.

Now a days strain gauges sensors are available in Integrated circuit (IC) formed with OP – AMP. This topic is discussed in Chapter 1 under the heading Microelectronics Instrumentation.

3.11 Temperature Transducers

Temperature measurement is widely useful technique in various industries and process plants. This is old technique but now a days various techniques and methods are in existence and under research. Temperature is fundamental quantity and totally different in nature from the quantities like mass, time, length and area.

The term temperature is defiined as **the degree of hotness or coldness of a substance or medium**. Following phenomenon is used for temperature measurements. Due to heating or cooling following effects are used for measurement (1) Change in physical or chemical state of the substance or medium. (2) Change in physical dimensions. (3) Change in electrical properties such as resistance. (4) If two dissimilar metal gets joined or welded, e.m.f. gets generated at the junction. (5) Change in the intensity of total radiation emitted. Temperature may also be defined as **degree of heat**. Above mentioned effects are used for temperature measurement transducers which are discussed in following sections.

(A) Temperature Scales

In this section different temperature scales are discussed but before that some primary discussion is necessary.

For defining standard temperature scales in numerical form some reference temperature should be selected or we must choose some reference temperature. With this, some rules should be defined for obtaining difference between reference temperature and other temperatures. So these reference temperatures are two fixed points which are (i) Lower fixed point or ice point (ii) Upper fixed point or steam point.

For lower fixed point the ice temperature is taken into account which is prepared from distilled water, when melting under a pressure of 760 mm of mercury.

The upper fixed point or steam point is temperature of steam from purely distilled water boiling under pressure of 760 mm mercury Temperature difference between ice point and steam point is called **fundamental interval**.

The temperature scales are (i) Fahrenheit and centigrade (Celsius) temperature scales. (ii) Kelvin and Rankine temperature scales.

(i) **Fahrenheit and centigrate scales** : Fahrenheit scale is indicated by °F and introduced in 1709 by a German philosopher Fahrenheit.

Centigrade scale is indicated by °C introduced in 1742 by Professor Celsius. Both scales are linear and can be converted from one scale to another using the equation. ;

$$\frac{°C}{1000} = \frac{°F - 32}{180} \quad ...(3.50)$$

For Fahrenheit temperature scales the melting point of ice is 32°F and 212°F is boiling point. Similarly for centigrade temperature scale melting point of ice is 0°C and boiling point is 100°C.

(ii) **Kelvin and Rankine temperature scale** : The Kelvin scale was introduced by Lord Kelvin in 1848 on this scale ice-point is 273·15°K and steam point is 373·15°K. For converting °C (centigrade) into °K (Kelvin) we use following equation.

$$°K = °C + 273{\cdot}15$$

The Rankine scale is indicated by '°R' on which ice point is 491·7°R and steam point is 671·7°R. The Fahrenheit (°F) can be converted into Rankine (°R) by the equation,

$$°R = °F + 459{\cdot}69 \quad ...(3.51)$$

DESIGNATION	METAL COUPLES USED		TEMPERATURE RANGE		MILLIVOLTS	
	Positive Wire	Negative Wire	Min (°C)	Max (°C)	Min	Max
B	Pt 70–RH 30	Pt 94–Rh 6	0	1860		
J	Iron	Constantan	–196	760	0·0527	0·05745
K	Chromel	Alumel	–190	1371	0·040	0·041
	Platinum 90% Rhodium 10%	Platinum	0	+1400		
T	Copper	Constantan	–190	399	0·04277	0·052
Y	Iron	Constantan	–129	982		
E	Chromel	Constantan	–280	+850		

Table 3.1 : (a) List of Industrial Thermocouples.

The Internationl Practical Temperature Scale (IPTS) is an international scale developed in 1929 and revised in 1948 and then in 1968. These scales are confirm, accurate and uniform for temperature measurements. IPTS is based on number of points such as boiling point, freezing point, triple point which can be used for calibration of temperature measuring instrument and devices. The calibration of given temperature-measuring device is carried out by—(1) Subjecting it to some established fixed-point environment such as melting or boiling points of standard substances. (2) By comparing its reading with more accurate, stable and calibrated temperature sensor or secondary standard.

(B) Thermocouple

This transducer is widely used in industrial applications for temperature measurement. Thermocouple is active transducer because there is no need of voltage source and transducer bridge circuitry.

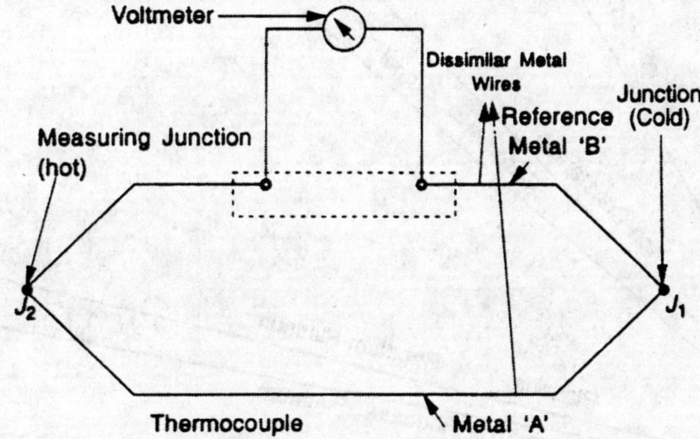

Fig. 3.25 : Thermocouple working principle

Thermocouple works on the following principle :

"When two dissimilar metals A and B are welded or joined together to form a closed circuit and the junctions (J_1 and J_2) are kept at two different temperatures (T_1 and T_2) then an *e.m.f.* is generated resulting flow of current in the loop or circuit". The two junctions in the loop are reference or cold junction which is generally kept at 0°C and the other is hot junction at which the temperature is to be measured as shown in Fig. 3.25.

The *e.m.f.* (E) generated is proportional to the difference of temperatures (T_1 and T_2), the materials used for thermocouple. This phenomenon is called as **Seebeck effect**. Thus the amount of *e.m.f.* generated is a function of temperature at hot and cold junctions.

Construction : In thermocouples the thermojunctions are formed by welding or soldering two different metals. The metals are welded by gas or electric welding. Thermocouples are available in different sizes, shapes, temperature ranges for different applications. According to media and the temperature range various combination of materials in fixed proportions are used for the manufacturing of thermocouples.

Generally, the important parameters to be considered for thermocouple materials are linear relationship between temperature and output voltage, to withstand highest temperature and temperature variations, corrosion resistance and sensitivity Figure (S).

The sensitivity figure is defined as,

$$S = dE/dt = \sum \alpha_n t^{n-1} \quad \text{...(3.52)}$$

where ; α_n = Seebeck co-efficient depends upon the material and temperature.

t = Hot junction temperature when cold junction temperature is at 0°C.

E = Fermi energy.

This sensitivity figure depends upon the material which is to be used for thermocouple. The thermocouple should generate high thermoelectric power.

Table 3.1 shows the materials used and other parameters for industrial thermocouples.

Fig. 3.26 shows the relationship graph between temperature and *e.m.f.* for different thermocouple metals.

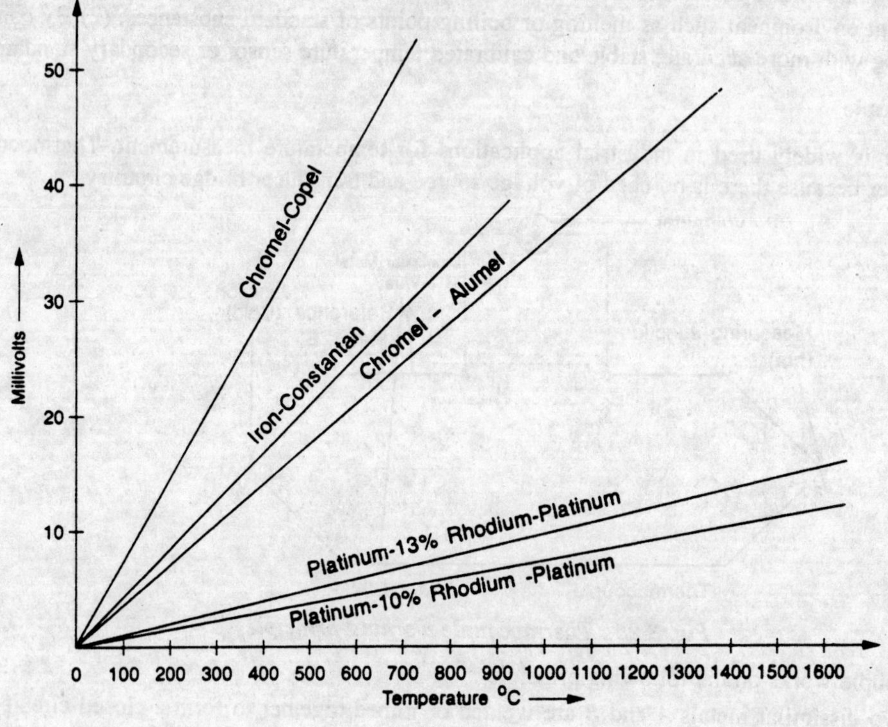

Fig. 3.26

Thermocouple probes : Thermocouples are used for the measurement of temperature ranges of different media in industrial processes and other applications. For protecting the thermocuples from operating environment, corrosion, shocks and vibrations a protecting metal sheath (metal tube) or well is used which is called **probe** as shown in Fig. 3.27.

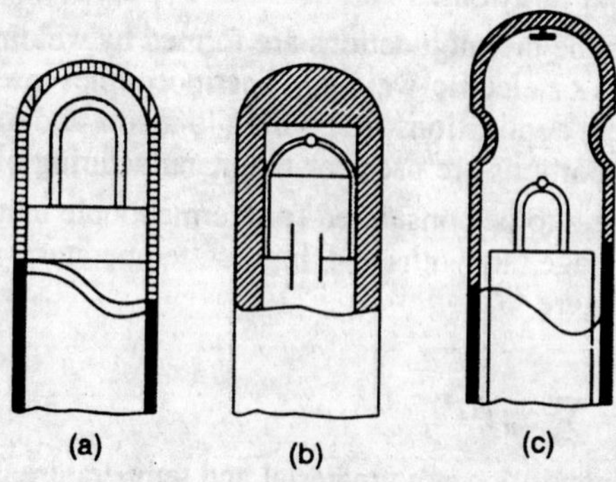

Fig. 3.27 Thermocouple probes (a), (b) and (c)

The probes are available as open element and closed element. In open element the hot junction is kept directly in contact with the media.

In closed element the hot junction is placed in sealed bulb and either the hot junction may be joined to the bulb or insulated. This system is used when the media is highly corrosive or at higher pressure.

Between the protective sheath and thermocouple wires proper separation is provided by using ceramic powder. Sometimes porcelain, fused silica can also be used as protecting sheath.

For the installation of thermocouple proper screw mountings, screws, clamps, housings and compensating cables are used.

Thermocouple junction compensation : The temperature indicating meters or display devices are placed at long distances from the place where the thermocouple is installed. These wires are long and unable to maintain constant temperature throughout their lengths which introduces the error signal. For eliminating such errors the same metal wires are connected throughout by which the thermocouple elemem is manufactured. Otherwise use such metal wires having same temperature – *e.m.f.* characteristics to that of thermocouple element material. These wires are called as compensating leads or compensating cables.

Sometimes the reference junction temperature changes because of ambient conditions under such circumstances, compensating arrangement is used to avoid the error signal generated due to the temperature variations. This error signal is eliminated or compensated by keeping the reference junction in ice bath or by keeping the reference junction in oven at fixed temperature. Such compensating arrangement provides the accuracy to the measurement systems.

Now-a-days thermistors or semiconductor sensors such as AD590 are used for compensation.

Thermocouples used for industrial applications has the arrangement such that it has one hot junction in the process and the reference junction is connected to meter with compensating wires to avoid the ambient temperature variations.

Thermocouples are simple in construction, rugged, operate on wide temperature. Using thermocouples maximum accuracy can be achieved by using compensating wires or cables. For protecting the thermocouple metal tube or probe (sheath) is essential.

Advantages of Thermocouple

(1) Construction is mechanically rigid and strong.

(2) Suitable for reading (measurement) of rapidly varying temperatures. (Fast response)

(3) Low cost.

(4) No need of bridge circuit.

(5) calibration and installation is easy.

(6) Suitable for temperature range – 270 to 2800°C.

Disadvantages of Thermocouple

(1) They require protective wall or sheath.

(2) Thermocouple needs compensation arrangement.

(3) Amplifier circuit is necessary to increase the output voltage level.

(4) For long distance temperature measurement compensating wires are necessary.

(C) Thermistor (semiconductor temperature sensors)

Thermistors are also called thermal resistors. For thermistors the absolute temperature – resistance relationship is given by,

$$R_T = R_{T_1} \exp\left[\beta\left(\frac{1}{T_1} - \frac{1}{T_2}\right)\right] \quad ...(3.53)$$

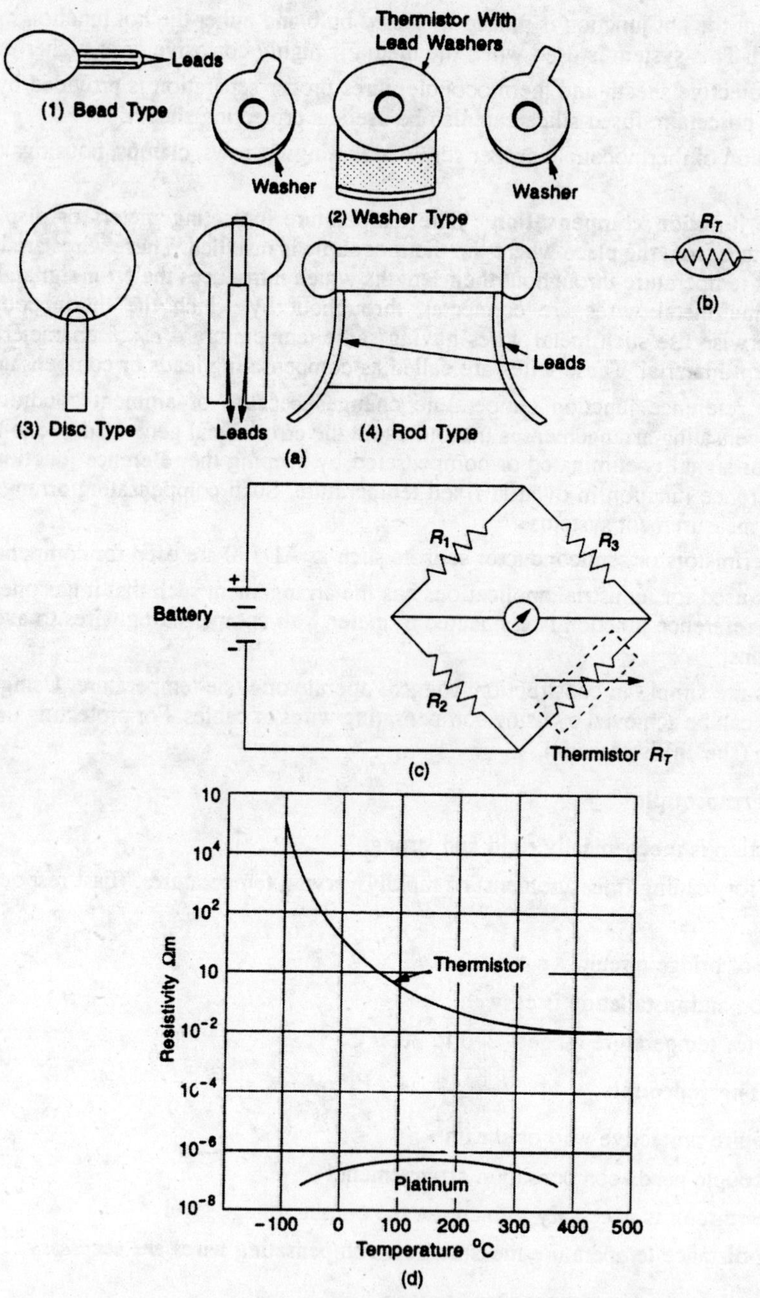

Fig. 3.28 : (a) Various configurations of thermistor. (b) symbol, (d) Resistance temperature characteristics of a typical thermistor (c) Wheatstone Bridge with Thermistor R_T

where ; R_T = Resistance of the thermistor at absolute temperature T. (°K)

R_{T_1} = Resistance of the thermistor at absolute temperature T_1. (°K)

β = Constant.

T_1 and T_2 = Absolute temperatures. (°K)

Fig. 3.28(b) shows the symbol of a thermistor.

Thermistors are made up of semiconductor materials. As temperature changes the resistance of materials also changes. The temperature coefficient of resistance for thermistors are mentioned below.

If the value of resistance increases with increase in temperature, it is positive temperature coefficient (PTC) thermistor.

If the value of resistance decreases with increase in temperature, it is negative temperature coefficient (NTC) thermistor.

Construction : Thermistors are semiconductors and made up of sintered mixtures of manganese, cobalt, zinc, aluminium, iron, copper, uranium oxides powder above 982°C in combination with binders. The mixture of these materials is pressed into different (types) shapes and size like beads, washer, rods, discs, flakes. All these types are represented in Fig. 3.28(a). The electrical connections or leads are embedded before sintering process.

These thermistors are placed in glass or plastic encapsulation so as to protect from damage, vibrations, flow rates etc. Thermistors are used to determine temperature, liquid level, rate of flow, etc.

The temperature range for thermistors is – 60° C to + 15° C. Its resistance varies from 0·5Ω to 0·75 MΩ. Thermistor is placed is contact with the media whose temperature is to be measured. As the temperature of the media changes the resistance of thermistor gets changed. This change of resistance can be measured by connecting the thermistor in any one arm of the Wheat stone bridge. According to change in the temperature that is increase or decrease in the temperature the resistance gets varied which causes to unbalance the bridge. It results to flow the current through galvanometer as shown in Fig. 3.28(c). The galvanometer indicates the readings calibrated as a temperature scale. Fig. 3.28(d) shows resistance temperature characteristics of Thermistor.

Advantages of Thermistor

(1) Small size, compact.

(2) Good response time from fraction of seconds to minutes.

(3) They can be installed in small areas like pipeline, inlet or outlet of tanks.

(4) No need of cold junction compensation.

(5) Good stability.

Disadvantages of Thermistor

(1) Non-linear performance *i.e.* temperature Vs change in resistance graph.

(2) Not suitable for higher temperatures.

(3) It requires bridge arrangement for converting change in resistance to voltage.

(D) Resistance Temperature Detectors (RTD) (Conductive temperature sensor)

This type of transducer is used for temperature measurement. Here the basic concept used is that electrical resistance of different metal changes in accordance with the temperature *i.e.* for temperature measurement. Principle used that is resistance of a conductor changes in proportion with the change in temperature. The unknown temperature is determined in terms of electrical resistance of the conductor which senses the temperature The change in resistance of this device is precisely determined either by bridge circuit or by

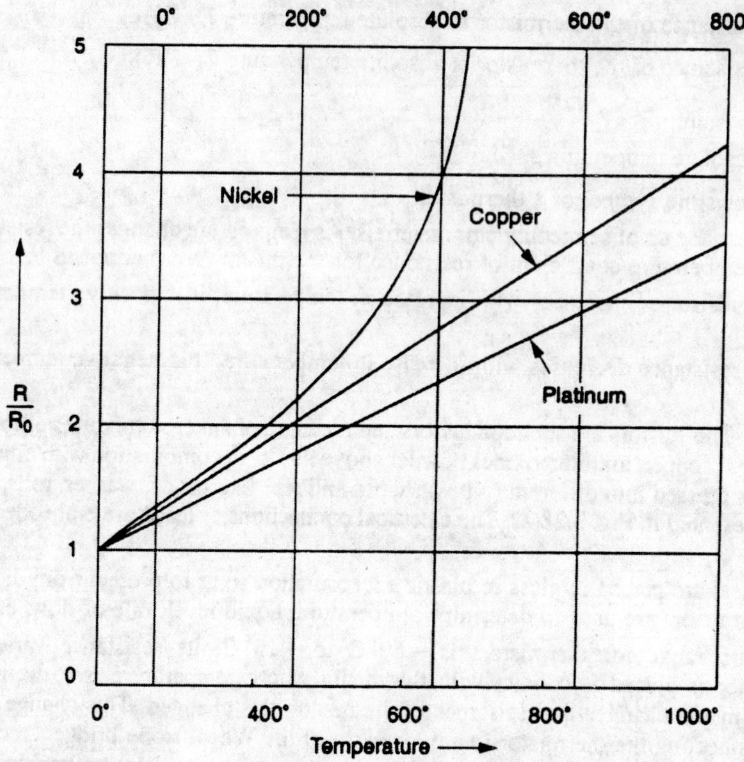

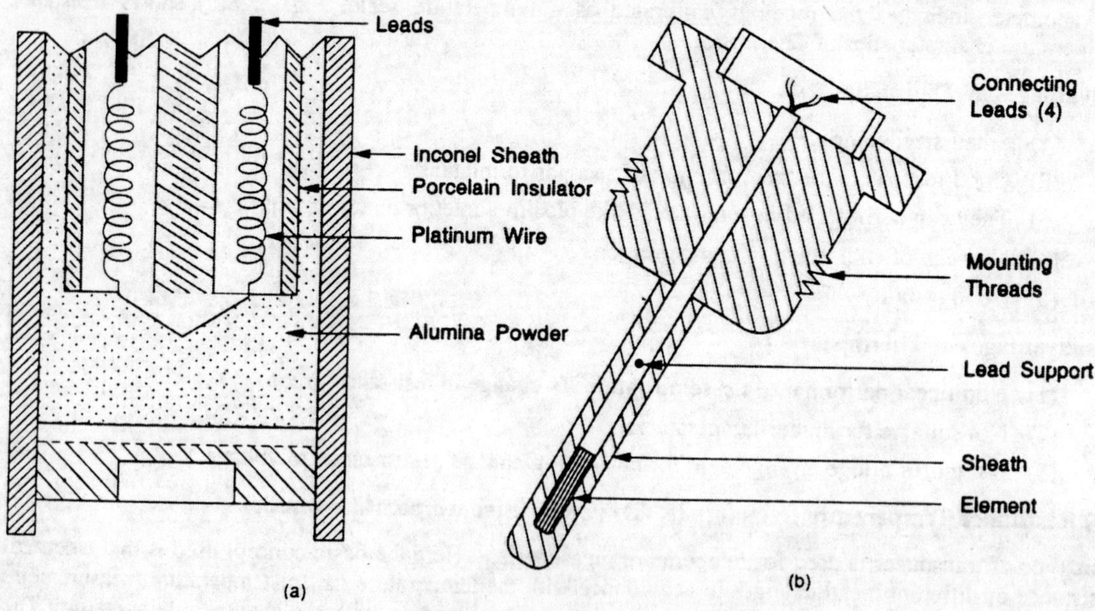

Fig. 3.29 : (a) Characteristics of materials used for resistance thermometers. (b) Resistance thermometer, (c) Industrial platinum resistance thermometer.

ohmmeter. Otherwise a separate circuit can be used to calibrate the value of resistance in terms of voltage or current.

For different metals the variations of resistance value (R) with temperature (T) is expressed as :

$$R = R_0 (1 + a_1 T + a_2 T^2 + \dots + a_n T^n) \qquad \text{...(3.54)}$$

where ; R_0 = Resistance of the conductor at temperature $T = 0$

$a_1, a_2, a_3, \dots, a_n$ = Constants

In metals the resistance increases with increase in temperature.

Metals used for RTD

(1) "Balco" handles the temperature range – 23°C to + 204°C having 2000 Ω resistance.

(2) Platinum has resistance temperature co-efficient 0·39 and temperature range – 260°C to 110°C.

(3) Nickel has resistance temperature coefficient 0·62 with temperature range – 220° C to 300°C.

(4) For tungsten resistance temperature coefficient is 0·45 and temperature range – 200°C to 1000°C.

Fig. 3.29(a) shows resistance versus temperature curve of different metals.

Generally, the metals used are platinum, copper and nickel for R.T.D. Most widely used resistance wire is "Platinum" because of the following features :

(1) Platinum provides good stability and accuracy.

(2) It can operate on wide range of temperature.

(3) It has good linearity over wide temperature range.

(4) Less errors during operation.

R.T.D. fabricated using platinum is called as Platinum RTD *i.e.* (PRTD).

Operation : Fig. 3.29(b) and (c) shows the construction of R.T.D. The important aspects to be considered for the designing of R.T.D. are (a) operating range (b) fast response (c) environmental conditions (d) ability to withstand corrosion, friction etc.

As R.T.D. is used for temperature measurements of fluids, solid and gases. R.T.D. are of probe type and immersed into the medium whose temperature to be measured. It consists of a platinum wire in spiral shape enclosed in a insulating material such as mica or ceramic. It is enclosed in a protective glass or porcelain tube. This assembly is placed at the tip of the probe. The R.T.D. is placed inside a well or tube to provide the protection to resistive element from corrosion, environmental conditions, mechanical damages etc. Mostly the well arrangement is used in liquid and gaseous medium at higher temperature *e.g.* power plants, pipelines etc.

EVALUATION COMPONENT

1. Why transducers are necessary ?
2. Define the term transducer. Classify them.
3. What is the difference between a transducer and a sensor ?
4. Compare :
   (i) Analog and digital transducers
   (ii) Active and passive transducers
5. Which are the important parameters to be considered while selecting a transducer for particular application ?

6. What is inverse transducer ? Where they are applicable ? State any practical example.
7. State the applications of potentiometric type transducers.
8. Give the constructional details of :
   (i) Wire wound potentiometers
   (ii) Non wire potentiometers
9. What do you mean by power rating of potentiometers ?
10. Elaborate the reasons due to which noise gets produced in potentiometers.
11. Explain how :
   (i) Translational (linear) POT measure linear displacements ?
   (ii) Rotary (angular) POT measure angular displacements ?
12. Define "sensitivity" of potentiometers.
13. Which materials are used for fabrication of both wire wound and non wire wound potentiometers ?
14. How linearity and sensitivity of potentiometers are co-related ?
15. How the linearity of potentiometers can be improved ?
16. On what principle inductive transducer works ?
17. Why residual voltage is present ir LVDT ?
18. State advantages and applications of LVDT.
19. With neat sketch explain the working of RVDT.
20. Compare LVDT and RVDT.
21. Explain the constructional details of synchrotransmitter.
22. Explain the construction of linear variable differential transformer. Write about its working.
23. What is electrical zero position of synchro ?
24. What is photoelectric effect ?
25. State the applications of photomultiplier tube.
26. Explain the working of photomultiplier tube with neat sketches.
27. Define photoconductive effect in semiconductors.
28. With V–I characteristics explain the working of photodiode.
29. List applications of photodiode.
30. What is piezoelectric effect ?
31. Classify piezoelectric materials. Elaborate about it.
32. Explain how output voltage gets developed across piezoelectric transducer ?
33. What are the characteristics of piezoelectric transducer ?
34. Define :
   (i) Strain
   (ii) Gauge factor
35. Explain working principle of **strain gauges.**
36. Calculate the Possion's ratio for a resistance wire strain gauge made up of platinum. Neglect piezoresistive effect. Gauge factor of platinum is 4·8 (Ans. $G = 1{\cdot}9$)
37. Explain the constructional details of :
   (i) Bonded strain gauge
   (ii) Bonded foil strain gauge
   (iii) Bonded semiconductor strain gauge
38. List the characteristics of strain gauges.

39. Which effects or phenomenon are used for temperature measurements ?
40. Define **temperature scales**.
41. State the working principle of **thermocouple**.
42. What is thermowell ? What is its significance ?
43. Write a short note on thermocouple covering following points :
    (i) Working principle
    (ii) Materials used
    (iii) Advantages
    (iv) Temperature range
44. Explain in detail **thermistors**.
45. What materials are used for fabrication of thermistor ?
46. With suitable diagram suggest a suitable method of temperature measurement using thermistor.
47. Write a short note on R.T.D.-

Chapter **4**

MEASUREMENT SYSTEMS (PART-I)

Learning Objectives

- To study temperature measurement methods.
- To study pressure sensing elements.
- To discuss working of different pressure measurement methods.
- To explain differential pressure measurement schemes.
- To define significance of thickness measurement.
- To discuss contact and non-contact type thickness measurement methods.
- To define terms : Humidity, relative humidity, dew point hygrometer.
- To state working principle of hygrometer.
- To study working of different hygrometers used for humidity measurement.
- To elaborate : Vibration and significance of vibration measurement.
- To discuss different vibration measurement methods.
- To define working principle of torque measurement.
- To discuss torque measurement methods.
- To explain working principle of flow measurement.
- To discuss characteristics of head flowmeters.
- To discuss construction, working, advantages, disadvantages and applications of different flowmeters.
- To explain thermal, ultrasonic, electromagnetic flowmeters.
- To explain velocity measurement using mechanical and electrical methods.

4.1 Introduction

Measurement of physical quantities is an important function in industries. In engineering field or applications various physical parameters such as temperature, pressure, velocity, thickness, flow, torque, etc. are involved. It is necessary to measure these parameters with high degree of accuracy and precisely. Measurement helps to control these parameters at a fixed value or set point.

In this chapter different measurement schemes of parameters like temperature, pressure, thickness, humidity, vibrations, flow, torque, velocity etc. are discussed. For measurement of these parameters their respective transducers are also covered. Different aspects such as advantages, disadvantages and applications of each measurement scheme is described.

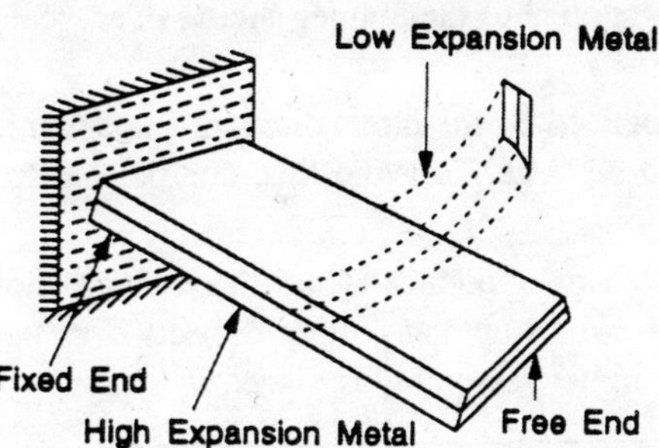

Fig. 4.1 : Principle of bimetallic thermometer.

For pressure measurement different transducers and pressure sensors such as diaphragms, bellows, bourdon tubes are discussed. Manometers, differential bellow methods of pressure measurement are also covered.

Thickness measurement which is important method in paper, plastic, textile, rubber industries is also covered in this chapter. Contact and non-contact thickness measurement schemes are discussed.

Various schemes of vibration measurement are discussed in this chapter. In engineering applications rotating machines, turbines, motors, AC/DC drives are used for driving different loads. For measurement of torque of these devices, different methods are suggested.

Flow measurement is an important task in power plants, process industries (paper, sugar), oil refineries, chemical industries so as to measure flow of liquids, gases, fluids, slurries, acids etc. Different flowmeters like venturi tubes, orifice plates weirs, thermal flowmeters, electromagnetic flowmeters, ultrasonic flow-meters etc. are covered in this chapter.

Tachometers (mechanical and electrical) are used for speed measurements which are described in the last section of this chapter.

This chapter helps to study how suitable methods can be implemented for measurement of different physical quantities like temperature, flow, velocity, torque etc. It also provides the knowledge and conceptual ideas regarding measurement of above stated parameters.

4.2 Temperature Measurement Methods

(A) Bimetallic Thermometers

It uses the concept of differential expansion of bonded strips of two metals. It uses the important aspect that some metal changes volume with temperature as well as the coefficient of change is not same for all the metals.

Working principle : If the two different metal strips are joined or welded together and these strips are heated, the resultant strip having lower expansion coefficient gets bend.

The deflection of metal strip is directly proportional to the square of the length, temperature and inversely proportional to the thickness of metal.

As the name indicates it consist of two different metal strips. According to the change (rise) in temperature the length of the metal gets changed with respect to the rates of thermal expansion. This expansion results in the bending of bimetallic strips towards the side having low coefficient of thermal expansion as shown in Fig. 4.1.

The deflection of metal strips is applied to the pointer sliding over scale, which is proportional to the temperature variations.

The deflection or movement produced by the bimetallic strips is small if the size of strip is small. For large deflections, size of strip should be large. To avoid this, the metal strip element is wound in spiral or helical shape.

At the outer part of the spiral strips a pointer is attached which moves over calibrated scale. As temperature increases the spiral strips winds up and pointer deflects over the calibrated scale in clockwise directon as shown in Fig. 4.2. The complete assembly is placed in a case.

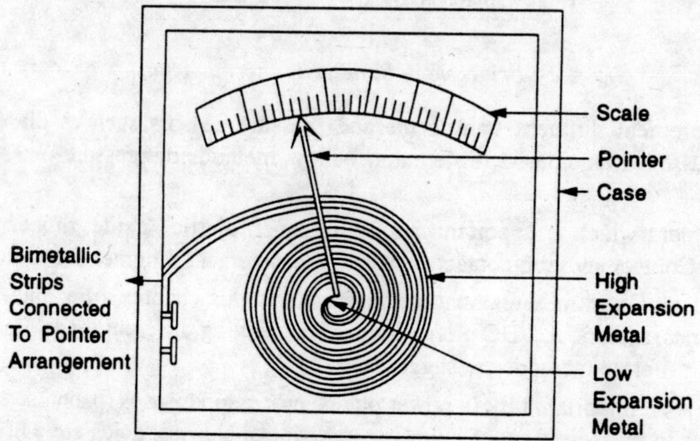

Fig 4.2 : Construction of bimetallic thermometer.

Metals used : Mostly the metals used are low-expansion invar (64% Fe, 36% Ni) and higher-expansion nickel-alloy with chromium or manganese.

This complete set (bimetallic strips) are sealed in stainless steel with plastic or glass crystals which is a protect tube or well.

(B) Bimetallic Thermometer Using Helical Elements

As name indicates the element is helical shaped as shown in Fig. 4.3 which is placed in thermo well for protection against corrosion and breakage. This type of thermometer is used for measurement of temperature of gas or liquid inside duct.

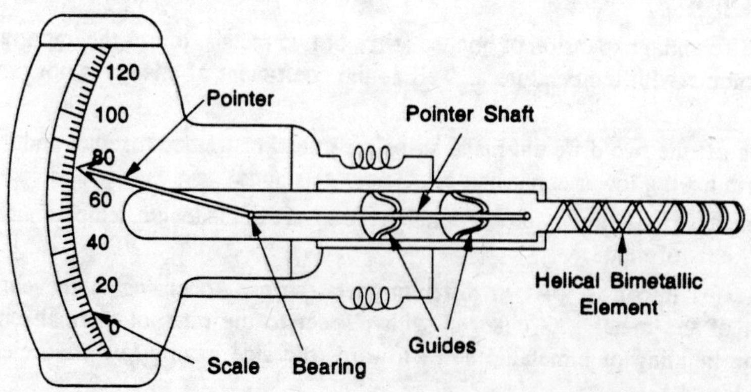

Fig. 4.3 : Bimetallic thermometer using helical element.

The helical bimetallic strips are placed inside the stem of thermometer. One post of the strips is connected to the shaft of pointer whereas other is fixed to the outer body. The shaft slides over a calibrated scale or dial.

When the temperature of whole stem changes the expansion of the bimetallic strip takes place resulting in the rotation of pointer shaft. This pointer shaft is attached to pointer which moves over the scale to measure the temperature.

Temperature ranges for bimetallic thermometers : Standard range 70 to 50°C up to 100 to 550°C.

Advantages :

(1) Easy for installation.
(3) Mechanically rigid and tough.
(2) Maintenance is not complex.
(4) Low cost.
(5) Wide operating temperature range.

Disadvantages :

(1) Calibration changes because of rough handling.
(2) Accuracy is not as good because of glass stem design.

Application :

Helical bimetallic thermometers are used in industrial and residential temperature measurements.

4.3 Pressure Measurement

(A) Diaphragm

Diaphragms are widely used for the measurement of pressure ranges from (0 to 0·7 kPa and 0 to 350 kPa).

Materials used : Important aspect to be considered for diaphragm is elasticity, the material should have good elasticity and low temperature co-efficient of elasticity. Phosphor bronze, Beryllium copper, Monel, brass, tantalum, Teflon, duranickel, Ni- span *C* materials are used for the manufacturing (construction) of diaphragms. In quartz diaphragm minimum hystersis and drift can be achieved.

Principle and construction : The diaphragm is a thin membrane made from sheet metal of precise dimensions. It is used for the measurement of pressure. When the pressure which is to be measured is applied to the diaphragm it deflect. The deflection (displacement) of diaphragm is proportional to the applied pressure.

This is the suitable transducer for sensing the pressure. The unknown pressure is applied to one side of the diaphragm. This unknown pressure is determined by measuring the deflection or displacement.

The thin membrane type diaphragms are applicable for the measurement of low pressure range. For higher pressure ranges circular flexible disc diaphragms are used. Types of diaphragms are :

(1) **Flat type** : It consists of a thin flexible circular metallic plate. These diaphragms provide low deflection because they operate on low pressure ranges. Fig. 4.4(a) shows flat diaphragm.

(2) **Corrugated diaphragm** : In this type of diaphragm surface area is increased due to which pressure handling capacity is increased, ultimately the deflection is also higher. Fig. 4.4(b) shows single corrugated diaphragm.

The pressure handling capacity of single corrugated can be a increased by bonding or welding two single corrugated diaphragms. Used for highly sensitive measurement when the two diaphragms are welded together

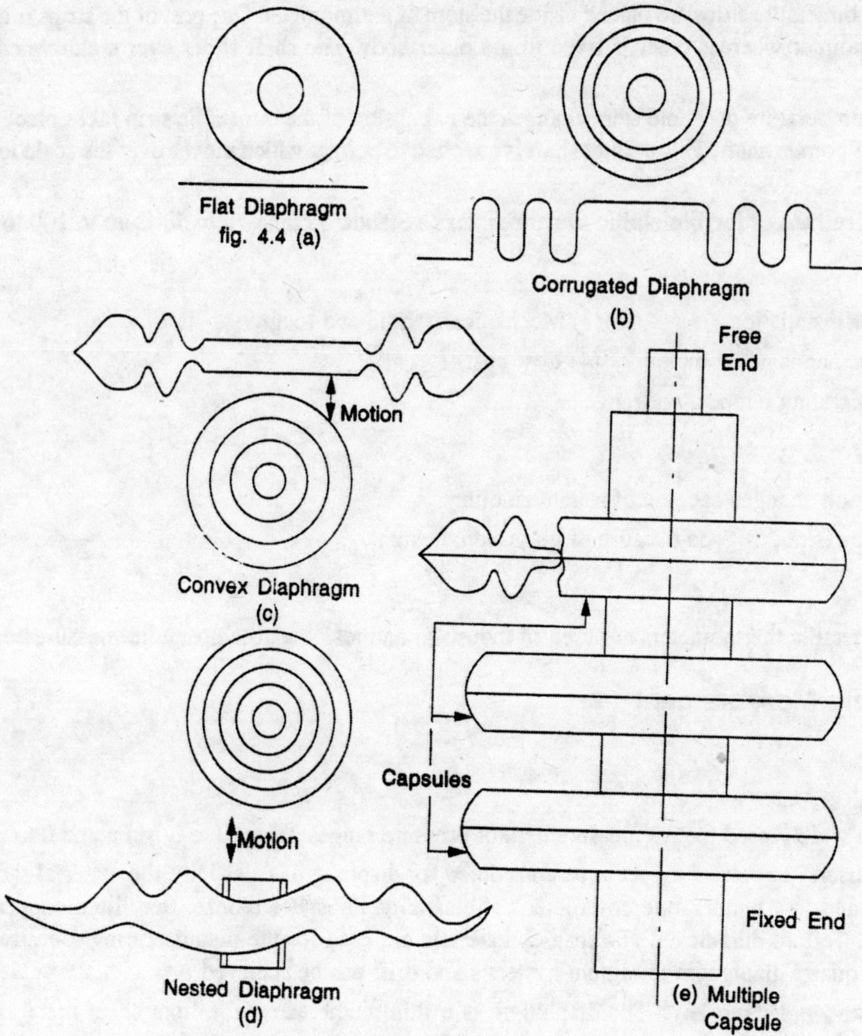

Fig. 4.4 : Types of diaphragms (a) Flat diaphragm, (b) Corrugated diaphragm, (c) Convex diaphragm, (d) Nested diaphragm, (e) Multiple diaphragm.

at the peripheries and the resulting diaphragm is called **Capsule**. The capsule diaphragms are again classified as :

(a) **Convex diaphragm capsule**

Fig. 4.4(c) shows convex diaphragm capsule.

(b) **Nested diaphragm capsule**

Fig. 4.4(d) shows nested diaphragm capsule.

(c) **Multiple diaphragm capsule** : When the number of convex or nested capsules are joined the resulting device is multiple capsule element. The capsule elements are connected axially with each other. Total deflection is sum of the deflection of individual capsules. This element provides increased output displacement in accordance with the change in applied pressure. Fig. 4.4(e) shows multiple diaphragm.

Pressure measurement using diaphragm

Fig. 4.5 shows pressure measurement using diaphragm. It consists of number of metallic diaphragms connected to each other on a metallic element. The metallic element is connected to a pointer through mechanism or spring arrangement. The pointer slides or moves over a calibrated scale as shown in figure.

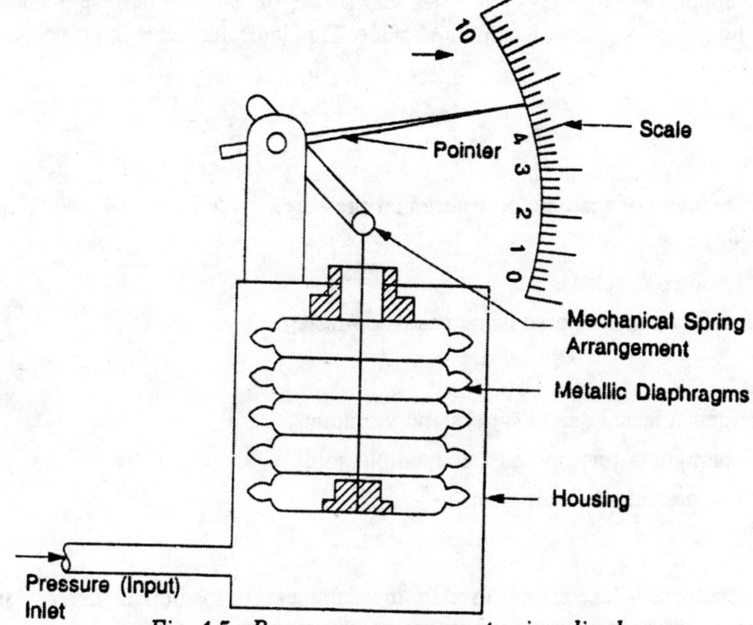

Fig. 4.5 : Pressure measurement using diaphragm.

When the pressure which is to be measured is applied through the pressure inlet due to which deflection of metallic diaphragm takes place. This deflection is attached to a pointer through spring or proper mechanism arrangements. Due to which the pointer slides over calibrated scale. The reading indicated by the indicating or recording instrument is proportional to input pressure signal which is to be measured.

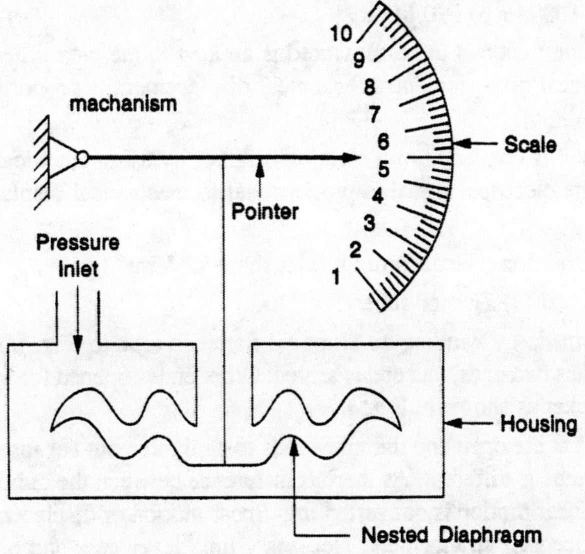

Fig. 4.6 : Pressure measurement using nested diaphragm.

Pressure measurement using nested diaphragm

Fig. 4.6 shows pressure measurement using nested diaphragm. It consists of a metallic housing in which nested type diaphragm is placed. Displacement of diaphragm is connected to a pointer through mechanism. This pointer slides over a calibrated scale and measures the displacement.

When pressure is applied through pressure inlet, deflection of nested diaphragm takes place. This deflection is indicated by the pointer over a calibrated scale. This indicated scale is proportional to applied input pressure.

Advantages of Diaphragm :

(1) Low cost.

(2) Useful for measurement of absolute or differential pressure.

(3) Linearity is good.

(4) Small size and good compactness.

(5) It can be fabricated or manufactured using corrosion resistive elements.

Disadvantages :

(1) Its performance gets affected due to shocks and vibrations.

(2) Repairing of diaphragm is very complex and tedious job.

(3) Applicable for low pressure measurements.

(B) Bourdon Tube

These are the primary pressure sensing elements used for measurements of medium to high pressure. Bourdon tubes are elastic members and converts the pressure into mechanical displacements.

Important aspect is that Bourdon tube is primary sensing element which senses pressure and converts into mechanical displacement, this displacement is converted into electrical signals using secondary transducers.

Materials used for construction of Bourdon tube are phosphor bronze, beryllium copper, monel, Ni-Span C, alloy steel. Pressure range is 100 kPa to 690 MPa.

Principle : When the pressure which is to be measured is applied to the pressure sensitive element, it deflects and resulting in mechanical displacement. Mechanical displacement is proportional to the applied pressure.

This mechanical displacement is converted into electrical signals by using electrical transducers. From this discussion it is noted that the electrical signal is proportional to mechanical displacement and so it is proportional to applied pressure.

These are various types of Bourdon tube according to its shape or form.

(1) C-type (2) Spiral (3) Helical (4) Twisted tube.

(1) **C-Bourdon tube** : It is made by winding an elliptical flattened tube to form a segment of a circle having an arc of 250°. This tube has two ends, one end is sealed. Other end is opened for applying the process pressure and it is fixed at the socket as shown in Fig. 4.7.

When the pressure is applied at the open end the tube tends to straighten out because the internal radius and outer radius of the Bourdon tube is different. As there is difference between the radius, it takes different areas for the pressure. The non linear motion is converted into linear motion or displacement by means of a pointer and calibrated scale (deflection) arrangement. Necessary link, lever gear and pinion attachment is

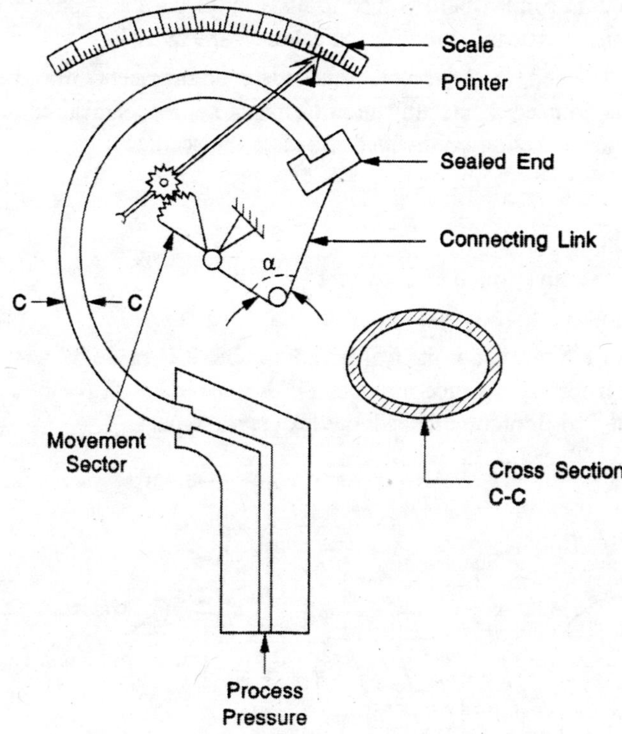

Fig. 4.7. C-Bourdon Tube

provided to the deflection system. Thus the applied pressure is measured by means of deflection over linear scale.

Pressure range for C-type bourdon tube : 0–1000 kPa to 0–140 MPa.

Application :

It is used for measuring fluid pressure.

Accuracy of Bourdon tube depends upon the diameter of tube, design quality, materials used for constructing calibrated arrangement. Accuracy is ± 0·5 to ± 2%.

(2) Spiral Bourdon tube : Spiral Bourdon tube is made by winding the elliptical flattened tube into spiral form of 2 or 3 windings around the same axis as shown in Fig. 4.8 one end of the tube is sealed and other is open.

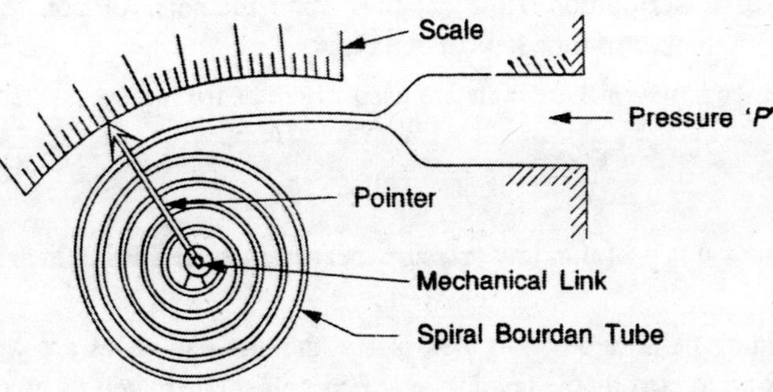

Fig. 4.8.

When the pressure is applied to spiral Bourdon tube it tends to uncoil and produces the deflection or displacement at free end. This displacement is used for indication or transmission.

Produced displacement is applied to the pointer and scale arrangements for measurement. The displacement is more so there is no need of magnification elements *i.e.* mechanical amplification. Due to this the gear and pinion arrangement is eliminated which increases the accuracy.

Advantage :

(i) More accuracy.

(ii) No friction because no gear and pinion arrangement.

(iii) Higher sensitivity because no friction and loss of motion.

(3) Helical Bourdon Tube : Fig. 4.9 shows helical Bourdon tube. It is made by winding the elliptical flattened tube into helical (helix) form. This tube provides higher movement of the free end. Mechanical movement or displacement of helical Bourdon tube is more than spiral type.

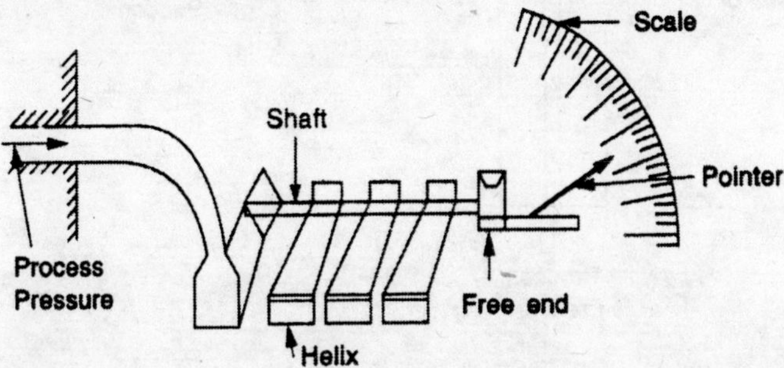

Fig. 4.9 : Pressure measurement using helical Bourdon tube.

A shaft is connected at the center of helical winding and a pointer is attached at the shaft using proper connecting link.

As the applied pressure changes the helical winding provides circular displacement which is indicated by the pointer connected to the shaft.

Advantage :

(1) It provides satisfactory operation over continuous fluctuating pressure signals.

(2) Higher capacity to handle the over range pressures.

(3) No necessity of mechanical amplification, gear and pinion arrangement.

Pressure handling range of this Bourdon Tube depends upon diameter of coil, thickness of tube, constructional material used and number of tube helical coils/tubes.

For low pressure measurements two or there coils are used where as for high pressure measurements 16 to 20 coils can be used.

(C) Bellows

This is pressure sensing element and is useful in low pressure measurement and vacuum or absolute pressure measurements.

Bellows are made by welding number of preformed plates, the circular plates are welded together so that they can be expanded or contradicted by the application of pressure. The materials used for construction of bellows are brass, stainless steel, beryllium copper, Monel, bronze, inconel, Ni-span. The material should be ductile, flexible, high strength, and should have low hysteresis etc. Fig. 4.10 shows basic bellows element.

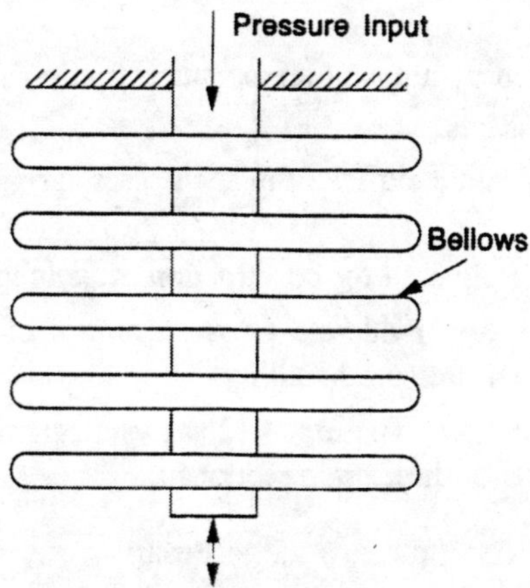

Fig. 4.10 : Basic bellows element.

Bellows are prepared by seamless tubes by hydraulic or mechanical roll formation or seamless bellow can also be formed. They can be directly attached with recorders or indicators.

(a) **Pressure measurement using single bellows** : Fig. 4.11 shows design of pressure measurement using single bellow element. It is exposed to atmospheric pressure which functions as reference pressure. At the bottom end, process pressure is applied. Calibrated spring is provided so as to counter balance the deflection of bellows and pointer.

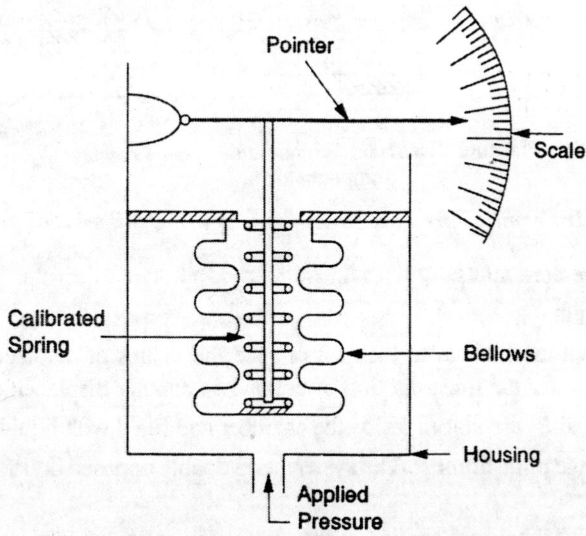

Fig. 4.11 : Pressure measurement using single bellows.

As the pressure is applied to spring loaded metal bellows it gets compressed, this forces the lower end of bellow in upward direction. This results in opposing the spring force. The pointer is connected to spring through suitable linkage and calibration. When pressure is applied in vertical movement the pointer moves over the calibrated scale. In this way the unknown pressure can be measured by this arrangement.

Limitations :

(1) Sensitivity may be affected by ambient temperature, friction and drift.

(2) Not suitable for high pressures.

(3) If the elasticity of bellow is not sufficient in such cases spring arrangement may be attached for accurate measurement.

(4) Work-hardening of the metals used for construction of bellows.

Advantages are—good for low and moderate pressure ranges of 9–13 kPa to 0–240 kPa. Easily applicable for absolute and differential pressure handling.

(b) **Differential bellows configuration** : The Fig. 4.12 shows diagram for differential pressure measurement unit. This system consists of three main sections :

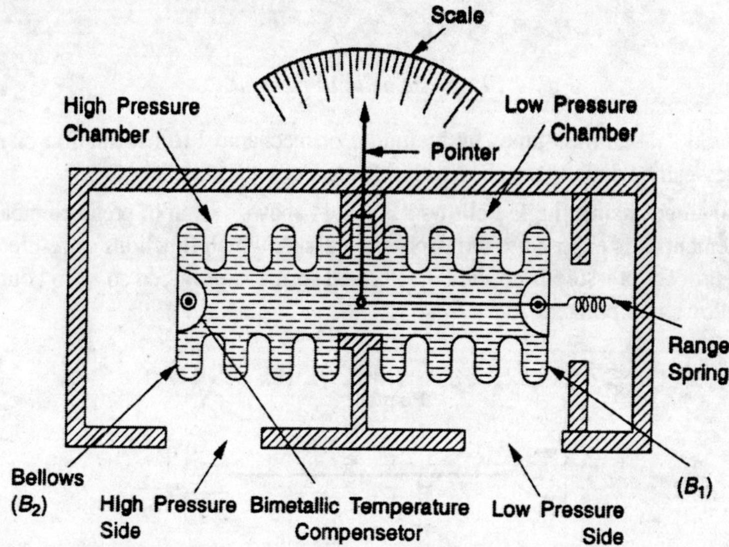

Fig. 4.12 : Differential pressure measurement using differential bellows.

(1) Low and high pressure inlets.

(2) Spring arrangement.

(3) Pointer attachment mechanism to indicate or read the values of displacement of bellow. (drive mechanism to transfer the linear motion of bellows in angular displacement).

Both bellow B_1 and B_2 are placed inside the chamber and filled with liquid. The function of spring is to determine the range of instrument. The system uses double opposed bellows placed in liquid filled chamber.

When the pressure signals are applied at the high pressure chamber it results in compressing the bellow due to which the liquid inside the chamber flows towards the low pressure bellows. So the pressure bellows gets expanded and applies a force against the range spring. The bellows gets displaced with linear motion.

The linear motion is converted to rotary (angular) motion using link, lever, gear and pinion arrangements with pointer and calibrated scale. The bimetallic temperature compensator is provided at the high pressure inlet so as to compensate the capacity of bellows with respect to change in the volume of liquid because of ambient temperature.

For protecting the bellows from over range of pressure nested diaphragms, diaphragm disks or metallic spacer rings can be used as a solid stop.

(D) Manometers

Generally the manometers are used for measuring the pressures of **liquid / fluid** *i.e.* it is a **wet medium** pressure detector. The manometer is based upon the principle of pressure excited by a liquid/fluid at the depth.

(a) **Float manometer** : It uses the concept of differential pressure *i.e.* the high pressure gets compared with the pressure. The Fig. 4.13 shows schematic of low pressure inlet. Float manometer consists of high and low pressure inlet, float and a pointer drive mechanism.

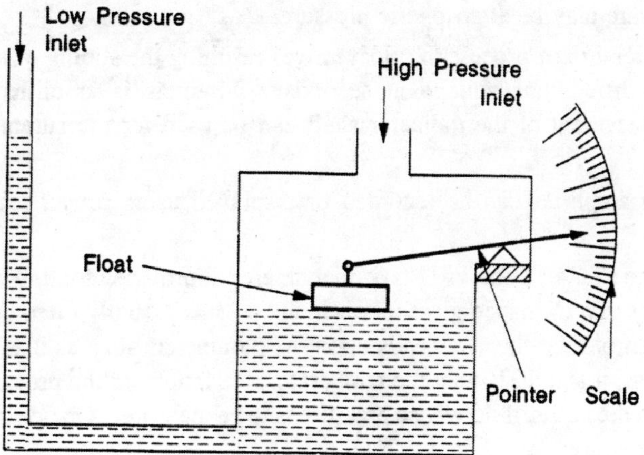

Fig. 4.13 : Float manometer.

Applications :

The float manometers are uses when the liquids are at high pressures or if the process liquids are hazardous.

Working :

This system uses differential pressure phenomenon which is detected and sensed by float mechanism.

When the pressure of liquid varies it results in changing the liquid level. Due to variations in liquid level the float moves. Float is connected to the pointer for indication or radiant through proper gear and lever drive mechanism.

This system is applicable for differential pressures of range 5 to 250 kPa.

(b) **Servo manometer** : Fig. 4.14 shows servo manometer. It uses principle of differential pressure. There is one inlet for process pressure placed inside the stainless steel or glass tube.

As the process pressure changes position of float varies. Float is connected to the core of differential transformer (LVDT). Float is placed inside the stainless steel or glass tube.

The core is generally made of high permeability ferromagnetic alloy. At primary winding '*P*' supply is applied whereas the secondary windings (S_1 and S_2) are in differential configuration and differential voltage is derived at secondary windings.

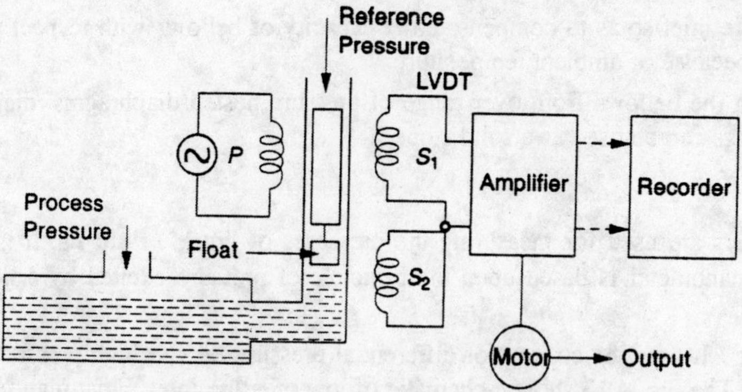

Fig. 4.14 : Servomanometer using LVDT.

The reference pressure may be atmospheric pressure.

Due to variations in the pressure of process the float moves resulting the sliding of artnature or core of differential transformer. The differential voltage at secondary windings is amplified and supplied to servomoter. The angular displacement of the indicator shaft can be used for measurement of level in the liquid manometer tube.

The gain or output of an amplifier can be recorded or displayed using proper recorder or indicator arrangements.

(c) **U-type manometer** : The Fig. 4.15 shows U-type manometer. It provides continuous measurement at high pressures. It consists of a U-shaped glass tube which contains partially filled liquid such as water or mercury. The selection of this liquid depends upon the parameters such as it should not be freeze or vaporize in the process, it should handle the temperature variations in the process if any, it should not react chemically with the liquid/fluid whose pressure is to be measured. Orifice is placed to develop high pressure and low pressure.

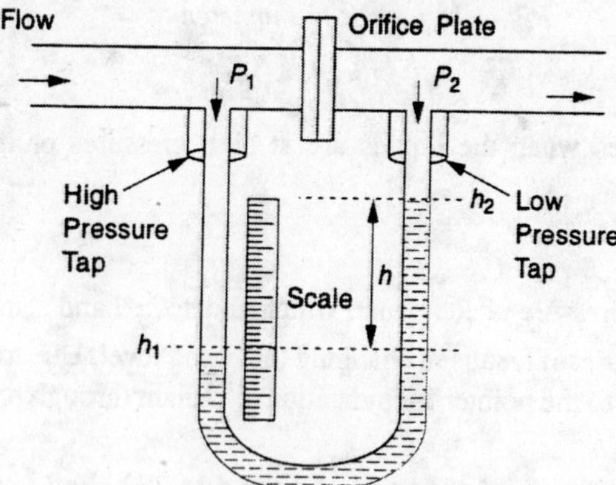

Fig 4.15 : U-type manometer.

When there is pressure difference at the two columns of the tube, fluid level decreases at one end and increases at the other end of the tube. The differential pressure is developed according to heights of liquid levels.

The height 'h' is proportional to the differential pressure of P_1 and P_2. The height 'h' is calculated as follows.

Analysis of Differential Pressure

The difference in liquid levels from one side to other indicates difference in pressure.

Let ; $P_1 \rightarrow$ Pressure at high pressure tap.

$P_2 \rightarrow$ Pressure at low pressure tap.

So, $$(P_1 - P_2) \propto (h_1 - h_2) \quad ...(4.1)$$

$\therefore$ $$(P_1 - P_2) = (\rho - \rho_1)(h_1 - h_2)\oint \quad ...(4.1a)$$

$\therefore$ $$P = (\rho - \rho_1)\, h \oint \quad ...(4.1b)$$

where; $P \rightarrow$ Difference in pressure $(P_1 - P_2)$

$h \rightarrow$ Difference in fluid levels $(h_1 - h_2)$

$\rho \rightarrow$ Density of the fluid in U–tube.

$\rho_1 \rightarrow$ Density of fluid whose pressure is to be determined.

$\oint \rightarrow$ Acceleration due to gravity.

Let, P_2 be atmospheric pressure, then height 'h' is direct measure of "P_1" as a process pressure.

Advantage : For indication purpose no requirement of pointer mechanism *i.e.* deflection system.

Limitation : Accuracy may be affected by visibility of the scale as it is read through glass tube.

4.4 Thickness Measurement

The thickness measurement techniques are used for measuring or monitoring the thickness of sheets of different materials, foils, films, cloths (textile industries), gears, papers and insulating sheets etc.

Thickness measurement transducers are widely useful for controlling the thickness during manufacturing processes. These transducers are useful in industries such as textile, steel, paper, plastic, rubber etc.

Methods for thickness measurement : These devices are (a) contact type and (b) non -contact type. Measurement methods are resistance, capacitance, inductance, pneumatic, reluctance, optical, ultrasonic, and X-ray. These transducers function on the aspect that comparison of sheet of reference (specific) thickness with the sheet under test.

(A) Contact Type Method

(i) **Differential roller gauge (LVDT)** : The Fig. 4.16 shows differential roller gauge method for thickness measurement. This method is applicable for continuous measurement of thickness. The thickness of sheet is sensed by linear variable differential transformer (LVDT). It uses two LVDT *i.e.* reference and measuring LVDT.

The sheet under test passes through a roller drum. Two separate wheels *i.e.* measuring wheel and reference wheels are used. Reference wheel is placed or adjusted over roller, whereas the measuring wheel is on the sheet.

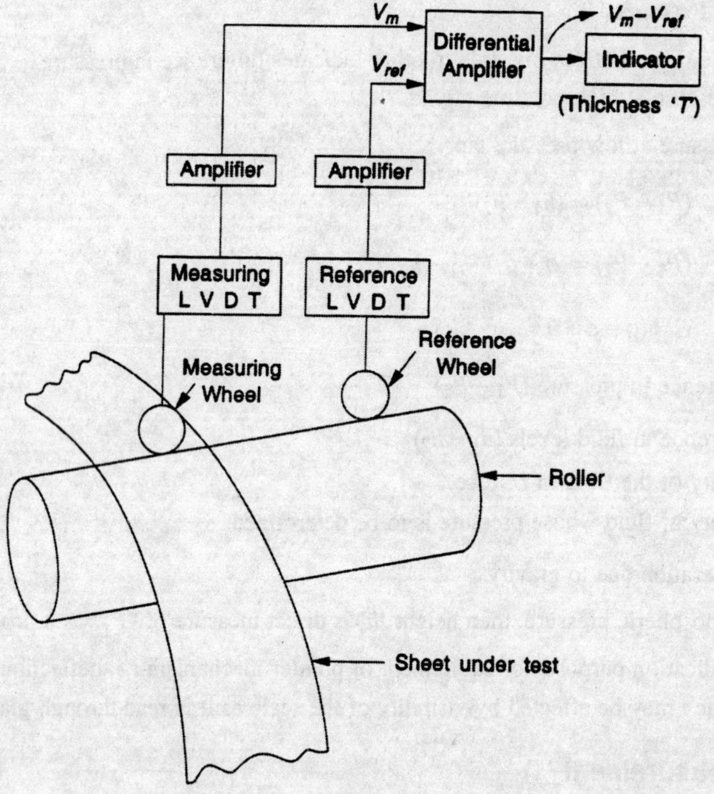

Fig. 4.16 : Differential roller gauge for thickness measurement.

The thickness is determined by the difference between secondary voltages of both LVDT due to linear displacements of core or armature positions. The thickness T of the sheet is determined by the relationship :

$$T = k\,(V_m - V_{ref}) \qquad \text{...(4.2)}$$

where ; T = Thickness of sheet.

k = Calibration constant.

V_m = Output voltage of measuring LVDT.

V_{ref} = Output voltage of reference LVDT.

This equation indicates that the core of LVDT displaces with respect to change in thickness. Difference between secondary voltage is due to displacement of core which is proportional to the displacement of measuring wheel with respect to thickness of the sheet. The final indicator will indicate the thickness.

Operating thickness measurement range is from 0·001 inche to several inches.

(B) Non-Contact Type

(i) Capacitance method :

Generally this method is useful for thickness measurement of insulating materials.

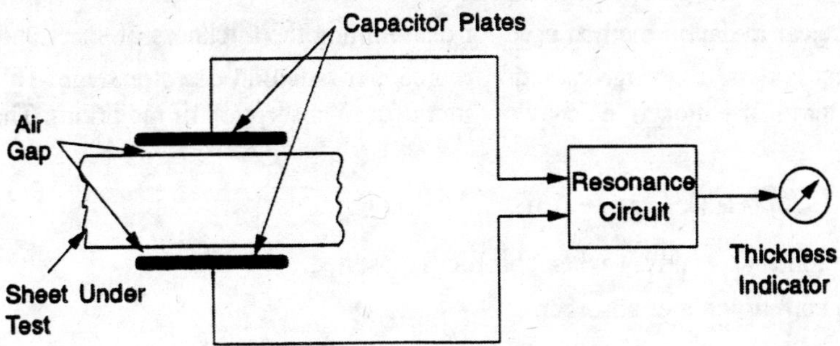

Fig. 4.17 : Capacitance method.

Fig. 4.17 shows capacitance method for thickness measurement. This system consists of two metal electrodes, on the two sides of the electrode insulating material sheet whose thickness is to be measured is placed. This construction functions as a parallel plate capacitor *i.e.* the two metal electrodes act as plates of capacitor and insulating material act as dielectric media.

The capacitance of this capacitor varies according to or in proportion with the thickness of insulating material (sheet under test). This change in capacitance is calibrated in terms of thickness by means of resonance circuit arrangements. Due to variation in the value of capacitor, the resonance ($X_L = X_C$) gets disturbed which is indicated by thickness indicator.

(ii) Nuclear radiation method :

As the name implies this system uses nuclear radiations (alpha, beta, gamma, X-rays) radiated or generated from radiation source for the measurement of thickness of sheets. It has wide operating range. This is a non-contacting type method, it is useful for the measurement of thickness of hot and cold bodies and thickness of plastic, rubber, paper sheets.

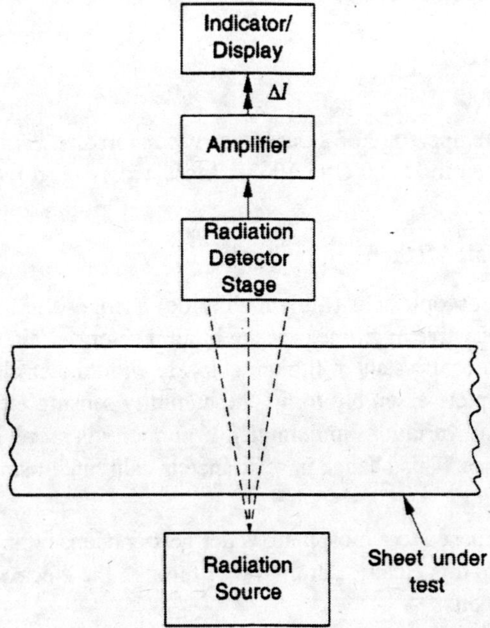

Fig. 4.18 : Nuclear radiation method.

Fig. 4.18 shows nuclear radiation method used for determining the thickness of sheet under test.

The sheet under test is passed through radiation source and radiation detector stage. Thickness of the sheet is determined in terms of radioactive decay or (attenuation) absorption of radiations. The relationship is :

$$\Delta I = I_0 [1 - \exp(-\mu t)] \quad ...(4.3)$$

where ; ΔI = Change in ionization current when absorber is inserted.

I_0 = Ionization current without absorber.

μ = Absorption coefficient ($cm^2/\mu g$).

t = Thickness ($\mu g/cm^2$).

The display or indicator is calibrated to read the value of thickness.

4.5 Measurement of Humidity

In various industrial process controls the amount of water vapor in air is determined. Humidity is defined as amount of water vapor present in a gas or air. In instrumentation systems these devices use following phenomenon for the measurement of humidity :

(1) **Relative humidity** : The ratio of mole fraction of moisture in a gas mixture to the mole fraction of moisture in a saturated mixture at the same temperature and pressure.

(2) Dew point (saturation) temperature. (°C/°F)

(3) Volume ratio or mass ratio (parts per million volume or weight).

(4) **Absolute Humidity** : The mass of water vapor present per unit volume.

Hygrometer is a transducer used for measuring humidity. This transducer is based on principle that some materials changes their electrical properties due to humidity.

4.6 Dew-Point Hygrometers

Dew point is defined as saturation temperature of a gas-water vapor mixture *i.e.* the temperature at which condensation of water starts over the range –80°C to 100°C. Most widely used methods to detect the dew point are as follows.

(A) Lithium Chloride Saturated Salt Method

It is based on principle that the hydroscopic salts (lithium chloride) changes their resistivity in accordance with humidity. Fig. 4.19 shows the resistive hygrometer using lithium chloride salt. It consists of thin walled hollow metal socket; in the hollow area, crystals of lithium chloride with binder elements are placed in the metal electrodes or wires. This complete assembly forms the humidity sensing element. The resistance of element changes when it is exposed to variations in humidity. If humidity is more, lithium chloride absorbs the moisture and lowers the resistance. This change in resistance is calibrated using Wheatstone Bridge. It is also known as resistive hygrometer.

When this sensing element is exposed to atmosphere, water gets condensed on the crystals resulting in formation of an ionic solution. Due to this electric current flows through the electrodes which is measure of resistance and hence of relative humidity.

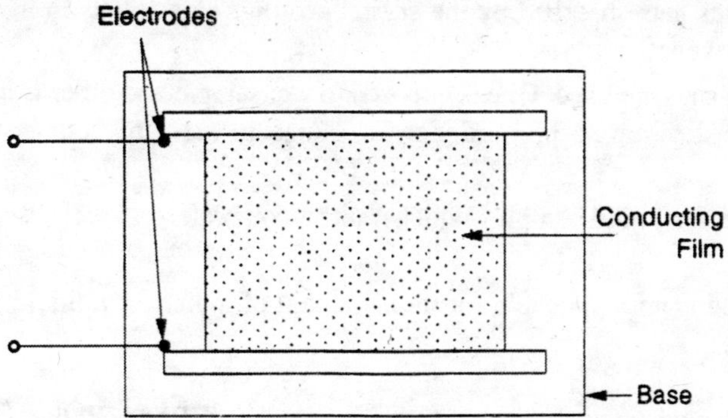

Fig. 4.19 : Resistive hygrometer using lithium chloride.

(B) Aluminium oxide capacitance hygrometers

The Fig. 4.20 shows capacitive hygrometer. It consists of a conductive aluminium substrate. Over this substrate a layer of porous aluminium oxide is deposited. The aluminium oxide is coated with thin film of gold. The aluminium substrate and gold film acts as electrodes.

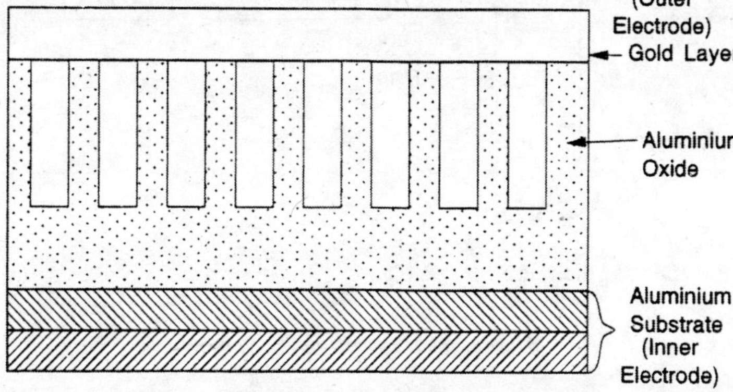

Fig. 4.20 : Capacitive hygrometer.

The capacitor is formed between the two electrodes of the capacitor (aluminium substrate and gold layer).

When this sensor is exposed to air, the aluminium oxide will absorb the water vapors through gold layer it changes the dielectric constant of aluminium oxide. Hence it results to change in the capacitance. This capacitance is measured by introducing frequency determining oscillator and other circuits. As the rate of water contents varies the capacitance of the sensor changes.

Now-a-days instead of aluminium oxide as a dielectric, hygroscopic polymer dielectric are used.

This sensor is small in size, wide humidity measurement range detects the low dew points. This type of hygrometer is not suitable for polar materials such as alcohol because the capacitor gets short circuited at frequency of 15 kHz.

(C) Piezoelectric Hygrometer

Piezoelectric crystals like quartz, barium titanate, Rochelle salt etc. has wider application area in electronic systems, communication systems and also in instrumentation systems, industrial applications for measurement of various parameters. These crystals are also useful for measurement of humidity. In humidity measurements the quartz crystals are coated with hydroscopic material. This crystal is exposed to the sample due to which the water contents gets absorbed by the crystal coating. This results in increase the mass and decrease in the oscillating frequency of crystal.

For measurement, two crystals are used. One is exposed to wet sample and other is to dry reference gas. Then frequency difference is determined proportional to change in mass of both crystals and moisture contents in the sample.

Application : Applicable for liquid samples which can be completely vaporized within the limits of instrument.

Limitations : This method is not applicable for measurement of moisture in liquid phase samples.

(D) Microwave Absorption Hygrometer

This system uses microwave frequencies of range 20 to 22 GHz *i.e.* K-Band. In this frequency range the water molecule produces molecular resonance.

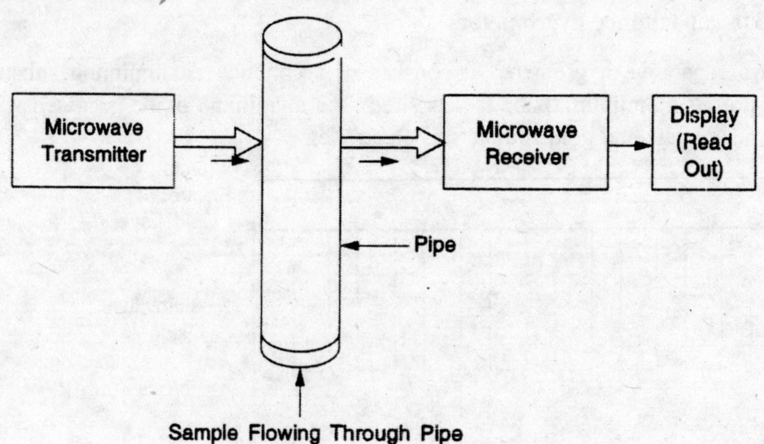

Fig. 4.20 (a) : Microwave absorption hygrometer.

Fig. 4.20(a) shows microwave absorption hygrometer. It consists of a microwave transmitter (Klystron, Magnetron) which transmits microwave frequency signal (beam) through the pipe in which sample is flowing. Receiver receives the reflected microwave beam. The receiving unit senses beam attenuation and phase shift due to moisture contents in the sample. The Readout is in terms of mass of moisture per unit volume.

Applications : Suitable for moisture content measurements in slurries and pastes.

Limitations : Not Suitable for moisture measurement in gases.

(E) Hair Hygrometer

Working Principle :

Some material such as Human Hair, Animal membranes, wood or paper changes their dimensions when they absorb moisture from atmosphere.

This property is used in Hair Hygrometer.

Construction :

The Fig. 4.20 (b) shows construction of Hair Hygrometer. It consist of a strand of Hair which is attached to the arm through mechanical linkage. The arm is connected to the pointer which slides over calibrated scale. The spring is connected to maintain the proper position of arm and the strand.

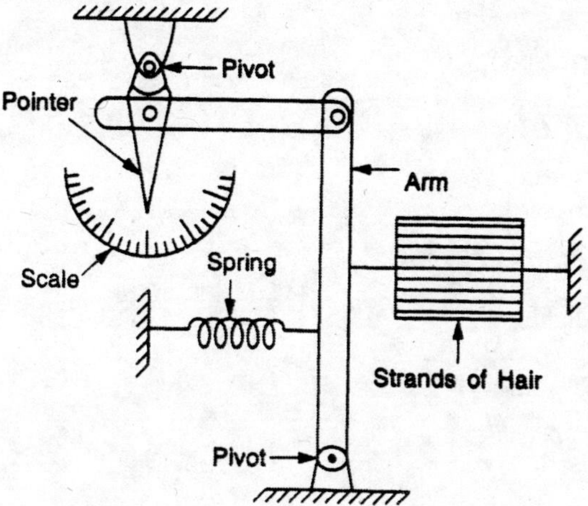

Fig. 4.20 (b) : Hair hygrometer.

Working :

According to the working principle *i.e.* when the hair absorbs the moisture their length gets changed. This change in the length is applied to the arm. Due to which the pointer slides over calibrated scale.

Application :

(i) In industrial applications.

(ii) It is used where high precision of measurement is not necessary.

Advantages :

(i) Small size

(ii) Low cost.

(iii) Easy installation

Disadvantages :

(1) Less accuracy.

(2) Periodic maintainces is necessary.

4.7 Measurement of Vibration

The term vibration relates with displacement, velocity and acceleration. So vibrations can be measured by using the transducers sensitive to displacement, velocity or acceleration. Measurement of vibration is generally useful in electric motor, power station turbines, generators, revolving machineries, installation of various structures, compressors etc.

Vibration is measured by its frequency and amplitude. Every vibrating body/element has mass, the frequency of vibration is a function of this mass. The amplitude of vibration is a function design of equipment *i.e.* it depends upon the designing parameters of the equipment whose vibrations are to be monitored (*e.g.* electric motor, compressor etc.).

The sinusoidal vibration displacement is given by :

$$x = A \sin \omega t \qquad ...(4.4)$$

where ; A = Amplitude.

ω = Angular frequency (rad/sec.)

Therefore velocity $V = \frac{dx}{dt}$ ...(4.5)

Therefore $V = \omega A \cos \omega t$ ...(4.6)

Maximum velocity $V_0 = A\omega$ ...(4.7)

Acceleration can be specified as :

$$\therefore \quad a = \frac{d^2x}{dt^2} \qquad ...(4.8)$$

Therefore $a = -A\omega^2 \sin \omega t$

Maximum acceleration $a_0 = -A\omega^2$ ...(4.9)

Thus the sinusoidal vibrations has frequency with amplitude or maximum velocity or maximum acceleration.

From the above discussion it is clear that displacement, velocity and acceleration are time related with each other.

If it is decided to measure any one parameter from displacement, velocity and acceleration; it can be determined by differentiation or integration.

Vibration is caused by any type of excitation which displaces some part of the mass or complete mass from its equilibrium position.

(A) (Self-generating) Piezoelectric Sensor

When mechanical force is applied to certain piezoelectric crystals (quartez, Rochelle salt etc.) an electric field appears across the surface plates of the crystal. This effect is called **Piezoelectric effect**. They do not require external power source so they are self-generating.

This sensor consists of a sensing element (piezoelectric crystal) which is sandwiched between the transducer body and seismic mass. The assembly is placed in housing.

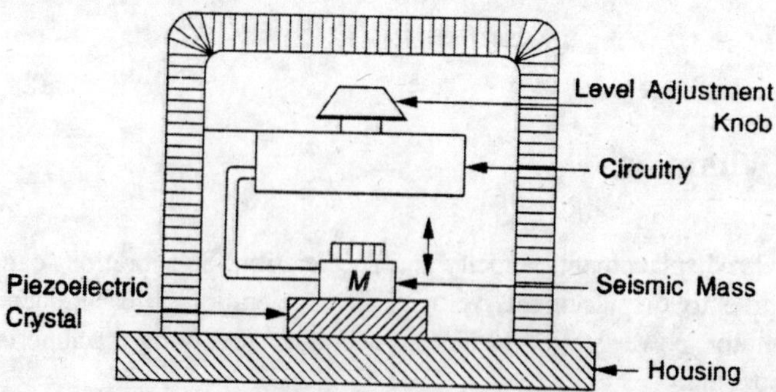

Fig. 4.21 : Piezoelectric crystal vibration sensor.

Fig. 4.21 shows piezoelectric crystal vibration sensor. The seismic mass is constant. Forces acting on sensing element are proportional to acceleration.

$$F \propto a$$

$$\therefore F = m \times a \qquad ...(4.10)$$

where ; m = mass

The quartz crystal produces an electric change which is proportional to force and therefore proportional to acceleration.

When the accelerating forces are applied to the transducer electrical voltage gets developed proportional to mechanical deformation of crystal and so proportional to acceleration. The developed electrical voltage is amplified in the amplifier stages and passed through control element. This amplified signal is provided to display devices or readout meters for monitoring or indication. Otherwise can be displayed on oscilloscope in the waveform manner for analysis.

Proper electrical isolation and protection is necessary for the device from noisy working conditions.

The operating vibration range is 19 to 1000 (g) Vibration Amplitude.

(B) (Non-self generating) Strain Gauge Sensor

This is a non-self generating sensor because it requires external power supply for operation. It senses the accleration.

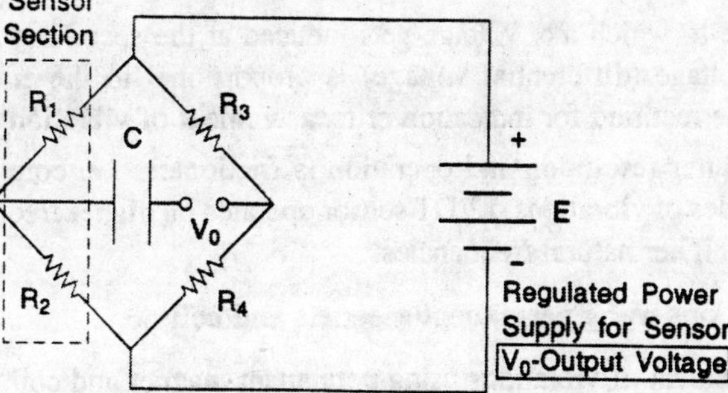

Fig 4.22 : Strain gauge vibration sensor.

The above Fig. 4.22 shows strain gauge vibration sensor. This sensor consists of a Wheatstone bridge having the arm arrangement such as semiconductor strain gauges are connected in two arms. In the bridge circuitry half bridge section senses tension and other half senses the compression. R_1 and R_2 are the strain gauges (sensor section).

Strain gauges measure a change in electrical resistance which is proportional to force because of acceleration. As stated above acceleration is related with vibrations, the meter is calibrated so as to read the voltage proportional to acceleration.

Because of accleration the resistance of strain gauge gets changed which unbalances the bridge. This voltage drop across bridge (output voltage) is recorded or indicated by the indicator.

Ranges : Range of this method is 5 to 1000 g. (Vibration Amplitude) 0 to 1000 Hz (Vibration frequency).

(C) LVDT Sensor

It measures the vibrations in terms of acceleration. LVDT produces an AC voltage proportional to displacement of movable ferromagnetic core. Fig. 4.23(a) and 4.23(b) shows LVDT sensor.

The core of LVDT acts as the mass. The spring action is produced by connecting the flexible reeds to the core of LVDT. Vibrations which are to be measured or monitored are applied to the reeds. These reeds are connected to the housing. When the vibrations are applied to the core, it moves in up and down directions

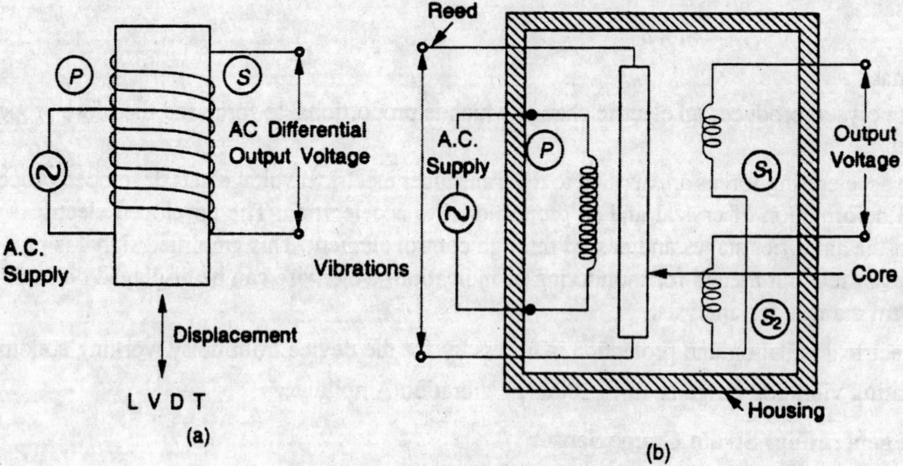

Fig. 4.23 : (a) and (b) LVDT sensor.

according to vibrations. Due to which a.c. voltage gets induced at the secondary windings of LVDT. The magnitude of output a.c. voltage (differential voltage) is proportional to the amplitude of vibrations. If necessary, this voltage can be rectified for indication or measurement of vibration amplitudes.

This system provides better resolution and operation is frictionless *i.e.* core moves in the air gap in accordance with the amplitudes of vibrations. LVDT sensor operates on higher frequencies because the mass of LVDT core is small with higher natural frequencies.

(D) Measurement of vibrations using permanent magnets and coil

The Fig. 4.24 shows measurement of vibrations using permanent magnet and coil. This method is used for measuring vibrations by determining it in the form of velocity.

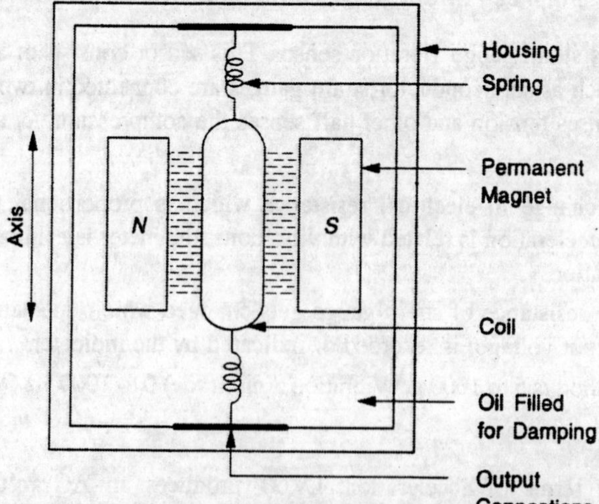

Fig. 4.24 : Vibration measurement using permanent magnet and coil.

It consists of an electrical coil suspended by a spring in the magnetic field of permanent magnets, placed in silicone oil. This complete assembly is mounted in a housing. The housing is placed in contact with the vibrating surface. The coil has low natural frequency by which it moves with housing.

When the vibration velocity is applied to the coil, it induces voltage in the coil. This amplitude of voltage is proportional to the vibration velocity, with the frequency equal to vibrational frequency. The induced voltage is monitored or displayed on a read-out.

4.8 Measurement of Torque

Torque measurement is widely useful in engineering applications. Specially it is used in rotating machines, electrical machines (AC and DC drives), pumps, turbines, generators etc. to measure the developed torque.

The term torque is defined as that force which tend to produce the rotations or to change the linear motion. It can also be defined as tangential force required to set a body into revolutions or rotations.

Mathematically torque is specified as :

$$T = F \times D \qquad ...(4.11)$$

where ; T = Torque

F = Force

D = Perpendicular distance from the axis of rotation or line of action of force.

Due to torque, angular displacement of the body takes place about its axis of rotation.

Torque measurement systems are classified as :

(a) In–line rotating torque sensors.

(b) In–line stationary torque sensors.

Transducers or sensors used for measurement of torque is designed with two basic stages :

(1) A structure which deforms microscopically when torque is applied to it. (Deformation stage).

(2) Conversion of this deformation into measurement signal. (Conversion stage).

(A) In–Line rotating torque sensors

For the measurement of torques of rotating shaft (motors or pumps) In–line rotating torque sensors are used. These type of sensors are useful for the measurement of large torques.

Fig. 4.25 shows In–line rotating torque sensor which consists of bonded strain gauges placed on a

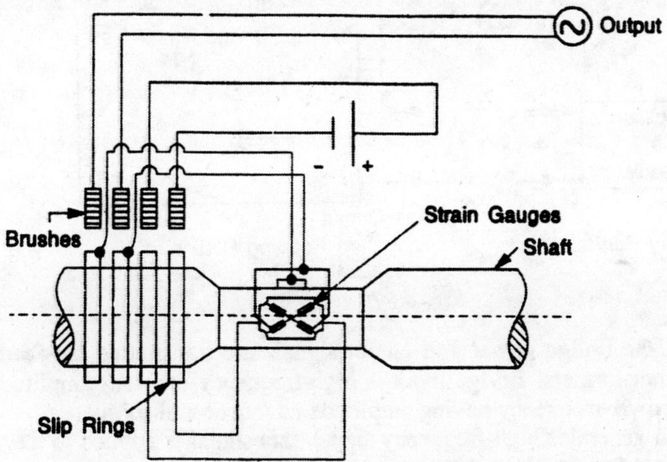

Fig. 4.25 : In-line rotating torque sensor.

rotating circular metal shaft. Strain gauges are connected in Wheatstone bridge configuration having electrical connections.

In one direction at a 45° angle to the axis, pure tensile stress exists and 45° in the other direction pure compressive stress is extant. The strain gauges are placed on the shaft of motor so as to sense the compressive and tensile deformation due to torsion.

The output of Wheatstone bridge is proportional to torsion and hence proportional to the torque.

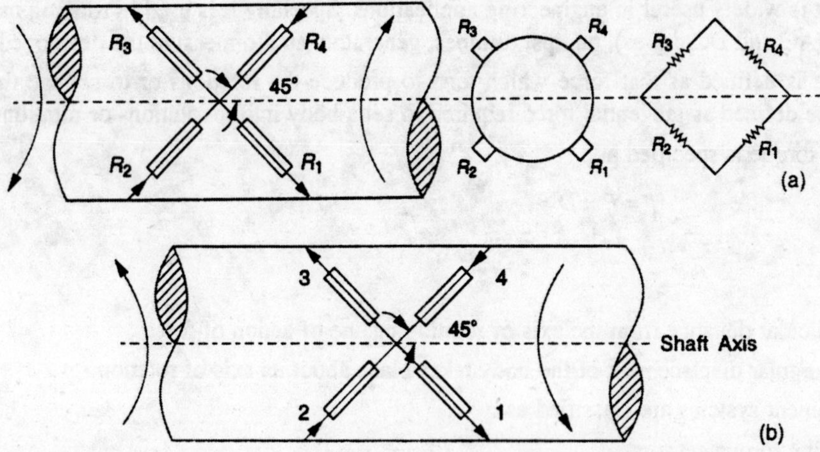

Fig. 4.26 : (a) Connections of strain gauges, (b) Slip ring arrangement with bridge

Fig. 4.26 (a) and (b) shows the connections of strain gauge elements using slip rings upon which the brushes are placed.

The slip ring arrangement produce noise due to changes in contact resistance so non-contacting type torque sensor is used which is discussed below.

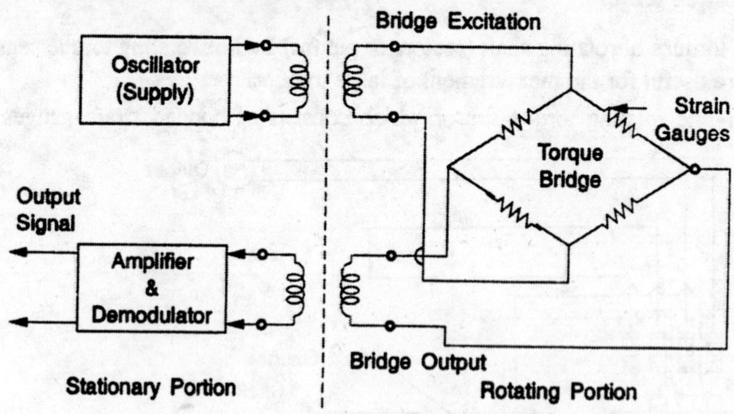

Fig. 4.27

Consider Fig. 4.27, here, the bridge power and output signals are transmitted between rotating and stationary members through transformers. Bridge input is high-frequency constant amplitude signal. The output signal is sine wave of power frequency having amplitude as function of torque.

The input oscillator circuit generates high frequency signal, this signal is applied to a.c. bridge where the modulation takes places due to torque. The output stage includes demodulator stage to suppress the carrier frequency of oscillator. All output stage DC signal in proportion to the peak of sine wave is produced which is shown in Fig. 4.27.

(B) Torsion meter

The relationship between torque 'T' and diameter 'd' of the solid shaft is specified as :

$$T = f_s\,(\pi/16 \times d^3) \qquad ...(4.12)$$

Therefore $$\gamma = f_s/G \qquad ...(4.13)$$

$$\varepsilon_{45} = \gamma / 2$$

$$\therefore \quad T = \frac{G\theta\pi d^4}{32L} \qquad ...(4.14)$$

L = Shaft Length

where ; f_s = Shear stress induced in shaft.

G = Shear modulus.

γ = Shear strain.

ε_{45} = Longitudinal strain in the shaft at 45° to the axis.

θ = Twist in the shaft (Angular twist).

From above equation if the angular displacement or twist 'θ' is measured the torque 'T' can be easily calculated. Following arrangement shows the measurement of 'θ' *i.e.* angular twist. The angular displacement between two sections of a rotating shaft is related to torque.

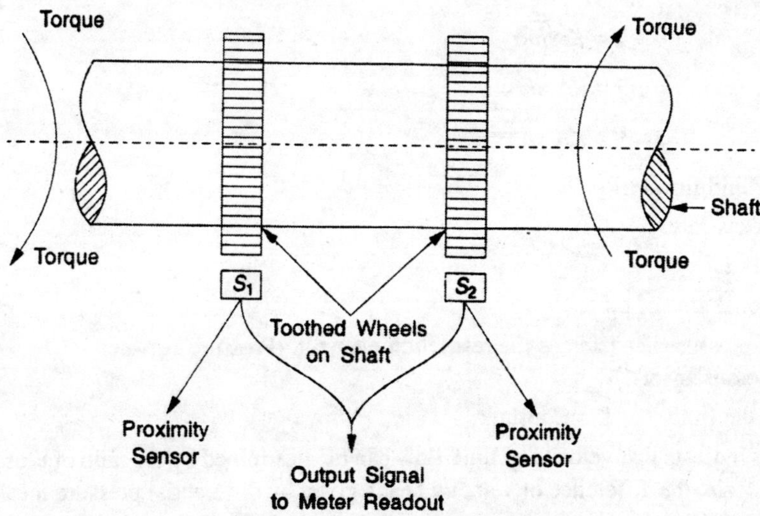

Fig. 4.28 : Torsion meter.

Fig. 4.28 shows torsion meter used for torque measurement. It consists of two identical toothed wheels which are fixed on the shaft at certain distance. The two proximity sensors (s_1 and s_2) produce output voltages whose phase difference is proportional to torque. A photocell and light source can be mounted on opposite ends of wheel. The amount of light falling on photocell is related to torque.

4.9 Flow Measurement

In instrumentation systems and process control flow measurement is an important method for the measurement of flow rates of different process variables. Flow measurement devices are widely useful in various industrial applications to measure the flow rates of liquids, different fluids, vapors, gases, slurries etc. The fluid matter may be clean or dirty, wet or dry.

By mechanical means the flow measurement system or device operates by placing an obstruction in the fluid path. Due to this obstruction pressure gets changed with respect to rate of flow. The differential pressure sensor is used to measure the pressure before and after the obstacle. This differential pressure provides the rate of flow.

Flow measurement can also be possible by electrical methods like allowing the liquid to flow in a magnetic fields, change in the resistance of an element placed in fluid path etc. The developed electrical voltage is calibrated so as to read the proportional rate of flow.

(A) Head flow meters

The head flow meter changes the velocity of fluid flow and creates differential pressure or pressure head in the fluid. These are the primary devices. This type of flowmeter uses Bernoulli's theorem. This theorem states that **in a fluid flowing stream, the sum of pressure head, velocity head and the elevation head at one point is equal to their sum at another point removed in the direction of flow plus loss due to friction between two points.**

The head flowmeters are orifice plates, venturi tubes, flow nozzles, weirs etc.

Velocity head is defined as the vertical distance through which a liquid would fall to attain a given velocity. The pressure head is the vertical distance which a column of the flowing liquid would rise in an open-ended tube as a result of the static pressure.

The basic equations for flow rates are ;

$$V = K\sqrt{h/\rho} \quad ...(4.15)$$

$$Q = KA\sqrt{h/\rho} \quad ...(4.16)$$

$$W = KA\sqrt{h\rho} \quad ...(4.17)$$

where ; V = Velocity of fluid flow.

Q = Volume of flow rate.

W = Mass flow rate.

A = Cross-sectional area of pipe.

h = Differential pressure head across the restriction element. (Pressure between points of measurement)

ρ = Density of the fluid. K = Constant

The above equations indicate that velocity of fluid flow can be determined by the ratio of cross-sectional areas of pipe and flow nozzle, the difference of velocity heads given by differential pressure measurements.

(B) Characteristic of Head meters

From the basic equation two important characteristics are studied (1) Square root relationship between flow rate and differential pressure (2) for the measurement of volume and mass flow consider the density of the fluid which is flowing. The characteristics are as follows :

(1) Square root relationship

(2) Density of the fluid which is flowing

(3) β (Beta) ratio

(4) Reynolds number.

(1) Square root relationship

The primary sensors such as orifice, venturi tube etc. develop the differential pressure head. A linear read-out of differential pressure expands the high end of the scale and compresses the low end in terms of flow.

Fifty percent of full flow rate produce 25% of full differential pressure. At this point a flow change of 1% of full differential. At 10% flow the total different pressure is 1% and a change of 1% of full scale flow results in only 0·2% full scale change in differential pressure. It suffers both accuracy and readability.

(2) Density of the fluid which is flowing

Consider basic equations. Head type meters does not provides the direct readings of mass of volume flow. Because density is square root and it is advantageous in mass flow measurements. The square root relationship reduces any error which may exist in the value of density used for measurement of mass flow.

(3) β (Beta) ratio

The head meters works on the principle that the restriction in flow path produces change in velocity. If the pipe and restriction are circular then 'β' is the ratio between diameter of circular restriction and internal diameter of the pipe. The ratio between velocity in the pipe and velocity at the restriction is equal to ratio of areas or β^2. 'β' is specified as it is a diameter ratio. Therefore

$$\beta = \frac{\text{Diameter of restriction element}}{\text{Internal diameter of pipe}}$$

β = Diameter ratio

If the pipe is non circular then 'β' is defined as the square root of ratio of area of restriction to the area of pipe.

(4) Reynolds number

It is a reference number used for the determination of flow. These numbers were discovered by Sir Osborne Reynolds in 1883. It is expressed as :

$$\text{Re} = \frac{VD\rho}{\mu} \quad ...(4.18)$$

where ; V = Velocity of fluid

D = Diameter of pipe (internal diameter)

ρ = Density of fluid

μ = Absolute viscosity

These numbers are used to determine the point at which the flow goes from viscous to turbulent stage.

Flow "noise" : The term flow noise represents variations in the **differential pressure**. These variations are developed at the differential pressure sensing devices.

The **noise** or variations are produced because of complex patterns of random velocities due to turbulent flow.

This noise or variations can be reduced by using two or more pressure sensing taps in parallel with high and low differential pressure connections.

(C) Measurement of Flow using Flowmeters.

Differential pressure flowmeters : The differential pressure flowmeter operates on the principle that if an obstruction is placed by means of devices such as orifice plates, venturi tubes, elbow, and the fluid flows through the pipe, differential pressure is developed at obstruction element. This differential pressure is proportional to the rate of flow. The different pressure can be measured by using pressure sensing elements like U-shaped manometers, Bellow, Bourdon tubes etc.

(a) **Venturi tube** : Venturi tube consists of three sections :

(i) **Convergent cone entrance** : At the inlet section there is convergent cone entrance having same diameter as the pipe. At this point high pressure tap is located, at which velocity head increases and pressure head decreases.

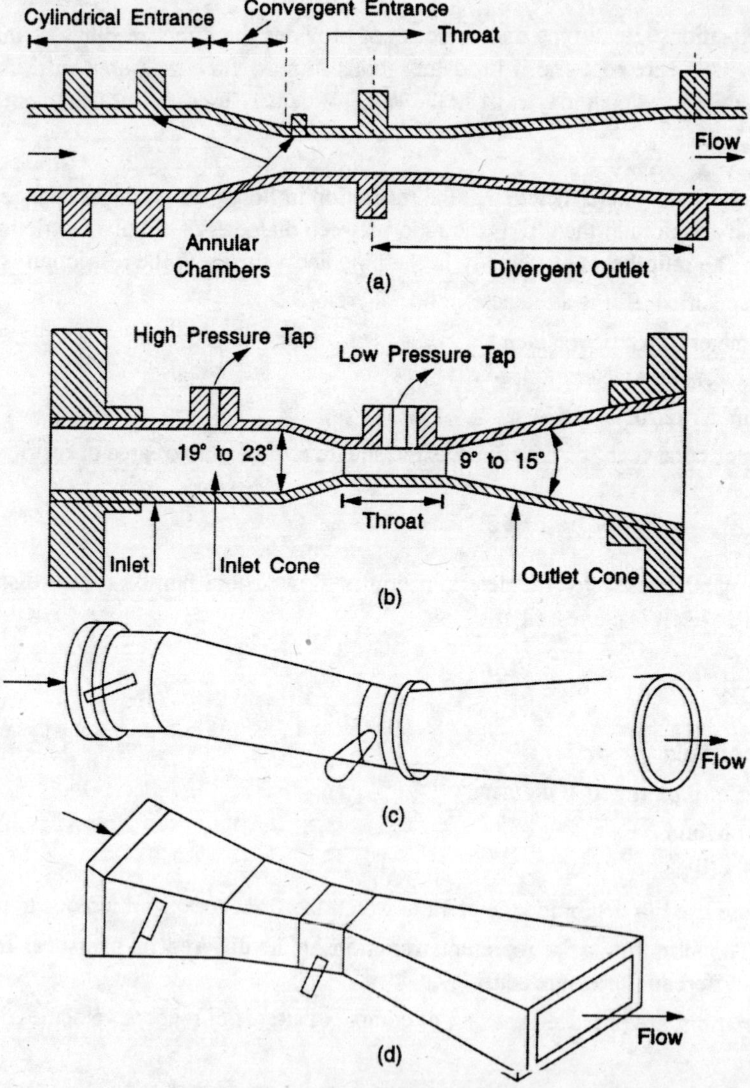

Fig. 4.29 : Venturi tube : (a) Classic type, (b) Short form venturi tube, (c) Eccentric venturi tube, (d) Rectangular venturi tube

(ii) **Throat section** : The throat section is cylindrical in shape. At this section the velocity of fluid is constant or flow rate is steady.

(iii) **Divergence outlet cone** : The diverging outlet cone expands from the throat section. In this section velocity of fluid decreases and maximum pressure head gets recovered.

Venturi tubes are classified as :

(1) Classic or long form

(2) Short form

(3) Eccentric type

(4) Rectangular type

(1) Classic or long venturi tubes : It is a differential pressure transducer used when high accuracy is necessary. Basically, venturi tube consists of cylindrical inlet section equal to the diameter of pipe. A converging conical section is provided in which cross sectional area decreases by increasing the velocity with increase of velocity head and decrease of pressure head. In cylindrical throat section velocity of fluid is constant and decreased pressure head is measured. Further, in the diverging outlet cone velocity of fluid decreases and original pressure head is recovered. The unrecovered pressure head is known as head loss.

Fig. 4.29(a) shows classical type venturi tube. Materials used for this type of venturi tube is cast iron body and throat is made up of bronze or stainless steel. At the mid point of throat six to eight pressure taps connect the throat to an annular chamber so that throat pressure is averaged. This tube senses static pressure because there is no measurement of fluid in annular chamber.

The classic venturi meter is applicable for the flow measurements of clean, non-corrosive fluids and gases.

(2) Short form venturi tube : Fig. 4.29(b) shows short form venturi tube. It consists of a straight inlet section of same diameter as the pipe over which a high pressure tap is installed whereas the low pressure tap is placed at throat. The pressure taps (high pressure and low pressure) are placed at one-quarter to one-half pipe diameter up-stream of inlet cone and at the middle part of throat section.

Short form venturi meters are applicable for flow measurement of slurries and dirty fluids because this tube has no sharp corners and does not projects inside the fluid stream. For dirty fluids or liquids it is important aspect to protect the pressure taps from plugging.

(3) Eccentric venturi tube : Used for handling mixed phases or to avoid build up of heavy materials. Fig. 4.29(c) shows eccentric venturi tube.

(4) Rectangular venturi tube : Fig. 4.29(d) shows rectangular venturi tube and it is used in duct work.

Advantages of Venturi tubes :

(1) Useful for high flow rates.

(2) Useful for pipes of small diameter as well as large diameters.

(3) More accuracy in measurement as compared to orifice and flow nozzles.

(4) It has low permanent pressure loss.

Disadvantages of venturi tubes :

(1) High cost.

(2) Difficult maintanance and inspection due to its construction.

(3) Limitation of lower Reynolds number of 150,000.

(4) Careful installation is necessary.

Size : Venturi tubes are available in following ranges :

(a) 1 in. up to 120 inches.

(b) 4 in. up to 48 inches.

(c) 1 in. up to 60 inches.

Reynold number range is Re > 100,000 with beta (β) ratio 0·3 to 0·75.

Applications of venturi tubes : Used for measurements of liquids, gases and steams.

(b) **Orifice plates** : Orifice plates are head type flowmeters.

They are also differential pressure flow meter. These devices change the velocity or direction of fluid, resulting in developing **differential pressure** or **pressure head**. This differential pressure can be measured by **pressure sensor** or transmitter.

Orifice plates are the simple and cheap devices for flow measurements, these are primary devices and placed in the flow line *i.e.* in pipe line and differential pressure is developed. The secondary elements will measure the pressure difference.

The orifice plate is a circular metal disc (nickel, Monel, stainless steel, phosphor bronze) of thickness 1/8 to 1/2 in. Thickness of metal disc depends on size, temperature of flow, the development of differential pressure across the orifice plate, diameter of pipe, Velocity of flow. This plate has a circular hole (orifice).

Different types of orifice plates are :

(1) **Concentric orifice plate** : This type of plate has a circular hole in the middle as shown in the Fig. 4.30(a).

(2) **Eccentric orifice plate** : In this type of plate the offset hole is bored which is tangent to the inside wall of the pipe or tangent to a concentric circle with diameter of 98% of internal diameter of the pipe. Fig. 4.30(b) shows eccentric type orifice plates.

These type of plates can be used for the measurement of flow gas containing liquid and liquid containing gases.

The pressure taps should be located in the quadrants directly to the hole.

Segmental orifice plates : Fig. 4.30(c) shows segmental orifice plate. Here the hole which is segment of a circle. The segment arc is parallel to pipe wall. It is used for flow measurement of vapor containing liquid and liquid containing solids. The pressure taps should be in line with the maximum dam height.

Quadrant edge orifice plates : These plates are useful for lower pipe Reynolds numbers. Fig. 4.30(d) shows quadrant edge orifice plate.

Integral orifice : This type of orifice plates are used for the measurement of small fluid flows. Basically they are flow restrictors and act as primary sensors for small flow rates. Integral type orifice plate consist of a plate with small hole which restricts flow, mountings, differential pressure sensor. Fig. 4.30(e) shows integral orifice.

Applications of Integral orfice :

(1) Used for laboratory work, additive measurements of higher flows.

(2) In general all the orifice plates are used for liquid, gas, vapor flow measurement.

Advantages of orifice plates :

(1) Low cost.

(2) Simple construction and higher reliability.

(3) Used with differential pressure sensors.

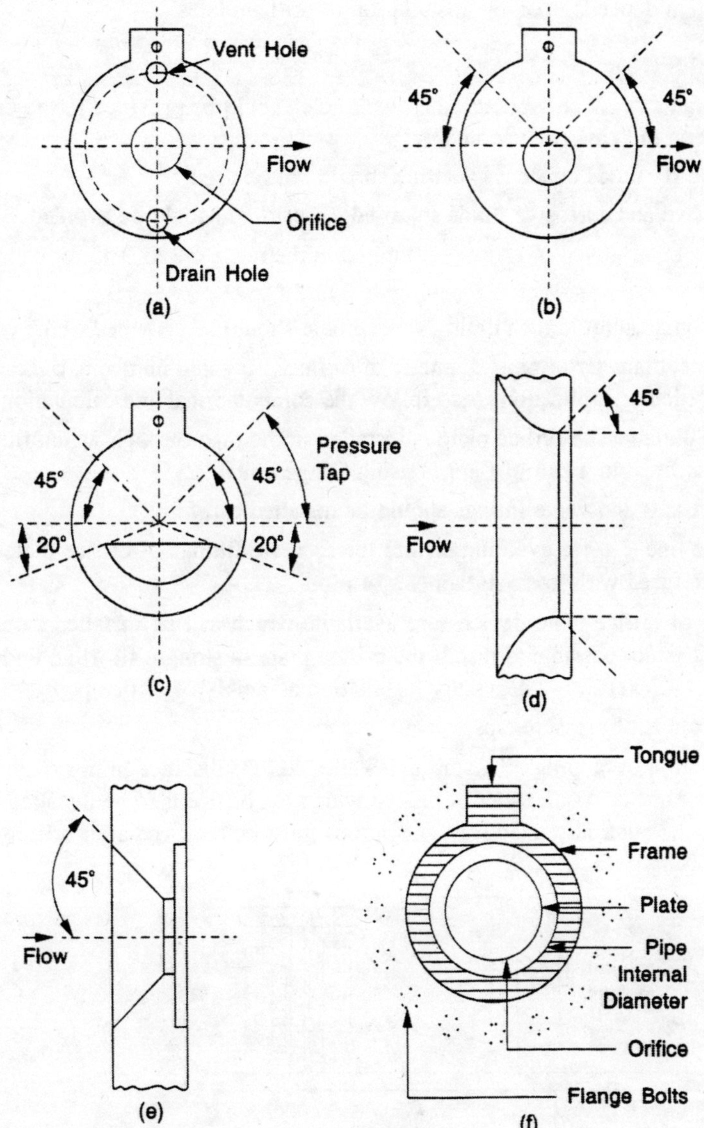

Fig. 4.30 : Orifice Plates : (a) Concentric orifice plate, (b) Eccentric orifice plate, (c) Segmental orifice plate, (d) Quadrant orifice plate, (e) Integral orifice plate (f) Installation of orifice plates.

(4) Easy installation and replacement.

(5) Maximum size is available according to the size of pipe.

Disadvantages :

(1) High amount of pressure loss.

(2) For slurries and dirty fluids some orifice (concentric type) is not useful because of accumulation of solids near the orifice plates.

(3) Orifice plates may be affected by corrosion, scaling, erosion.

(4) Accuracy of system depends upon the installation of orifice plates.

Limitations of orifice plates :

(1) If the liquid contains "gas" or "vapour" then vent hole should be provided to the orifice so as to avoid accumulation of "gas" or "vapours".

(2) For slurries and dirty fluids concentric orifice should not be used.

(3) For strongly erosive and corrosive fluids sharp-edged orifices should be avoided.

(4) The accuracy can be achieved if 'β' ratio is limited in the range 0·2 to 0·65 for concentric orifice plates.

(5) If the gas flow contains liquid then drain or weep hole should be provided to the orifice.

(6) Consider flow rate, diameter of pipe, diameter of orifices, Reynold numbers, β etc. for designing the orifice for particular application. Also follow the computational and calculation approach.

Installation of orifice plates : The orifice plate is installed in the pipe through which fluid is flowing. For installation the orifice is fitted in a pair of flanges using flange bolts.

Necessary mounting gaskets and valve fittings should be installed carefully.

Prefer straight run pipe line (*i.e.* for avoiding errors) for accurate fittings of orifice plates. The orifice hole should be properly interfaced with internal diameter of pipe.

Now-a-days such type of orifice plate devices are available which can be installed carefully without disturbing the flowline. If it is not possible to install the orifice plate in straight flowline under such cases use straightening vanes. Fig. 4.30(f) shows necessary installation assembly of orifice plate.

The important installation is of pressure taps.

Fig. 4.31 shows installation of tappings. The tap is installed at "*D*" distance in upstream and installed at "*D*/2" distance in downstream. "*d*" is diameter of pipe at which the orifice is to be installed. For accurate installation of pressure taps the burrs at the tap fitting locations must be removed after drilling process.

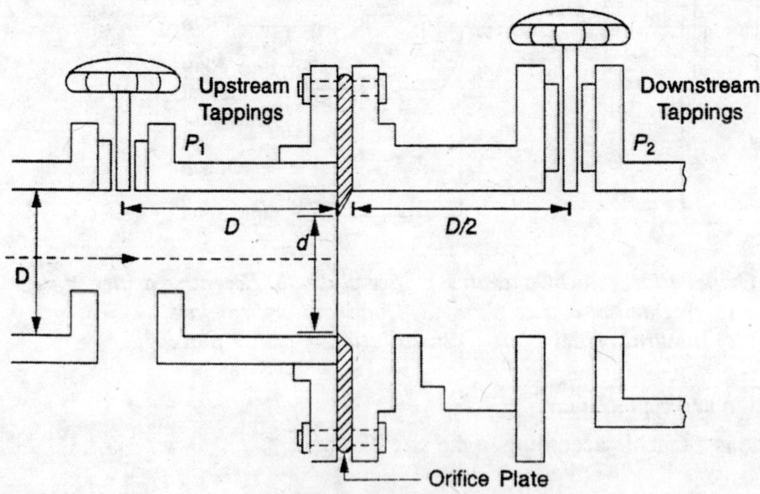

Fig. 4.31 : Installation of orifice.

(c) **Flow nozzle** : It is also a differential pressure developing primary sensor. It consists of a quarter ellipse convergence section and cylindrical throat-section as shown in Fig. 4.32.

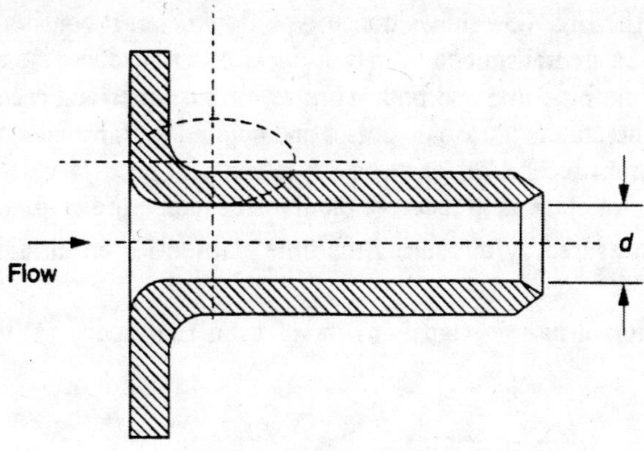

Fig. 4.32 : Flow nozzle.

Flow nozzles are installed in the pipe line by weldiing or placed in holdiing ring, mounted between flanges. Materials used for flow nozzles are aluminium, stainless steel, chrome-moly steel. The differential pressure measurement taps are installed at one pipe diameter in upstream and one-half pipe diameter in downstream from inlet.

Flow nozzles can be used for Reynolds numbers 50,000 or above, high co-efficient of discharge upto 0·99 and wide Beta range of 0·2 to 0·8, low pressure loss than orifice plates.

Applications :

Useful for slurries, conductive liquids, chemicals, gases, corrosive fluids, non conductive liquids.

Applications :

Suitable for measurement of steam flow and wet gases, fluids containing solids, high velocity fluids.

Flow nozzles can be installed in any position with pipeline. For wet steam, fluids containing solids vertical down flow position is used.

(d) **Ultrasonic flowmeters** : The ultrasonic flowmeters are classified into two types :

(i) Transit–time design for clean fluids.

(ii) Doppler reflection type for dirty liquids and slurries.

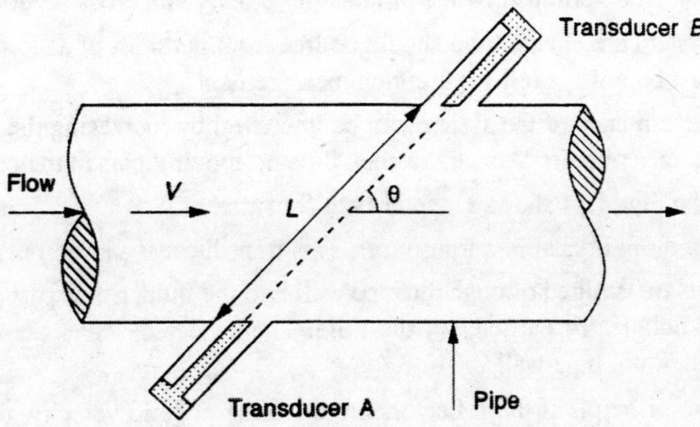

Fig. 4.33 : Transit-time flowmeter.

(i) **Transit–time flowmeter** : The Fig. 4.33 shows transit type flowmeter. It consists of two transducers A and B. The ultrasonic waves are transmitted from transducer A to transducer B and vice-versa. These transducers are placed into the pipe line and both work as a transmitter and receiver. An electronic oscillator is connected to generate the ultrasonic pulses and applied to transducer 'A' or 'B' which act as a transimitter. Similarly, transducer A or B acts as a receiver with necessary detector and amplifier stages. The function of detector stage is to measure the transit time period elapsed.

The flow of fluids is measured by measuring the time elapsed for an ultrasonic pulse to travel through pipe section.

Let, the time elapsed for ultrasonic signals to travel from transducer 'A' to transducer 'B' is 'tAB' which is given by :

$$tAB = \frac{L}{(C + V . \cos\theta)} \quad ...(4.19)$$

Also, the time period elapsed from 'B' to transducer 'A' is given by :

$$tBA = \frac{L}{(C - V . \cos\theta)} \quad ...(4.20)$$

where ; C = Velocity of sound in the fluid.

L = Acoustic path length between A and B.

θ = Angle of flow path with respect to pipe axis.

V = Velocity of fluid in pipe.

Therefore the time difference between tAB and tBA is calculated as follows :

$$\therefore \Delta t = tAB - tBA = \frac{2 . L . V \cos\theta}{C} \quad ...(4.21)$$

$$\therefore V = \frac{\Delta t . C}{2L \cos\theta} \quad ...(4.21a)$$

$$\text{Therefore } V = \frac{\Delta t . L}{2 . \cos\theta . t_A^2} = K . \Delta t / t_A^2 \quad ...(4.22)$$

Where t_A is the average transit time between the transducers.

Thus the flow rate can be determined by the product of velocity and cross-sectional area of pipe.

Limitation for this system are (1) the fluid should be free from particles of solid materials and air bubbles. (2) The pipe must be always full for accurate measurement.

The accuracy and performance of the system can be improved by increasing the transducers like two, four, six or more number of pairs. Also there should be no moving part in the operation.

(ii) **Doppler flowmeters** : The Fig. 4.34 shows Doppler-type flowmeter.

It consists of a transmitting crystal in a transducer. This transducer is placed outside the pipe.

The ultrasonic wave is transmitted through the pipe wall into the fluid. Some part of the ultrasonic wave is reflected by the bubbles or particles in the fluid. The reflected waves are detected by the receiver transducer placed at the pipe wall.

According to Doppler principle if the reflectors are traveling at fluid velocity, the frequency of reflected wave is shifted in proportion to the flow velocity.

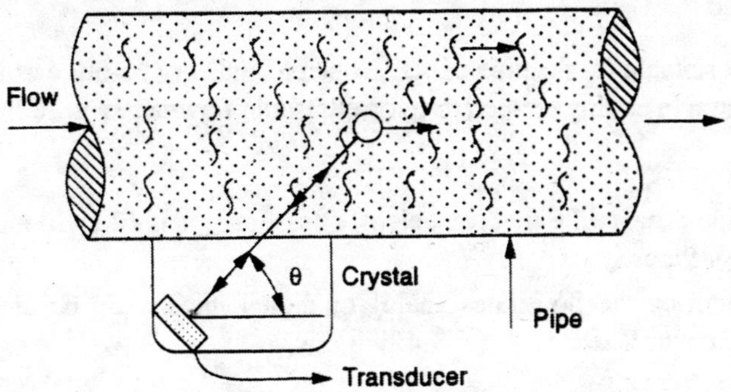

Fig. 4.34 : Doppler type flowmeter.

Therefore, the velocity of fluid is given by,

$$V = \frac{\Delta f \,.\, C_t}{2 \,.\, f_0 \cos \theta} = \Delta f K \qquad ...(4.23)$$

where : Δf = Difference between transmitted and received frequency.

C_t = Velocity of sound in transducer.

f_0 = Transmission frequency.

θ = Angle between transmitter and receiver crystal with respect to pipe axis.

K = Constant.

For both transmit-time and Doppler effect flowmeters horizontal pipes are suitable because easy and proper installation of transducers on circumference of the pipe is possible.

Advantages of ultrasonic flowmeters :

(1) More accurate.

(2) No moving part is involved in process or system so no chance of sparking or arcing.

(3) Installation is possible without shutting down the process.

(4) Performance does not get affected by environmental parameters such as moisture, temperature, dust, corrosion and oxidation.

(5) Design can be altered or changed according to flow rates.

(6) Suitable for corrosive, erosive, slurries and viscous fluids.

(7) Highly reliable.

(8) Operating speed is more.

Disadvantages of ultrasonic flowmeters :

(1) High cost.

(2) Maximum operating temperature for ultrasonic flowmeters is 100°C.

(3) The (construction) pipe line should be tight.

(4) Pipes of the materials like cement, clay and very porous cast iron are not suited for Doppler-type flow meters because they absorb the ultrasonic energy.

(5) Air bubbles or solid particles present in the liquid may affect the accuracy of system.

Limitations :

Doppler flowmeter is not suitable for the pipe of clay, concrete and very porous cast iron as they absorb the ultrasonic energy. Maximum operating temperature for ultrasonic flowmeters is 100°C.

For Information

The necessary straight pipe distance from valves, elbows, tees, pumps is 10 to 20 meters upstream and 5 meters downstream for clean fluids.

For displaying and monitoring the flow rates analog or digital meters can be used, simultaneously alarming function can also be controlled.

(e) **Thermal flowmeters** :

Theory of thermal flowmeters : In these type of flowmeters the temperature of fluids or liquids is used for measurement of flow rate. These flowmeters can be used to measure flow rates in terms of mass. They are classified into two types :

(1) If a known amount of heat is added to the fluid which is flowing, due to which its temperature increases. Such instruments are called as **Heat Transfer Flowmeters.**

(2) If the fluid or liquid is passed through a hot body or heated element the effect of temperature on a hot body is observed or measured. Such instruments are known as **Hot-wire Probes** or **Hot-wire Flowmeters.**

Working Principle :

(a) **Heat transfer flowmeters** : These instruments operate on following equation.

$$Q = WC_p (T_2 - T_1) \qquad ...(4.24)$$

where ; Q = Heat transferred (cal/hr).

W = Mass flow rate of fluid. (kgm/hr)

C_p = Specific heat of fluid. (cal/kgm°C)

T_1 = Temperature of fluid before heat is transferred to fluid (°C) (initial temp.)

T_2 = Temperature of fluid after heat is transferred (°C) (final temp.)

$$\therefore \qquad W = \frac{Q}{C_p (T_2 - T_1)} \qquad ...(4.25)$$

In this type of flowmeters the heat is added to the fluid flow with the help of heating element.

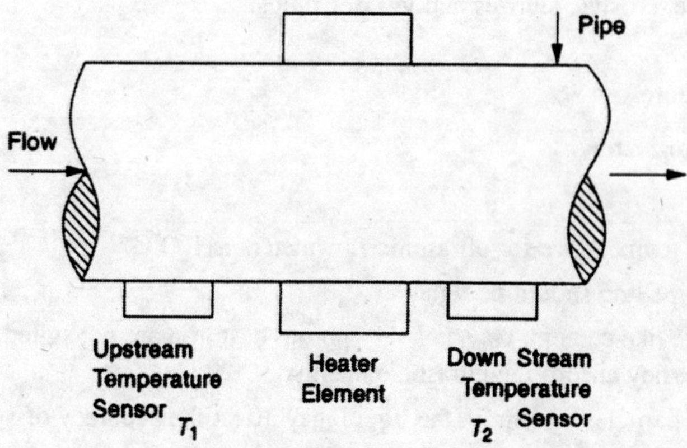

Fig 4.35. Heat transfer flowmeter.

Fig. 4.35 shows arrangement of heat transfer flowmeter. It consists of two temperature sensors. One is placed before the heater element so as to read the temperature in initial condition *i.e.* (T_1) at upstream end.

Another sensor element is placed after the heater element so as to measure the temperature of heated fluid at downstream flow *i.e.* T_2.

Thus the temperature difference of T_1 and T_2 is measured and the value of W *i.e.* mass flow can calculated proportional to flow rate.

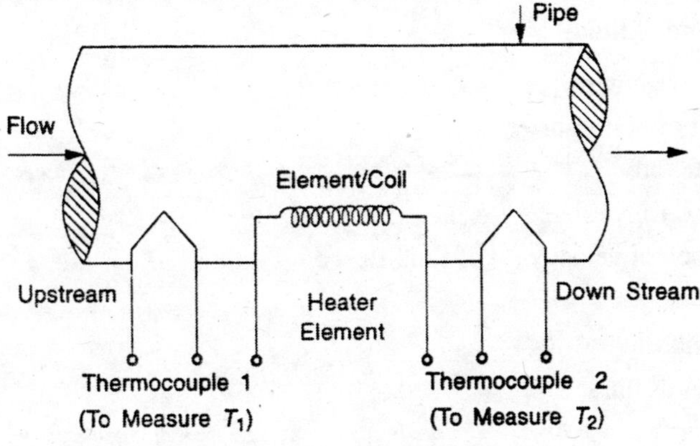

Fig. 4.36.

The heater element used may be electric immersion heater or transformer. The temperature sensors used may be either thermocouple or thermistor as shown in Fig. 4.36.

(b) **Hot wire instruments** : Hot wire anemometers are widely applicable for the measurement of flow of gases and liquids. It consists of an electrically heated wire, which is placed in fluid flow. Due to fluid flow the heat of the wire changes. So the relationship between heat loss and fluid flow is determined. Simultaneously because of heat loss, temperature changes results to change in the resistance of hot wire. By measuring the change in resistance or change in current the flow rate can be measured.

(i) **Hot wire anemometer** : Fig. 4.37 shows arrangement of hot wire anemometer. It consists of thermoelement wire made up of platinum having diameter 0·0005 to 0·03 cm and length of 0·1 to 1 cm. The dimensions of the wire depends upon size of the pipe and fluid flow, generally the length of wire is half of the pipe diameter. The wire should sustain fluid pressure, forces and shocks. This thermoelement or wire is placed in the flow stream.

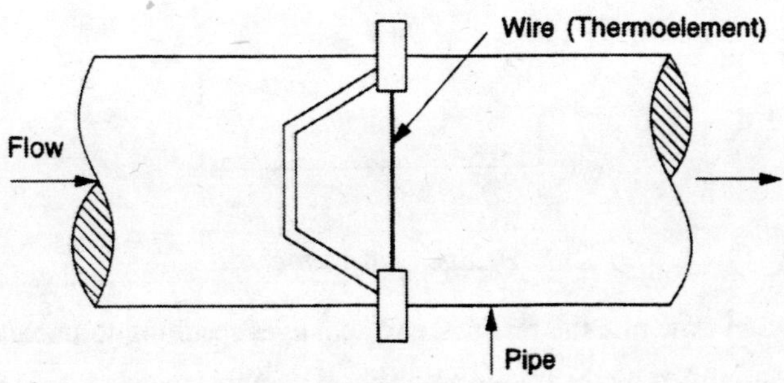

Fig. 4.37 : Hot wire anemometer.

When the fluid stream flows through wire, the heat losses from the surface of wire by convection phenomenon is balanced against heat generated in the wire at equilibrium.

The heat transfer from thermoelement or wire to the fluid is determined as.

$$I^2 R_w = K (t_w - t_g) h\alpha \qquad ...(4.26)$$

where ; K = Conversion factor

t_w = Temperature of wire

R_w = Resistance of wire

t_g = Temperature of fluid

I = Current through wire

h = Coefficient of heat transfer

α = Heat transfer area

Therefore, $h = \lambda + \zeta \sqrt{v}$ ...(4.27)

where ; h = Is the function of velocity (v) of fluid flow

ρ = Density

λ = Thermal conductivity

C_v = Specific heat of fluid

where ; $\zeta = (2\pi\rho C_v)^{1/2}$ ...(4.28)

Bridge circuit arrangement for hot wire anemometer : Fig. 4.38 shows circuit arrangement for bridge circuit. It consists of excitation supply and detector (D) for measurement. Resistances R_1, R_2 and R_3 are the bridge arm resistances, R_w wire resistance. R_i is internal resistance which has a large value.

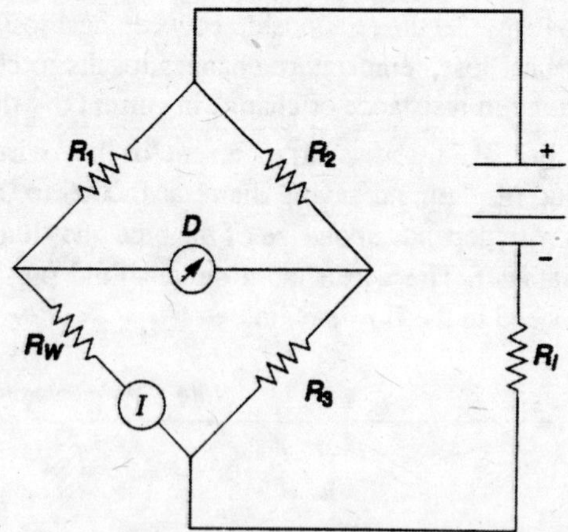

Fig. 4.38 : Bridge arrangement.

Because of change in flow rate the resistance R_w changes resulting to unbalance the bridge. The unbalanced voltage is amplified by the bridge amplifier. The unbalanced voltage is used as supply for bridge. The bridge current necessary to balance the bridge indicates the change in flow velocity.

(ii) **Hot wire flowmeter using thermocouples** : The Fig. 4.39 shows hot wire flowmeter using thermocouples. This meter consists of two thermocouples '*A*' and '*B*' which are connected in series so as to form a thermopile. This series arrangement of '*A*' and '*B*' thermocouples is supplied with AC supply through transformer for heating.

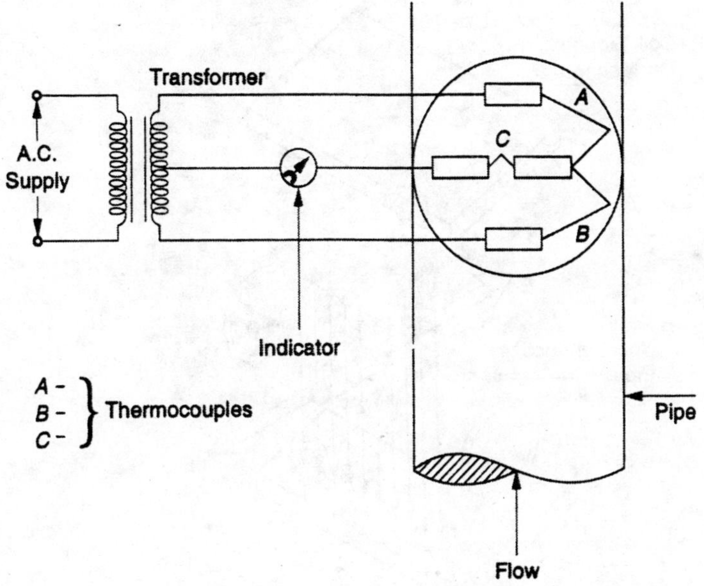

Fig. 4.39 : Hot wire flowmeter using thermocouple.

Third thermocouple element '*C*' is connected in d.c. output circuit, it is not supplied by AC supply hence not heated. This method is used for measurement of gas flow.

This complete assembly is then placed in gas stream, due to gas flow the heated thermopile cools according to convection phenomenon. Thermocouples are heated by constant input AC power supply it will attain maximum equilibrium temperature and produces an e.m.f. as a function of temperature of gas, velocity of gas, its thermal conductivity, density and specific heat.

The thermocouple '*C*' produces an e.m.f. proportional to gas temperature, it cancels the ambient temperature effect on output signals from thermocouples *A* and *B*.

The output voltage signal is a function of flow of gas which is given by.

$$E = \frac{C}{2\,(\pi\, K\, C_p\, \rho\, d\, V)^{1/2} + K} \qquad ...(4.29)$$

where ; E = Error voltage generated

C = Instrument constant

K = Thermal conductivity of fluid

C_p = Specific heat of fluid

d = Diameter of heated thermocouple element

V = Velocity of fluid

ρ = Density of fluid

(f) **Electromagnetic flowmeters** : This method is applicable for the flow measurement of fluids which are electrically conducting. Electromagnetic flowmeters uses Faraday's law of electromagnetic induction for determination of flow. Faraday's law states that when a conductor moves through a magnetic field of given field strength, voltage is produced in the conductor and it depends upon relative velocity between conductor and field. Consider Fig. 4.40 for studying working principle.

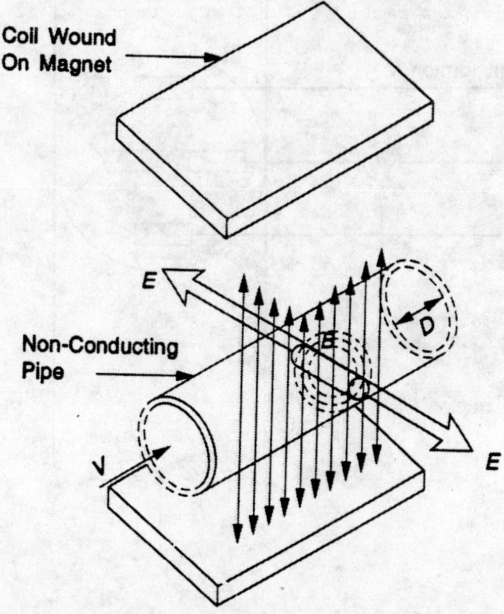

Fig. 4.40 : Working principle of electromagnetic flowmeter.

The Fig. 4.41 shows electromagnetic flowmeter. It consists of a permanent magnet or electromagnet, the non-conducting pipe (fiber glass) is placed in the magnetic filed. The coils are wound on the magnet. Two electrodes are mounted on the opposite side of each other and mounted around the pipe to pick up the induced e.m.f.

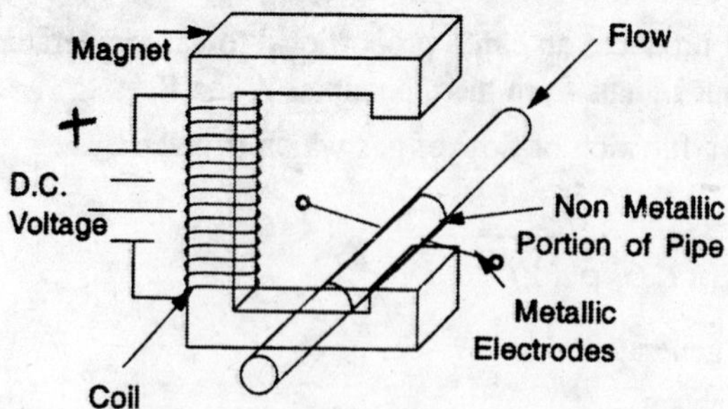

Fig. 4.41 : Electromagnetic flowmeter.

The liquid is the conductor having equivalent length to that of internal diameter of pipe, (*D*), this liquid conductor moves with the velocity "*V*" through the magnetic field strength "*B*". So, the induced voltage '*E*' is expressed as below :

$$E = BDVC \qquad ...(4.30)$$

Therefore $$V = \frac{E}{BDC} \quad ...(4.31)$$

where ; E = Induced voltage in volts

B = Magnetic field strength in wb/m^2

D = Length of conductor (m) (fluid)

V = Velocity of conductor or fluid (m/s)

C = Constant (dimensional)

When the fluid (liquid) enters through the pipe section it passes through the magnetic field produced by the magnetic coils. This magnetic coils are supplied or excited by AC or DC excitation source. Due to this magnetic field, voltage gets induced in the electrodes which are mounted on the walls of pipe. This voltage is 'E' which is expressed in previous discussion.

If the flow rate increases the liquid conductor moving through the magnetic field also increases the instantaneous output voltage "E".

The generated voltage signal 'E' is equal to average velocity of the fluid. The magnetic flowmeters senses the linear velocity of fluid and measures the volumetric flow rate which is expressed as :

$$Q = VA \quad ...(4.32)$$

where ; Q = Volumetric flow rate

V = Fluid velocity

A = Cross-sectional area of flowmeter

Therefore $$Q = \frac{EA}{BD} \quad ...(4.33)$$

For specified size of flowmeter A, D, B are constants, therefore $Q = KE$

$$K = \text{constant} = \frac{A}{BD} \quad ...(4.34)$$

Therefore, the induced voltage is directly proportional and linear with volumetric rate of flow.

Advantages :

(1) No moving parts, so frictionless operation.

(2) Obstructionless operation.

(3) Reduced power consumption.

(4) More accuracy and wide operating range.

(5) Suitable for acids, bases, and slurries.

(6) Possible to design for any size of pipe.

(7) Can be used as bi-directional meter.

Disadvantages :

(1) This method is useful only for conductive fluids, but for gases, pure fluid's (substances) it is not applicable.

(2) Careful installation of magnetic flowmeters is necessary.

(3) Cost is more *i.e.* expensive.

(4) Large and heavy in size.

(g) **Weirs** : These are variable head, variable flow meters. Weirs are used for the measurement of large volume of liquids in open channels. Special applications of these devices are in dams, open channels such as irrigation systems, waste water channels, rivers, dams and other civil engineering works. Weirs works on the principle that "If a restriction of a specified shape and size is placed in the path of flow, the liquid level at upstream end increases, this increase in the liquid level is a function of rate of flow through the restricted section". The head is measured as the difference in the level of liquid pool at particular distance upstream from weir in comparison to the horizontal crest of particular weir *i.e.* V-shaped, rectangular, trapazoidal.

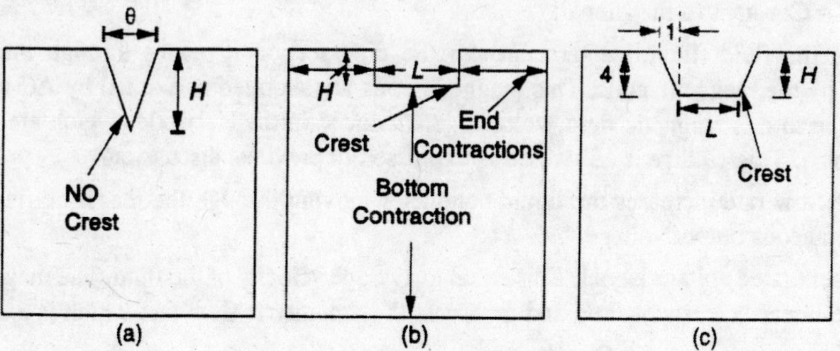

Fig. 4.42 : Types of weirs : (a) V-Notch weir, (b) Rectangular weir, (c) Cippoletti weir.

Weirs have different shapes of the notch or opening. Depending on this shape of notch, weirs are classified as follows :

(1) *V*-notch or triangular weir.

(2) Rectangular weir.

(3) Trapezoidal or Cippoletti weir.

(1) ***V*-notch weir** : The Fig. 4.42(a) shows *V*-notch weir. It has a *V*-shaped or triangular notch having an angle θ, which varies in the range of 30 degrees to 90 degree depending upon the flow capacity.

V-notch weirs are used for the measurement of smaller flow rates.

Here, for *V*-notch weir the relationship between flow and measured head is,

$$Q = 2{\cdot}48 \tan 1/2\ \theta H^{2{\cdot}5} \quad ...(4.35)$$

where ; Q = Rate of flow in cubic feet/second.

θ = V-notch angle in degree.

H = Head of the fluid flow.

L = Length of crest in feet.

The head is measured as the difference in the level of liquid pool at particular distance upstream from weir in comparison to the horizontal crest of particular weir *i.e.* *V*-shaped, rectangular, trapezoidal.

(2) **Rectangular weir** : The Fig. 4.42(b) shows rectangular weir. It has rectangular notch of length '*L*'.

Rectangular weirs are used for larger flows. The weir of 2 ft. crest long develops 0·2 ft head.

For rectangular weir the relationship between flow and measured head is as :

$$Q = 3{\cdot}33\ (L - 0{\cdot}2H)\ H^{1{\cdot}5} \quad ...(4.35a)$$

In this type of weir rate of flow is directly proportional to crest length and to the three-halves power of the head *i.e.* $L > 3H$.

(3) **Cippoletti weir (trapezoidal)** : The Fig. 4.42(c) shows Cippoletti weir. It has trapezoidal notch having slide slopes of 1 : 4. These type of weirs are used for larger flow measurements.

The relationship between flow and measured head for trapezoidal weir is;

$$Q = 3{\cdot}367LH^{1{\cdot}5} \quad \text{...(4.35b)}$$

Installation of weir : The weir plate may be placed in a weir box or can be installed directly in the fluid channel *e.g.* : dam. Sometimes the weirs may be installed in a steeling basin with flow line.

Advantages :

(1) Simple construction.

(2) Easy installation.

(3) Used for measurement of dirty fluids or liquids containing muddy particles, sewages discharge etc.

(4) Accuracy is more for large liquid flows.

Disadvantages :

Non-linear performance.

4.10 Measurement of Velocity and Speed

Fig. 4.43 shows velocity sensor. It consists of an electrical coil placed in permanent magnetic field. This assembly is placed inside the silicone oil. This method is discussed in measurement of vibration section. The output voltage is a function of vibration velocity and frequency equal to the vibration freauency.

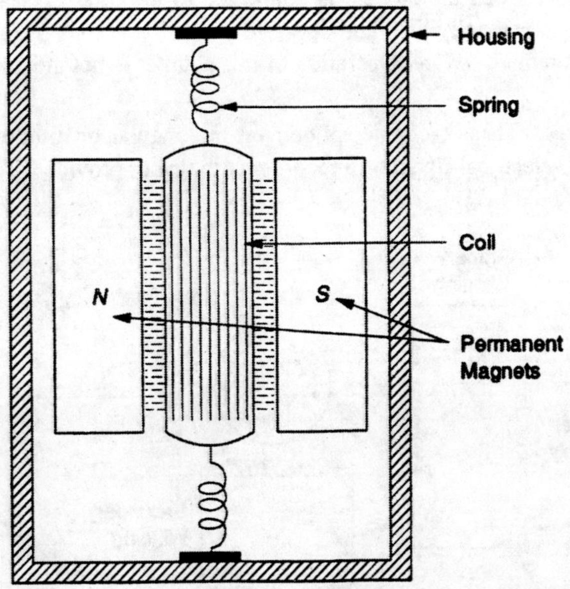

Fig. 4.43 : Velocity sensor using permanent magnet.

The velocity may be linear or angular. The linear velocity is along the axis of movement where as the angular velocity is around the axis. The angular speed is measured using tachometer. The speed is a scalar quantity equal to the magnitude of velocity. Tachometers are classified as :

(A) Mechanical Tachometers

(B) Electrical Tachometers

(A) **Mechanical Tachometers** : These type of tachometers consist of mechanical assembly and mechanical movements for the measurement of speed. Type of mechanical tachometer is :

Revolution Counter : Is a very simple method for the measurement of **rotational** speed. It consists of a worm gear connected to a spindle. The worm gear meshes with spur gear moves a calibrated dial to indicate the revolutions. The spindle or shaft is driven by the source speed. Fig. 4.44 shows revolution counter.

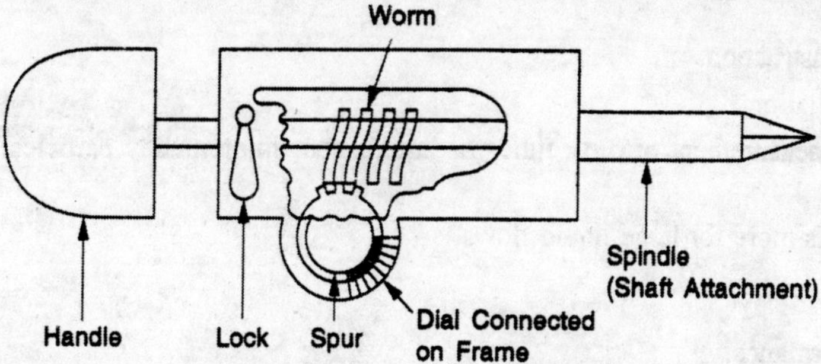

Fig. 4.44 : Revolution counter.

The counter has two dials–inner dial and outer dial to indicate the (revolutions) speed of spindle or shaft.

A stopwatch is attached to the assembly for indicating time, the revolution counter measures both revolutions and time.

For measurement of speed the spindle is connected to the rotating shaft whose speed is to be measured. This device is manually held and operated manually to start the counter and stopwatch. It is a direct speed measurement device. Operation of this counter is not affected by any environmental variations.

(B) **Electrical Tachometers** : These tachometers convert the angular or rotational speed into electrical signal and indicates this electrical signal on a recorder or indicator provided. The electrical tachometers are classified as—

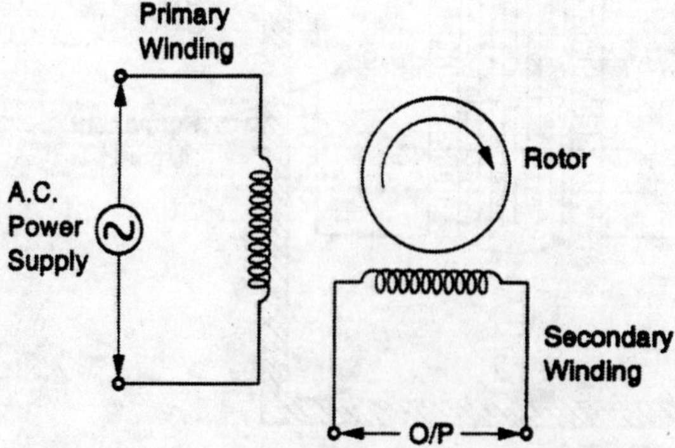

Fig. 4.45 : A.C. Tachometer generator.

(i) **AC tachometer generators** : (For rotary-velocity measurement) The A.C. tachometer generator is similar to a two-phase induction motor, it is an electromechanical device. It consists of primary winding as shown in Fig. 4.45. A secondary winding is placed 90° to that of primary winding.

The primary winding is exicted by sinusoidal voltage. As the two windings are placed at 90° positions the voltage across secondary winding is zero. As the rotor starts to rotate at the speed which is to be measured, sinusoidal voltage gets induced in the secondary winding. The amplitude (magnitude) of the voltage induced across secondary winding is proportional to the speed of rotation.

Output voltage signal is recorded or indicated by using proper indicator or recorder instrument.

(ii) **DC tachometer generators** : (For rotary-velocity measurement) : The d.c. tachometer generator consists of a permanent magnet which provides the magnetic flux. An output winding is placed on a rotor called as armature winding as shown in Fig. 4.46.

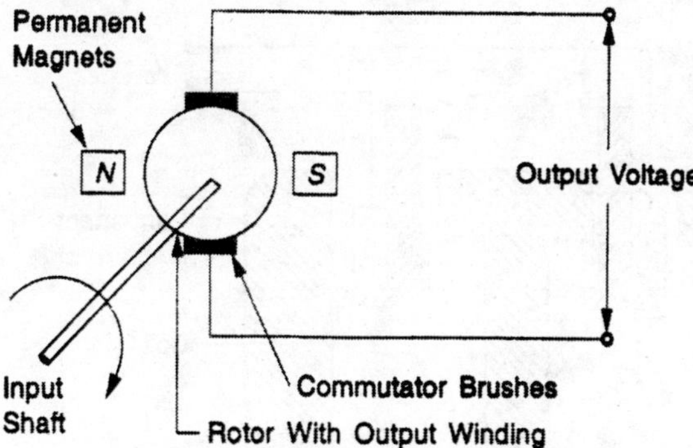

Fig. 4.46 : D.C. tachometer generator.

When the rotor speed is zero there is no relative motion between the magnetic field and winding hence, output voltage is zero. As the rotor speed increases the relative motion between magnetic field and winding increases resulting in increase of the output voltage.

As the rotor speed increases output voltage also increases, this output voltage is sinusoidal, which is converted into DC voltage using commutator and brushes connected on the motor as shown in Figure.

(iii) **Drag-cup rotor tachometer** : The working of drag-cup tachogenerator is shown in Fig. 4.47(a).

It consists of two stator field windings, mounted at right angles to each other *i.e.* two windings are placed quadrature in space. The rotor is a thin aluminium cup which rotates in the air gap of fixed magnetic structure. The speed which is to be measured is applied to the rotor.

The reference winding is supplied with the voltage V_r. The speed under measurement is angular rotations θ as indicated in the Fig. 4.47(a)

(C) Moving-Coil type translational velocity transducer : This type of transducer uses following principle :

$$e_0 = BlV_i \qquad ...(4.36)$$

where ; e_0 = Output voltage (V)

B = Flux density. (wb/m^2)

l = Length of coil (m)

V_i = Relative velocity of coil & magnet (m/s)

As the 'B' and 'l' are constants the output voltage linearly following the velocity and its polarity reverses when velocity reverses sign.

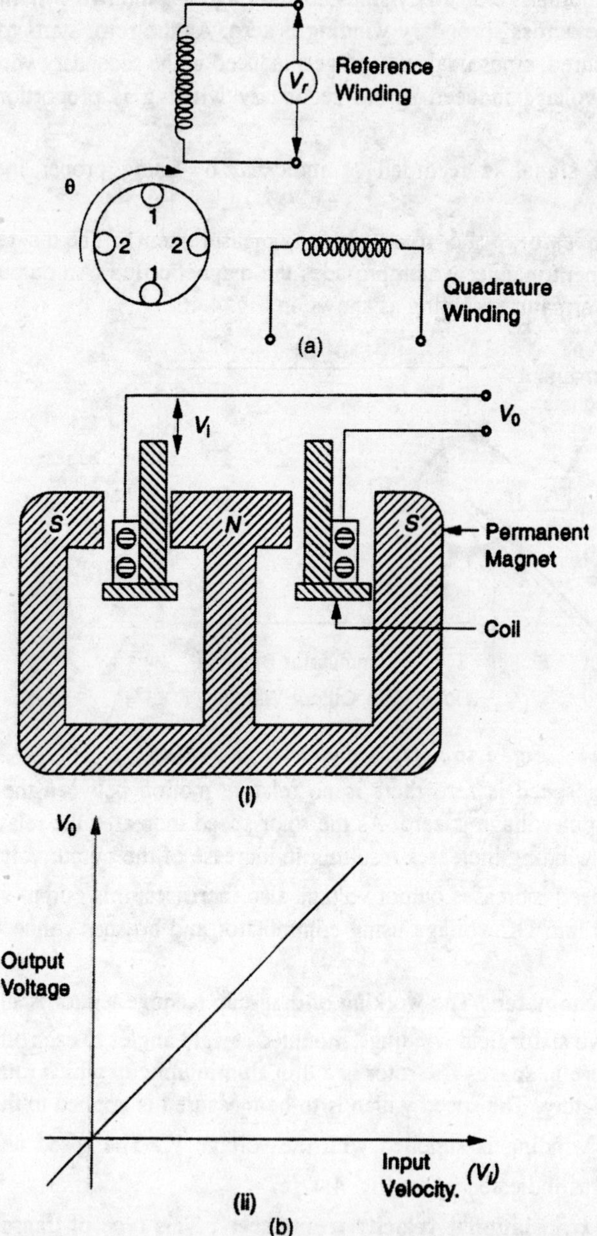

Fig. 4.47 : (a) Drag-cup rotor Tachogenerator, (b) Moving-coil type translational velocity transducer (i) Construction, (ii) Representation of V_0 and V_i

Fig. 4.47(b) shows the constructional details of moving-coil type translational velocity transducer. The coil moves in a permanent magnetic field. Due to this the output voltage is induced in the coil according to the principle as discussed above.

The output voltage is linearly proportional to the input velocity as shown in Fig. 4.47(b) (ii) This device is also applicable for the measurement of vibratory velocities.

Sensitivity can be increased by increasing length of coil. High resistance coil is preferable for avoiding the loading effects.

(D) Moving magnet type translational velocity transducer : Fig. 4.48(b) shows cross-section of

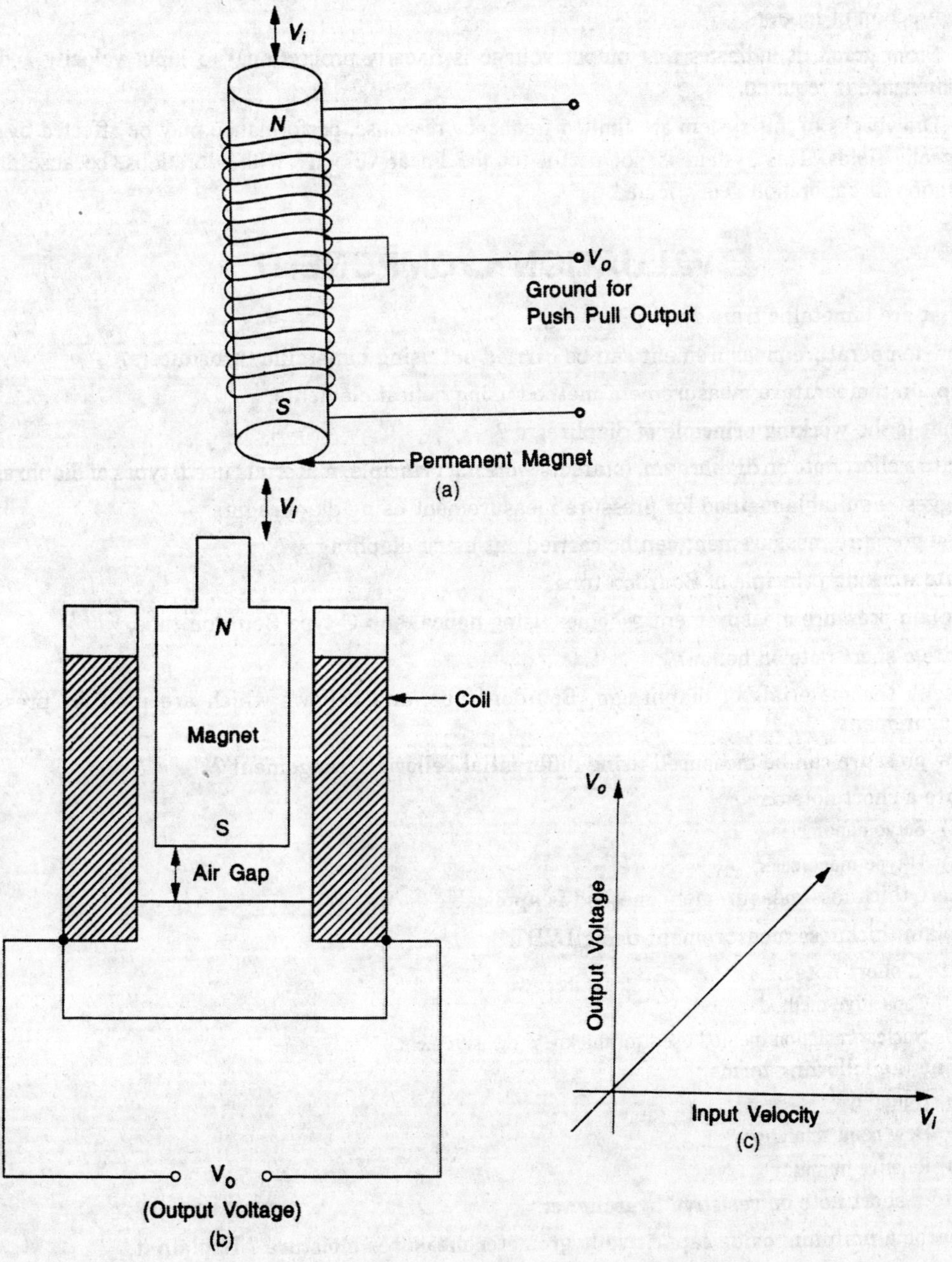

Fig 4.48 : (a) Circuit connection, (b) Cross-sectional view, (c) Representation of V_i and V_0

moving magnet-type transducer. The velocity sensing element is rod and this element is connected to the shaft or device whose velocity is to be measured. Velocity sensing element is permanent magnet rod and coil is wound on the surface of this magnet as shown in Fig. 4.48(a).

When the input velocity under measurement is applied to the magnet, voltage (V_0) is induced proportional to the input velocity 'V_i' as shown in Fig. 4.48(c). The polarity of output voltage indicates the direction of motion.

From graph, it indicates that output voltage is linearly proportional to input velocity and less maintenance is required.

Drawbacks of this system are limited frequency response, performance may be affected by stray magnetic fields. This system is not useful for the linear velocity with vibrations, because due to vibration its calibration gets affected.

EVALUATION COMPONENT

1. What are bimetallic transducers ?
2. How temperature measurement can be carried out using bimetallic transducers?
3. Explain temperature measurement method using helical elements.
4. What is the working principle of diaphragm ?
5. Write a short note on diaphragm. (state its working principle, materials used, types of diaphragm).
6. Suggest a suitable method for pressure measurement using diaphragm.
7. How pressure measurement can be carried out using diaphragm ?
8. State working principle of Bourdon tube.
9. Explain pressure measurement schemes using helical and C-type Bourdon tube.
10. Write a short note on bellows.
11. Specify the materials of diaphragm, Bourdon tube, and bellows which are used for pressure measurement.
12. How pressure can be measured using differential bellows arrangement ?
13. Write a short note on :
    (i) Servomanometer
    (ii) U-type manometer
14. Where thickness measurement method is applicable ?
15. Explain thickness measurement using LVDT.
16. Write a short note on :
    (i) Capacitive method
    (ii) Nuclear radiation method used for thickness measurement.
17. Define the following terms :
    (i) Humidity
    (ii) Dew point saturation
    (iii) Relative humidity.
18. Write a short note on resistive hygrometer.
19. How the aluminium oxide capacitive hygrometer measures moisture ? Explain it.
20. Write a short note on vibration measurement using piezoelectric crystal.
21. Explain the working of hair hygrometer.

22. How strain gauge can be used for vibration measurement ?

23. Explain how LVDT sensor measures vibrations as a function of acceleration ?

24. Which arrangement is suitable for measurement of vibrations in terms of velocity ? Elaborate it.

25. Define torque, classify torque measurement systems, where torque measurement can be carried out ?

26. With neat sketches write short notes on :

(i) In-line rotating torque sensor

(ii) Torsion meter

27. Why flow measurement is necessary ? Discuss it.

28. What do you mean by head flowmeters ?

29. What are the characteristics of head meters ?

30. Define :

(i) Flow noise

(ii) Reynolds number

31. Write a short note on differential flowmeters.

32. Explain suitable method for flow measurement of non-corrosive fluids using venturi tube.

33. For flow measurement of slurries and dirty fluids which method is useful, suggest it ? Explain the same.

34. Describe with neat sketches all types of venturi tubes.

35. On which factors Reynolds number depends ? Elaborate it.

36. Write a short note on orifice plates covering points such as :

(i) Operating principle

(ii) Materials used for orifice plates

(iii) Constructional details like, size, diameter etc.

(iv) Applications

37. What are the advantages and disadvantages of orifice plates ?

38. State the advantages and disadvantages of venturi tubes.

39. Write a short note on flow nozzle.

40. List the limitations of orifice plates.

41. Explain the procedure regarding installation of orifice plates. List the precautionary measures which are to be considered while installation.

42. Specify the applications of flow nozzle.

43. How ultrasonic flowmeters can be used for flow measurement ?

44. Describe :

(i) Ultrasonic transit-time flowmeters

(ii) Ultrasonic Doppler flowmeters

45. Write a short note on thermal flowmeters.

46. State and explain the types of thermal flowmeters with their working.

47. Explain the working of mass-flowmeters.

48. With neat sketch explain the working of flowmeters using thermocouples.

49. How hot wire anemometer works ? Explain it.

50. State the working principle of electromagnetic flowmeters.

51. State the advantages, disadvantages and applications of electromagnetic flowmeters.

52. How weirs can be used for flow measurement ?

53. Explain all the types of weirs.

54. Write in detailed manner about the installation of weirs.

55. Why tachometers are used ?

56. Explain with neat sketches :

(i) Mechanical tachometer

(ii) DC tachometer

(iii) Drag-cup-rotor tachometer

57. Write short notes on :

(i) Moving coil velocity transducer

(ii) Moving magnet velocity transducer

Chapter **5**

MEASUREMENT SYSTEMS (PART-II)

Learning Objectives

- To study the significance of biomedical instrumentation.
- To discuss man-instrument system.
- To study phenomenon of biopotentials.
- To study electrical phenomenon through Bioelectric potentials.
- To study function of biomedical electrodes.
- To explain measurement methods of bioelectric potentials.
- To define requirements of transducers used in biomedical applications.
- To discuss different blood flowmeters.
- To explain direct and indirect blood pressure measurements.
- To understand EEG, EMG and ECG systems.
- To state the significance of non-destructive testing. (NDT).
- To compare destructive and non-destructive testing.
- To study different non-destructive testings.
- To examine ultrasonic testing.
- To specify the significance of level measurement.
- To study different mechanical methods of level measurement.
- To examine various electrical methods of level measurement.
- To list advantages, disadvantages and application areas of various level measurement schemes.
- To define the terms : Sound, SPL, sound power level, microphone.
- To define function of electroacoustic transducers.
- To discuss construction, working and characteristics of different microphones.
- To define pH value and to study pH measurement.

5.1 Introduction

This chapter is divided into following sections :

(1) Biomedical instrumentation

(2) Non-destructive testings (NDT)

(3) Level measurement

(4) Sound measurement

Biomedical instrumentation is related with monitoring and measurement of parameters related with human body. These parameters are blood pressure, blood flow, body temperature, pH value of blood etc. These measurements are possible because the bioelectric potential gets developed in human body. This section covers the transducers like biopotential electrodes and their types, blood flow measurement transducers, transducers for body temperature measurement, blood pressure measurement methods (direct and indirect methods). Introduction of EMG, EEG, ECG is also highlighted. It helps physicians and nurses for diagnosis and treatment of various diseases.

Second section is non-destructive testing (NDT) which is useful for testing of metals, alloys, metal parts and finished products manufactured in industries and metallurgical industries. NDT is used for detection of flaws, internal defects, dislocations, tears, scabs, internal shrinkage etc. in metals and finished product. Non-destructive testings are quite advantageous over destructive testing, these points are also elaborated in this section. Different non-destructive testings such as visual inspection, magnetic particle detection, radiography, ultrasonic flaw detection and eddy current testings etc. are also discussed.

Level measurement is the oldest method used for level measurement of liquid, solid (particles) and gaseous matters. This method is applicable in power plants, process industries such as sugar, plastic, glass, oil refineries, petrochemical industries etc. This section covers different mechanical methods such as bubbler, float, thermal etc. used for level measurement. Electrical methods like resistive, capacitive, ultrasonic, microwave methods are also discussed.

Sound is the phenomenon due to which communication is possible. It is an important aspect of human life. Sound measurement is useful in audio recording and reproduction equipments, TV and radio broadcasting stations, noise pollution control fields. Characteristics of sound like sound intensity, loudness of sound, SPL are elaborated. Various microphones like piezoelectric, condenser, carbon, electromagnetic, velocity ribbon microphones are discussed in this chapter.

5.2 Biomedical Instrumentation

Biomedical instrumentation relates with the measurements of different parameters from a living human being. The parameters involved with human body are variables associated with the physiological process of human body. These involved parameters are body temperature, electrical activity of the heart *i.e.* ECG, electrical activity of the brain *i.e.* EMG, bloodflow, respiratory system *i.e.* air flow etc. In medical instrumentation system all these parameters are monitored and analysed for diagnosis and evaluation purpose.

This section *i.e.* biomedical instrumentation is widely useful in clinical fields for diagnosis and monitoring of patients under treatment or surgery. It helps the physicians and nurses to make the clinical decisions and to operate the patient in skillful manner.

In short, we can elaborate the biomedical instrumentation that it is an aid to the physicians for on line evaluation and monitoring of the physical conditions of patient, to monitor different activities of patient body under surgery, pH value of blood, it provides the information of internal and external body parameters *e.g.* temperature of body.

Important aspect is that the biomedical instrumentation system should be accurate, linear and have wide operating range, stable, good frequency response, it should not be affected by any noise signals or stray electric or magnetic fields. Essential aspect is that it should have proper isolation *i.e.* the instrument should not produce direct relation between human body and ground, also it should be error free and properly calibrated.

This instrumentation system involves following sections.

(1) Human body on which the measurements or monitoring is carried out.

(2) Stimulation arrangement.

(3) Transducer.

(4) Signal conditioning elements (Devices).

(5) Display devices, recording and data transmission elements.

(6) Control element.

5.3 Block Diagram of Man-Instrument System

Fig. 5.1 shows block schematic of bio-medical instrumentation or man-instrument system. This system is called as man-instrument system because it consists of both parts *i.e.* man and instrument.

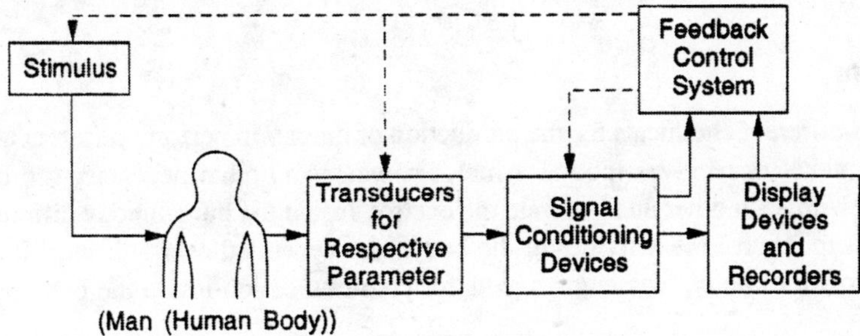

Fig. 5.1 : Block schematic of biomedical instrumentation.

Block schematic of biomedical instrumentation (man-instrument) system

This system is same as general block schematic of instrumentation system. The only difference is that the parameters under measurement are related to human body or biological parameters, chemical parameters etc.

Initial part of this system is human body it is called as subject because all the parameters are measured from human body using various instrumentation systems.

Second stage is transducer, the function of this stage is to convert the one form of energy into another. Transducer produces an analogous output electrical signal proportional to the quantity under measurement. Transducer converts various parameters like temperature; pressure, airflow, blood flow etc. related to human body into electrical signals. This system uses both active and passive transducers. At the same time two or more transducer can be connected to the human body for measurement of different variables.

Next stage of instrumentation system is signal conditioning elements or signal processing element. Sometimes the output of transducers are weak (low) under these circumstances, low signals should be conditioned *i.e.* amplified, which is done by signal conditioning devices (SCD). The function of SCD is to modify or alter the output of transducers so that they become usable at sufficient level to drive the next stages *i.e.* indicating or displaying (recording) devices and control elements.

The output of signal conditioning elements is to be displayed for the meaningful measurements of the parameters under study. In biomedical technology or medical fields the parameters should be recorded so as to get the permanent record of the data. The recorders used may be (1) analog recorders *i.e.* graph paper and stylus (pen) to produce analog waveforms. Different analog meters may be used.

(2) Digital devices such as computers and microprocessors which handles the on-line measurement

process of varying parameters accurately, rapidly, simultaneously. It also stores the data. Computer system has good feedback control in the process.

For human body sometimes external stimulus is essential, which is generated by the instrumentation system. Feedback control system is provided so as to control the complete functioning of instrumentation system automatically. It uses a feedback loop *i.e.* closed loop control system in which the operation is controlled according to the output of display devices or recorders.

Now a days the biomedical instrumentation systems are becoming sophisticated, accurate and precise. These systems use analog-to-digital (ADC) converters which facilitate to achieve good computer control and storage of information.

The human body has different sub systems such as electrical, hydraulic, pneumatic, chemical, thermal etc. which are discussed as below.

5.4 Sub Systems of Human Body

(A) Bio-Chemical System

The human body contains different chemicals for the production of energy to perform different activities and functions. Human body intakes or receives food as a fuel, air, water and other necessary source materials. All these interact or react with each other and generate the energy. Inside the human body different chemical reactions takes place so as to control the activities of the body. The generated energy is used for controlling the different activities, growth of body, messenger agents for communication inside the body.

(B) Hydraulic System

The cardiovascular system of human body acts as a closed complex hydraulic system. It has four-chamber pump *i.e.* heart connected with elastic tubings (blood vessels). The pressure of system is controlled by change in the diameters of tube.

The four chamber pump acts as isolated two stage pumps. First of these pumps collects the blood from the first stage and pumps it into second stage. First stage is atrium while second stage is ventricle. The blood flows in the laminar fashion and acts as communication and supply network for all the parts of systems.

In hydraulic system the variables or parameters are pressure of blood, flow rate of blood, blood volume at different locations in the body. These parameters are measured in the biomedical instrumentation system.

(C) Pneumatic System

The respiratory system of human body acts as pneumatic system. The air pump develops positive and negative pressure in a sealed chamber. It results to suck the air inside and force it out through the elastic bags within the compartment. The air pump is diaphragm and chamber is thoracic cavity. The elastic bags are lungs.

The system variables are respiration rate, respiratory volume, respiratory air flow and concentration of CO_2 *i.e.* carbon dioxide.

(D) Communication Network

The nervous system is equivalent to communication network. Brain is the control part which acts as computer and controls different functions and activities of the body. Brain performs different functions such as decision making, feeling of emotions, to sense different conditions such as heat (temperature), touch, cold etc.

The network is assembled by means of different nerves and muscles, the signals transmitted through this network are nerve action potentials *i.e.* the network lines are nerves and muscles where as the action potentials are information signal thus it constitutes the communication network. Somewhere the network is complicated so as to sense the signals such as light, cold, heat, sound etc.

Thus the system acts as computer *i.e.* brain with linked communication network of lines *i.e.* nerves and muscles.

Brain is assigned or compared as computer because it has computational power, ability of decision making and the prime important part is memory used for storing the information.

5.5 Electrical Phenomenon Through Bio-Electric Potentials

While functioning the various systems of human body generates different signals and contains the information regarding the function. These generated signals are nothing but the bio-electric potentials developed at nervous system, brain functioning, heart beats *i.e.* cardiovascular system performance, muscle activities etc.

Human body contains certain type of cells. The bio-electric potentials are ionic voltages produced because of electrochemical activity of special type of cells. These ionic voltages are converted into electrical voltages by means of transducers. Transducers are placed at various parts of the body so as to monitor and measure the developed voltages which will be helpful for the physicians for diagnosis and treatment of various diseases and surgery. The generation of bioelectric potential takes place at cellular level. This idea of generation of electricity in the body has been developed by the Luigi Galvani. The practical application of these potential is further developed by Dutch physician Willem Einthoven.

In the body there are certain type of cells (i) nerve cells (ii) muscle cells. These cells are surrounded by the body fluids which contain charged atoms called as ions. Ions are sodium (Na^+) potassium (K^+) and chloride (Cl^-). The membrane cells allows to enter potassium and chloride ions but blocks sodium ions. The various ions seek a balance between the inside and outside of the cell according to the concentration of electric charges the inability of sodium ions to enter into membrane causes into two conditions. Fig. 5.2(a) shows polarized cells.

(1) The concentration of sodium ions inside the cell becomes lower than the outside fluid. The sodium ions are positively charged, they try to make more outside positive than inside.

(2) To maintain the balance of the electric charge the additional positively charged ions of potassium enters the cell resulting in higher concentration of potassium inside than outside.

From above discussion it is clear that the charge balance does not occur but the equilibrium is achieved because of ion concentrations and hence potential difference is developed across the membrane *i.e.* negative inside and positive on outside. The membrane potential is called **resting potential.** Range of resting potential is − 60 to − 100 mV. Human cells vary from one micron to 100 microns in diameter and 1mm to 1m in length. The typical thickness of membrane is 0·01 micron.

The condition of negative charge is inside the surface of the membrane and positive charges outside the surface of the membrane. The unequal charge distribution creates some electrochemical reaction and during this reaction in the living cells potential is developed which is called resting potential. This condition of cell is **polarized condition**. The above discussed phenomenon is represented or denoted as a tiny biological battery.

Fig. 5.2 (a) shows polarized cell with resting potential.

When a section of cell membrane is excited externally by energy source the outer side of membrane becomes momentarily negative with respect to inner polarities (ions). This movement of ions constitute the ionic current such a phenomenon is called **depolarization** and the cell potential changes to approximately

+ 20 mV. The cell has slight positive potential on the inner side due to imbalance of potassium ions. This potential is known as action potential which is approximately + 20 mV as stated above.

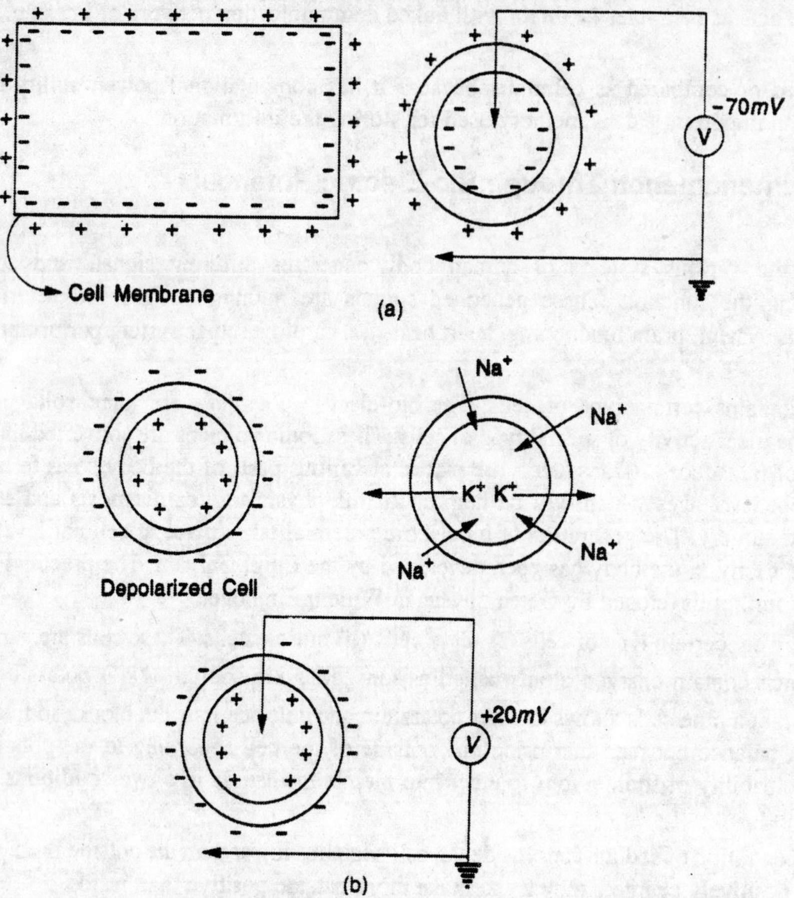

Fig. 5.2 : (a) Polarised cell with resting potential. (b) Depolarized cell

Change of the resting position of the membrane in polarized position *i.e.* resting potential to action potential is called depolarization which is shown in Fig. 5.2 (b).

For the measurement of biopotentials, transducers are used for converting the ionic potentials and ionic currents into electric currents and voltages. These transducers are biopotential electrodes.

5.6 Biopotential Electrodes

Electrodes are playing vital role in biomedical instrumentation, they are used for recording bioelectric signals. Electrodes are electrical contacts or terminals through which the electrical voltage can be picked up or obtained from the skin surface of human body. Application of electrodes are recording ECG, EEG, EMG, respiration curves etc. These electrodes are placed in contact with the skin surface of human body. Sometimes electrode paste, electrolyte or jelly is interfaced between electrode and surface of skin so as to reduce (lower) the input impedance of the system. The biopotential electrodes are classified into three main types.

(1) Micro Electrodes

These electrodes are used for the measurement of bioelectric potentials near to a single cell or within a single cell. In bioelectric potentials the electrode potential occurs at the interface of metal and an electrolyte.

(2) Skin surface Electrodes

These type of electrodes are used to measure the bioelectric potentials such as ECG, EMG and EEG from the surface of skin.

(3) Needle Electrodes

Needle electrodes are inserted or penetrated in the skin for recording EEG and EMG potentials. The electrodes should be (1) Properly connected to the body of patient (2) Correctly selected for particular application (3) Comfortable to the patients for long time spans. (4) Accurately interfaced with the surface of the skin.

(A) Measurement scheme :

For the measurement of bioelectric potentials two electrodes are essential which are placed on the skin surface of patient body. Across each electrode the potential gets developed. The output voltage measured is the difference between instantaneous potentials of the two electrodes. For higher accuracy in the measurements two same type of electrodes can be used. If the two electrodes are different types then electrode offset voltage may be developed. Because of this offset voltage current flows through the electrodes and input stage of the amplifier.

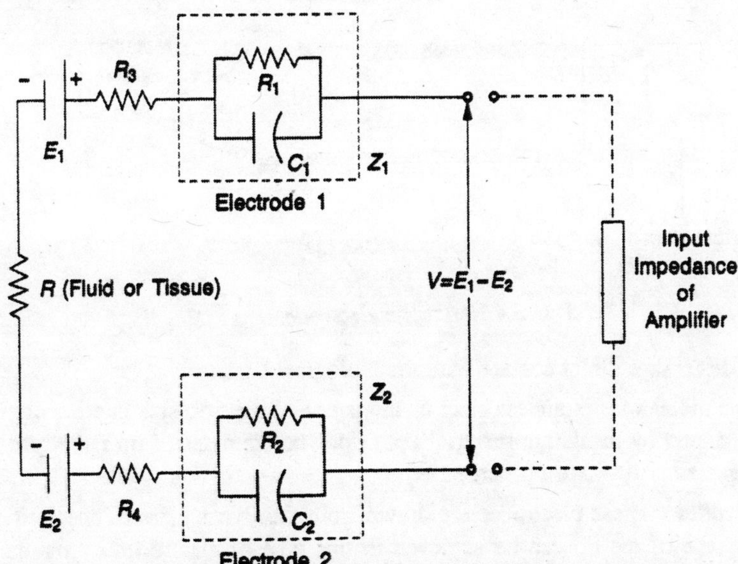

Fig. 5.3 : Equivalent circuit of two electrodes.

The Fig. 5.3 shows equivalent circuit of two electrodes which are to be connected or attached to the skin surface. The equivalent circuit consists of two electrode sections having a voltage source in series with resistance capacitance network. The difference voltage *i.e.* $V = E_1 - E_2$ depends upon the difference of ionic potential between the two points of body where the electrodes are to be attached.

Z_1 is the impedance of electrode 1 which is resistance-capacitance network of R_1 and C_1 similarly Z_2 is impedance of electrode 2 which is network of R_2 and C_2. Z_1 and Z_2 are the skin contact impedances. R is the tissue or fluid resistance from the patient's body.

In the low frequency region the impedance of electrode should be high and decreases with increasing frequency. For measurement of bioelectric potentials the potential drop across electrode impedance should be low. This condition can be satisfied by making low skin contact impedance and by increase in the impedance of measuring device at higher value.

For accurate and precise measurements of potentials the input impedance of amplifier stage should be several times more than the electrode impedance.

(B) Microelectrodes :

Microelectrodes are used for analysing the performance of single cell. The size of this electrode is very small so as to penetrate easily in the cell without damaging it. Generally the tip of microelectrodes ranges from 0·5 to 5 micron and strong enough to penetrate through the tissues.

There are two types of microelectrodes :

(a) Metallic microelectrodes (b) Glass or micropipet.

(a) **Metallic Microelectrodes :** These type of electrodes consist of a fine needle of stainless steel wire having a fine tip. For high input impedance tungsten metal can be used. Fig. 5.4 shows internal construction of metallic electrode. It uses tungsten metal because it has good mechanical strength and appearent inertness.

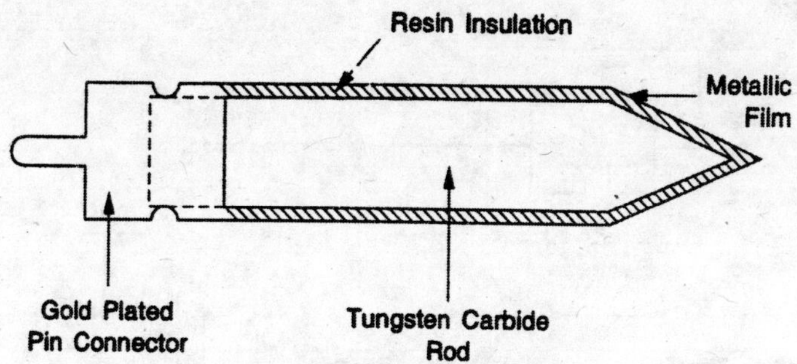

Fig. 5.4 : Metallic electrode.

Metal electrodes makes direct contact with the biological tissues.

The tip of fine metal wire (stainless steel or tungsten) of specific size is electrolytically etched. This wire is then coated by insulating material upto tip. The diameter of tip is 0·4 μm for tungsten. Now a days the tip size is 1 to 6 μm (micropipet).

(b) **Glass Microelectrodes :** These electrodes are drawn from the Pyrex glass of special type. Accurate desired size and shape of the tip can be achieved in this type of electrodes by maintaining proper timings, temperature and direction of pulling the tip at the time of fabrication. Usually its size is 1μm in diameter.

Inside the pipet electrolyte is filled, which is compatible with the fluid of cells. Micropipet has two types of contacts (1) Metal wire in contact with the electrolyte filled inside the micropipet. (2) Electrolyte in the micropipet and fluid inside the cell or fluid outside the cell.

(C) Skin surface electrodes

These type of electrodes are used to sense or record ECG, EEG and EMG potentials from the surface of the skin. There are various old type of skin surface electrodes like immersion electrodes, plate electrodes etc. In immersion type electrodes the subject like hands and feets is immersed in the saline solution. Plate electrode consists of metal plates having rectangular or circular shape. Materials used for plate electrodes are german silver, nickel silver, nickel plate. Other metal alloys can also be used. Their contact impedance is 2 to 5 kΩ. Plate electrodes are connected to the surface of skin with electrode jelly and attached using elastic strips.

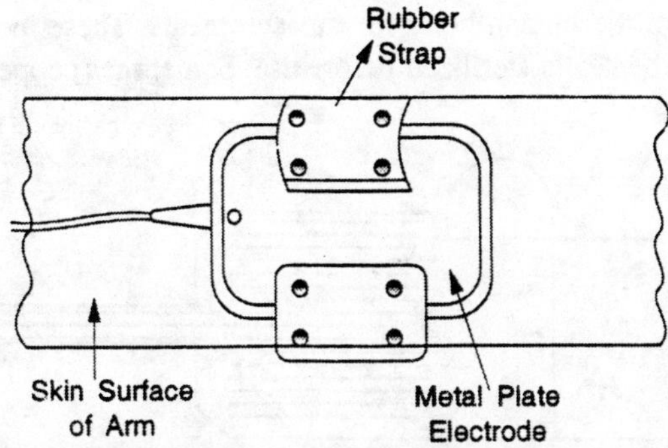

Fig. 5.5 : Skin surface electrode.

Fig. 5.5 shows plate electrodes. Plate electrodes are also called limb electrodes.

Drawbacks of plate electrodes are :

(a) Not suitable for long period patient monitoring.

(b) Possibility of electrode slippage.

(c) These electrodes are sensitive to movement. Due to small movement, thickness of the electrolyte between metal and skin changes the electrode potential and impedance. Sometimes, the bioelectric potentials may be blocked.

For avoiding this drawbacks adhesive tapes and nutmeg grater is used to eliminate the effect of movement *i.e.* movement artifact to reduce interfacing impedance.

New type of skin surface electrode is floating electrode as discussed below.

Floating electrode : In floating electrode direct contact between skin and electrode is avoided which eliminates the movement artifact.

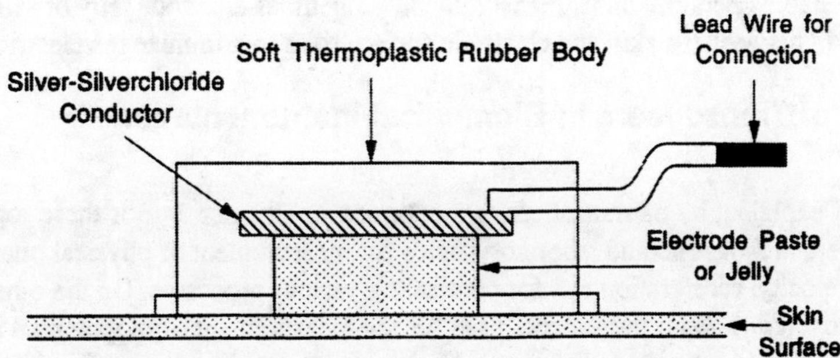

Fig. 5.6 : Cross-section of floating electrode.

Fig. 5.6 shows cross-section of floating electrode. It consists of a rigid plastic cup containing electrode jelly with metal electrode. The electrolyte or jelly is inserted in the electrode with the help of syringe. This jelly or electrolyte forms the conductive path between metal electrode and skin, constituting the bridge arrangement. Floating electrodes are used to record the ECG potentials satisfactorily.

(D) Needle Electrodes :

Fig. 5.7 shows the needle electrode. The needle electrodes consists of fine metal wire of material such as stainless steel coated with teflon coating but the tip is opened so as to insert in the surface of skin. As stated, these are insulated wires of metals and bare (exposed) only at the tip so this tip makes contact with the tissues, nerves and muscles of the human body for measurements. These electrodes can be reused for number of times, but they should be neatly sterilized before use. Sometimes copper platinum wires may also be used.

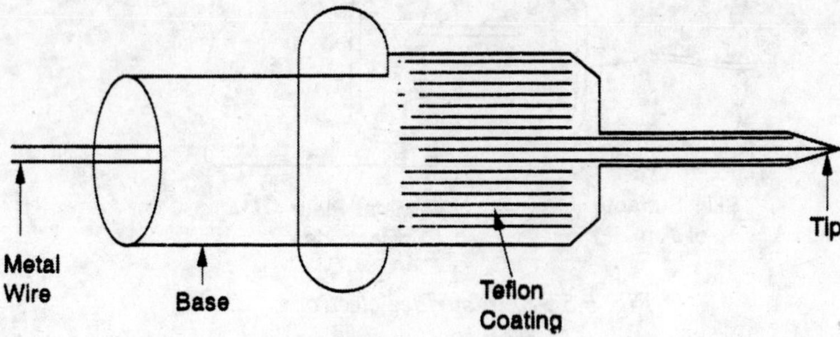

Fig. 5.7 : Needle electrode.

Needle electrodes are used in EMG, neurography, EEG measurements for this they are inserted into a section of skin just beneath the surface and parallel to it. Important types of needle electrodes are :

(1) **Unipolar needle electrodes** : It consists of a single metal coated with teflon, which is bare at the tip. This type of electrode is also called monopolar needle electrode. It measures the biopotentials at a contact point with reference to any indifferent point.

(2) **Bipolar needle electrodes** : It consists of two metal wires placed in a metal cannula, where the cannula acts as ground of the electrode. The tips are bared and makes contacts at two points for measurements.

(3) **Multielement needle electrodes** : These type of electrodes receive the biopotential signals from individual fibres of muscle tissues. They consists of number of fine metal wires and can be inserted at different depths for the measurement of biopotentials.

The needle electrodes are simple in construction and have good resistance to the thermal and chemical effects and also provide stable operation under measurement. Sometimes electrode jelly or saline soaked felt pads may be introduced in between the skin and electrode surface so as to minimize the electrical impedance.

5.7 Applications of Transducers in Biomedical Instrumentation

Detailed discussion and explaination of transducer is mentioned in Chapter 3. But these topics cover the applications of transducers in industrial and laboratory fields for measurement of physical quantities such as temperature, pressure, velocity, acceleration etc. for controlling various processes. On the other hand in this chapter the discussion covered is about transducers used for the measurement of biopotential signals. These measurements are applicable for monitoring the conditions of patient, diagnosis of disease, to monitor different parameters at the time of surgery or treatments. Also used in patient monitoring systems.

5.8 Requirement of Transducers used in Biomedical Applications

(1) To pick up the ionic potentials and currents from the surface of skin and convert them into electric voltages and currents.

(2) These received voltages and currents are monitored or displayed or recorded for the purpose of analysis and measurements.

(3) This analysis and measurements help the physicians to diagnosis the patient diseases, treatment and to monitor the patient at the time of surgery.

(4) The transducer should be accurate and can be easily interfaced with the patients body or skin surface.

(5) Transducer should be properly grounded or shock proof so as to protect patient from electrical shocks.

(6) Select the transducer according to factor of monitoring *i.e.* short time monitoring or long time monitoring.

(7) Range of the quantity which is to be measured.

(8) Transducers should be properly implemented according to the biological parameter which is to be recorded.

(9) Impedance of the transducer, body surface and measurement system should be properly matched.

Table 5.1 indicates some important biomedical (physical) parameters.

Parameter and respective transducer used for converting it into equivalent electrical (voltage/current) signals is highlighted in this Table 5.1.

Resistive, capacitive, inductive, piezoelectric, photoelectric, chemical transducers are used in measurement and conversion of physical (biomedical parameters) quantities into electrical signals. Respective working principle of stated transducers is discussed in Chapter 3.

Table 5.1

	Biomedical Parameter		Respective or Necessary Transducer(s)
1.	Blood flow	(a)	Ultrasonic blood flow meters
		(b)	Electromagnetic blood flowmeters
2.	Blood pressure	(a)	Strain gauges
		(b)	LVDT
		(c)	Low frequency microphones for Korotkoff sound
3.	Respiration rate	(a)	Thermistor
		(b)	Strain gauge
4.	Temperature	(a)	Thermocouples
		(b)	Thermistors
		(c)	Electrical resistance thermometer
5.	Heart rate	(a)	ECG method

In this chapter above measurement systems and their respective transducers are discussed which are important by the view of biomedical measurements. Other parameter measurement systems and transducers are beyond the scope of this book.

5.9 Blood Flow Measurements

Human body contains muscles, nerves, tissues, cells, bones etc. but through all the sections or parts of body blood flow is necessary. Blood flows through all the organs of the body. If the blood flow is irregular or defficiency it causes for different diseases. Thus, the blood flow measurement is very important because it helps to diagnose the disease. The blood flow or rate of blood flow is measured or expressed in liters per minute or milliliters per minute.

(A) Ultrasonic Blood Flowmeters

Most widely and commonly used type is Doppler shift ultrasonic blood flow meter which is explained as below.

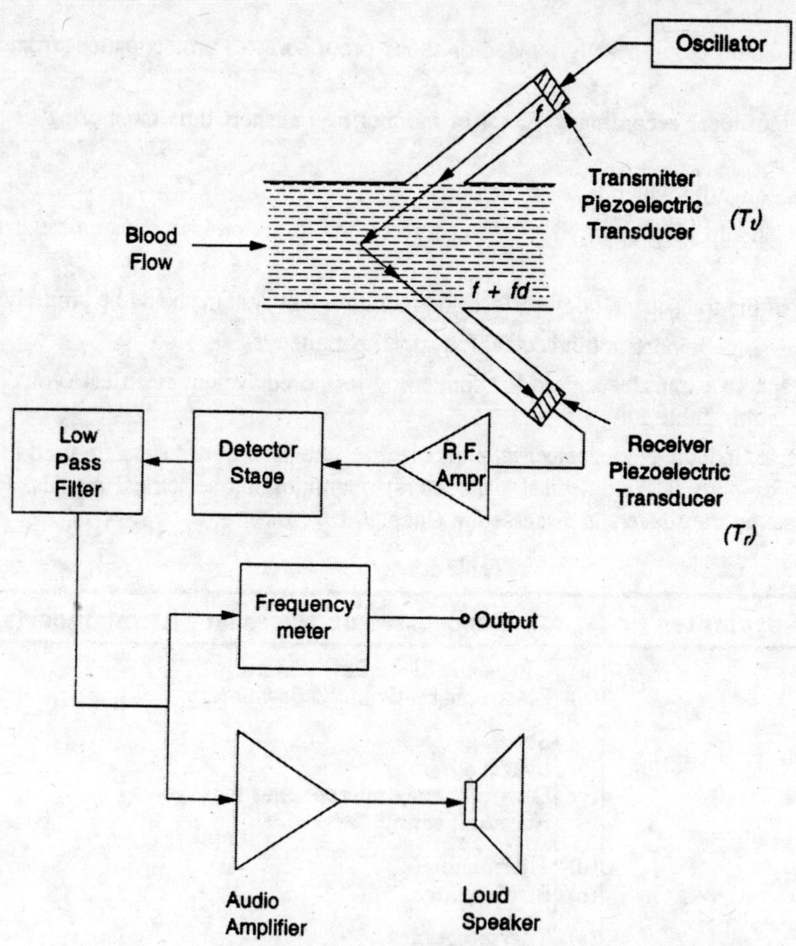

Fig. 5.8 : Ultrasonic blood flowmeter. (Doppler Effect)

Fig. 5.8 shows Doppler effect ultrasonic blood flowmeter. It consists of an oscillator which generates the frequency range of 5 MHz. This frequency is used to excite the piezoelectric transducer at transmitter end *i.e.* 'T_t'. This piezoelectric transducer 'T_t' will transmit an ultrasonic pulse or beam of frequency 'f' into the flow of blood.

Transmitted ultrasonic signal gets scattered from the moving blood cells and excites the piezoelectric transducer receiver 'T_r' which is opposite to the transducer.'T_t' at transmitter end.

Because of scattering phenomenon due to moving blood cells the reflected signal has different frequency component because of Doppler effect. These frequency components are $f + f_d$ and $f - f_d$ which depends upon the direction of blood flow. The Doppler frequency component is directly proportional to the velocity of blood flow.

The function of detector is to produce sum and difference of Doppler frequency *i.e.* $f + f_d$ and $f - f_d$. The function of low pass filter is to select the difference frequency signals and applied to the audio amplifiers. Same frequency signals are indicated by the frequency meter at the output stage.

The Doppler frequency is narrow-band signals having low frequency range *i.e.* audio frequency range so it can be heard by earphone or loud speaker. Thus the blood flow can be heard using loudspeaker or can be measured in terms of frequency.

(B) Electromagnetic Blood Flowmeters

Electromagnetic blood flowmeter uses the principle of Faraday's law of electromagnetic induction. Basic operating principle of this flowmeters is explained in Chapter 4 under the section measurement of flow using electromagnetic flowmeters.

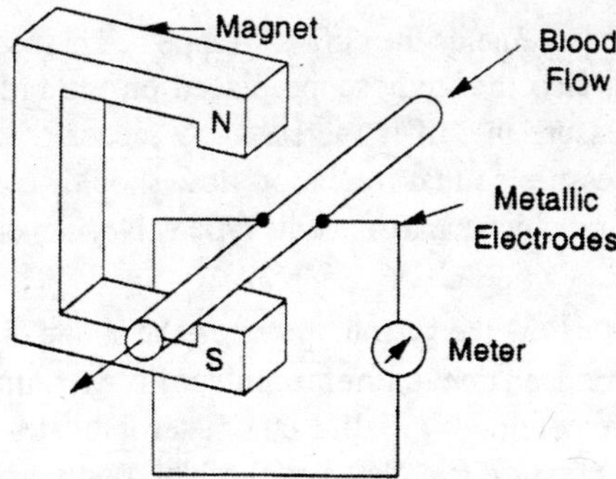

Fig. 5.9 : Electromagnetic blood flowmeter.

Fig. 5.9 shows construction of electromagnetic blood flowmeter. It consists of a permanent magnet placed around the blood vessel. This magnet produces magnetic field around the vessel, perpendicular to the direction of blood flow. Two electrodes are introduced to get the induced voltage.

Here, the blood stream cuts the magnetic field and voltage gets induced in the stream which is picked by the electrode. The induced voltage is given by :

$$E = BDVC \qquad ..(5.1)$$

where ; E = Induced voltage in volts

B = Magnetic field strength (wb/m^2)

D = Diameter of blood vessel

V = Velocity of blood flow (m/sec.)

C = Constant

Thus the blood flow rate can be determined in terms of voltage using electromagnetic blood flowmeter.

5.10 Blood Pressure Measurement

Blood pressure is one of the most important parameter of cardiovascular system. Blood pressure measurement helps the physicians to provide the idea of high blood pressure (Hypertension) and extreme low blood pressure (Hypotension). Due to these measurements physician can provide treatment to patient. This measurement helps to avoid sudden or untimely death of patient.

Blood pressure measurements can be carried out by either indirect or direct methods. The advantages of direct methods are accurate measurements and continuous or on line recording of blood pressure waveforms.

(A) Indirect Method

This method is useful for determination of systolic and diastolic blood pressure. Blood pressure is measured by a device sphygmomanometer, (The word sphygmos is a greek word which means pulse). This device consists of an inflatable rubber bladder enclosed in an inelastic fabric called as cuff, a rubber bulb and valve assembly, a manometer which may be mercury column type or aneroid type pressure measuring unit and a stethoscope. Fig. 5.10 shows performance of the system.

At the time of blood pressure measurements the cuff is wrapped or placed around the patients upper arm (area between shoulder and elbow), also the stethoscope placed on the branchial artery close to the skin surface near elbow. Initially, the pressure in cuff is increased by means of a rubber bulb. This pressure is increased well above the systolic pressure so that the blood flow should be stopped completely. Next is to reduce the pressure in cuff at gradual rate by means of needle type valve. Generally mercury in the manometer falls at the rate of 3 mm/sec.

When the cuff pressure falls just below the systolic pressure the sound is generated, this sound is called as Korotkoff sound. This sound is produced due to brief turbulent flow terminated by a sharp collapse of the vessel and persist as the cuff pressure continues to fall. Manometer indicates the pressure of cuff when first Korotkoff sound is heard that blood pressure is called systolic blood pressure. The Korotkoff sound can be heard by using stethoscope which is placed on branchial artery or microphone can also be used to hear the sound.

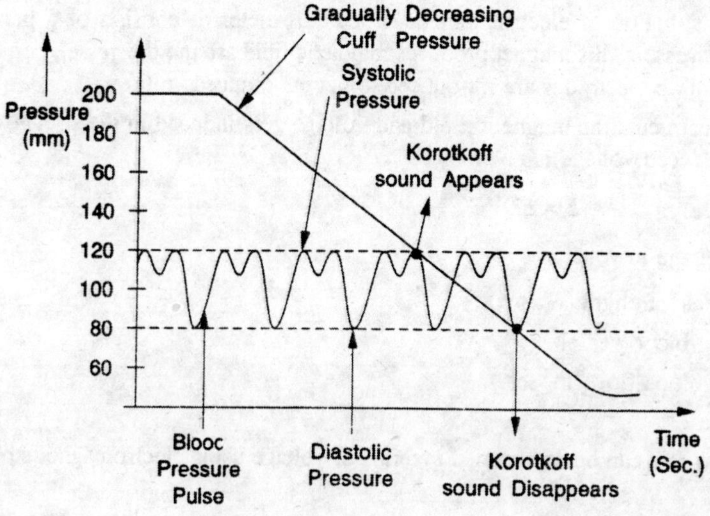

Fig 5.10 : Performance of the system.

For recording the blood pressure the operator or physician has to note the pressure at the on shot of Korotkoff sounds *i.e.* systolic blood pressure and also note the pressure when sound just disappears *i.e.* diastolic blood pressure. Generally these blood pressures are expressed in the ratio manner *i.e.* systolic pressure to diastolic pressure *e.g.* 120/80 as shown in Fig. 5.10.

Advantages of sphygmomanometry :

(1) Simple construction.

(2) No necessity of surgical operation for insertion of transducer or catheter. (It is non invasive method).

(3) Systolic and diastolic pressures can be measured.

(4) Useful in intensive care units (ICU's) and coronary care units (CCU's) if the stethoscope is replaced by microphone. The output of microphone can be conditioned so as to drive the alarm or lamp units.

(5) Suitable for hospital and clinical purposes.

(B) Direct Method for Blood Pressure Measurement

In direct method a catheter is inserted in the heart or major vessel through a superficial vein or artery *i.e.* it is an invasive type method. It provides continuous and permanent record of blood pressure variation measurements. This system uses suitable pressure transducer for the measurement of blood pressure. The pressure transducer is connected to patient by means of thin tube called catheter as stated above.

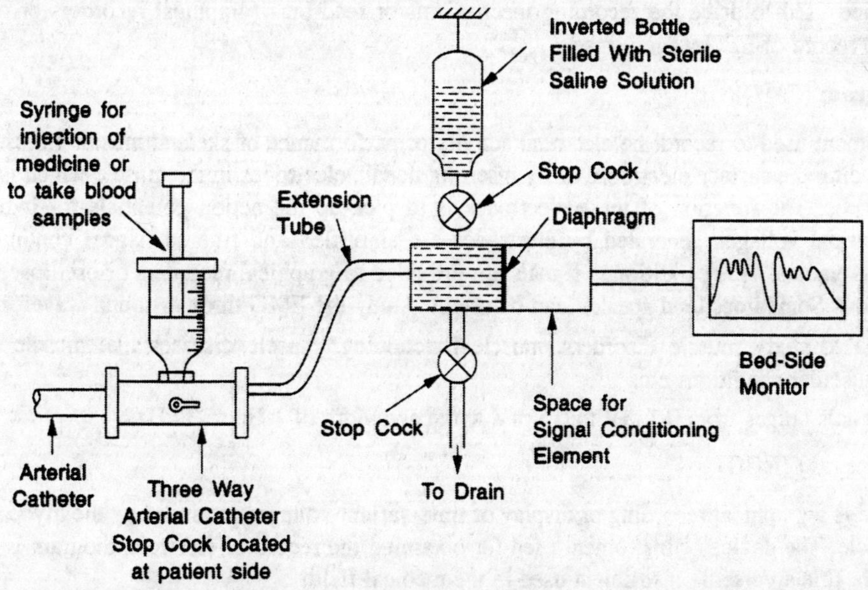

Fig. 5.11 : Direct method for blood pressure measurement.

Fig. 5.11 shows direct method for blood pressure measurement.

Through the catheter sterile saline solution is circulated to transmit the fluid pressure to the transducer outside the body. At the other end of catheter a three way stop cock is placed which provides some facilities and support to the patient at bedside. From the other *i.e.* second terminal of the stop cock, syringe is connected to introduce the medicine in or to take out some blood samples for laboratory testing from patient's body. Third terminal of stop cock is connected to the transducer which may be a diaphragm, strain gauge. LVDT may also be used but initially the blood pressure should be sensed by primary element so that the pressure gets converted into displacement. Transducer which may be used should be linear and it should have stable frequency response so as to record the pulsating nature of blood pressure variation between systolic blood pressure and diastolic blood pressure.

5.11 Study of (a) EEG (b) EMG (c) ECG

(A) Electroencephalogram (EEG)

This device or instrument records and provides the measurements of electrical activity of brain. It records the bioelectric potentials generated by neuronal activity of brain. The bioelectric potential gets recorded or EEG can be obtained by placing electrodes on the surface of scalp *i.e.* skull of the patient. It represents summention of action potentials of neurons in the brain.

EEG records the biopotentials in the range of 10 μV to 100 μV *i.e.* peak to peak amplitude with band width of d.c. to 100 Hz.

Some times needle electrodes can also be used to record EEG. For improving electrical contacts jelly or elecctrolyte is used. EEG analysis is used to trace the brain disorders, damaged sections of brain, mental diseases.

In recording process the electrical output of electrodes is amplified by preamplifier and amplifier stages *i.e.* conditioned and used to drive the recording mechanisms or read-outs (graphical recorders or oscilloscopes) to get the record of EEG waveforms.

(B) Electromyogram (EMG)

EMG is an instrument used to record the electrical activity or performance of skeletal muscle fibres. EMG is recorded by placing the surface electrodes or by inserting needle electrodes in the muscle which is under diagnosis or analysis. The function of these electrodes is to pick-up the action potentials for individual muscle. The electrical voltages generated by electrodes are amplified and filtered (signal conditioning) according to requirements. This conditioned signals are provided to graphical recorders or oscilloscopes to plot the wave forms. Sometimes loud speaker can be used to study the EMG through sound waveforms.

EMG is used to study muscle disorders, muscle functioning, muscle diseases like muscle injury, paralysis, neuromuscular conditions etc.

The EMG signals ranges from 0·1 mV to 0·5 mV with bandwidth of 1 Hz to 3 kHz.

(C) Electrocardiogram (ECG)

Electrocardiogram is a graphical recording or display of time-variant voltages produced by the myocardium during cardiac cycle. The device or instrument used for obtaining the record of electrocardiogram is called electrocardiograph. It is a versatile instrument used in the medical field.

It is useful for determination of heart disorders, positions (condition) of heart, diagnostic information regarding arrhythmias and myocardial infarction. This system provides the measurements of electrical activity of heart. As stated above that ECG is graphic recording or display of the electrical activity of heart muscle myocardium, which is represented in the form of time varying voltages in cardiac cyde.

The ECG recording technique consists of five electrodes which are to be connected to patient's body. It is represented by lead systems as indicated below (Fig. 5.12a). Four electrodes are used to record the ECG and only one electrode is connected on right leg as reference electrode. The lead connections are;

Lead I : Left Arm (LA) and Right Arm (RA)

Lead II : Left Leg (LL) and Right Arm (RA)

Lead III : Left Leg (LL) and Left Arm (LA).

"These electrodes are connected to the ECG machine through same number of electrical wires (cables) which are called Leads". These electrodes (leads) makes the physical contact with patient body. Inside the machine (instrument) these electrodes are connected to differential amplifier through lead selector switch. The stated three leads are also called as bipolar because ECG is recorded using two electrodes, the third electrode is not used.

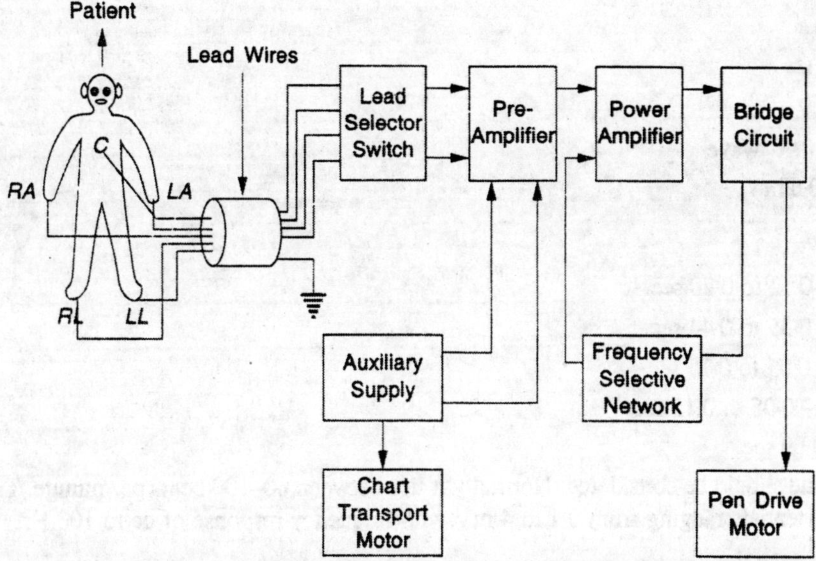

Fig. 5.12(a). Block diagram of ECG machine.

Fig. 5.12(a) shows block diagram of ECG machine. It consists of four (electrodes) lead wires which are connected to lead selector switch whereas the electrode of right leg is grounded. The potentials picked up by the electrodes are connected to lead selector switch. Function of lead selector switch is to select the leads in two by two manner as per lead systems.

The potentials are applied to preamplifier which is differential preamplifier. It provides better stabilization and impedance matching. This amplified signal is applied to power amplifier which is push-pull amplifier.

The frequency selective network is R-C network which provides necessary damping to the pen drive motor.

Auxiliary supply is used to drive chart drive motor and preamplifier. The base of transistors in power amplifier *i.e.* push pull amplifier are driven by the feedback signal frequency selective network from bridge through circuit.

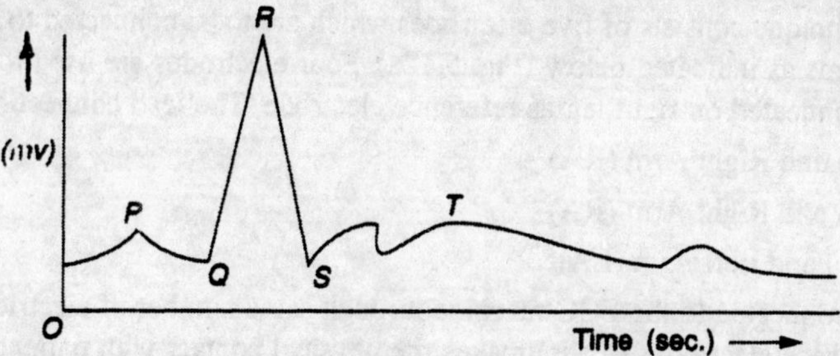

Fig. 5.12(b) Typical ECG waveform.

Fig. 5.12(b) shows typical waveform of ECG. It has some standard values of amplitude and time duration in seconds which are listed below.

Amplitudes :

'*P*' wave 0·25 mV

'*R*' wave 1·60 mV

'*Q*' wave 25% of '*R*' wave

'*T*' wave 0·1 to 0·5 mV

Duration :

P – *R* interval— 0·12 to 0·20 sec.

Q – *T* interval— 0·35 to 0·44 sec.

S – *T* segment— 0·05 to 0·15 sec.

P wave interval— 0·05 to 0·15 sec.

QRS interval— 0·09 sec.

For ECG, heart rate should be considered. Normally it lies between 60–100 beats per minute. The ECG should measure the potentials ranging from 0·1 to 4 mV with frequency response of dc to 100 Hz.

5.12 Non-Destructive Testing (NDT)

Engineering work involves manufacturing and fabrication of different products and items. For fabrication purposes various metals and alloys are used as well as the processes like welding, casting, forging, cutting, drilling etc. are carried out. During these processes defects like cracks, porosity, scabs, tears, cavities, internal shrinkage etc. may takes place or occur. So it becomes necessary in metrology to locate or trace these defects. Such defects can be traced or analyzed using non-destructive or destructive testings. This section deals with non-destructive testing, destructive testing has some disadvantages and limitations which are discussed in the next section.

Now-a-days, non-destructive testing of materials (metals and alloys) is widely used in industries and manufacturing units, workshops etc. Important aspect of this testing is to check and verify the parameters like size, shape, hardness etc. of the material so as to provide good quality and safe product to the customer for particular applications.

5.13 Comparison of Non-Destructive and Destructive Testing

Non-Destructive Testing	Destructive Testing
1. The part or section under test will not be damaged so the manufacturer can inspect or check the product which is to be sold.	1. The part or section under test gets damaged so it has no commercial value *i.e.* part under test becomes unusable.
2. These testings are carried out directly on product items without considering quantity and cost of product.	2. These testings are carried out on the samples of products.
3. No scrap losses occurs in this testing except bad parts.	3. Samples under test may be damaged which results into scrap.
4. Same test or different tests can be carried out on same product or material. Different tests can be carried out simultaneously or sequentially.	4. No same or different tests can be carried out because the sample under test gets damaged or may be damaged.
5. Number of tests can be carried out on the same sample.	5. For different tests different samples may be necessary.
6. Requires costly and complex equipments for testing.	6. Requires equipments of moderate cost for testing.
7. This testing requires skilled operators and technicians to check.	7. As these tests are operated on different samples it requires high manpower.
8. This method is suitable for materials having higher costs.	8. This testing is not suitable for the materials having higher costs.

The term non-destructive testing is defined as **testing of materials *i.e.* metal and alloys without destroying them**. These tests can be carried out in situ manner. NDT detects the discontinuities in the structure and surface of metals or parts which are in rest positions or in continuous process (on line operations). These discontinuities takes place because of creeping, swaging, wear and/or erosion, cooling effects, friction etc.

The important function of NDT is to detect and inspect the discontinuities to avoid the accidents, hazards and great losses. There are some defects which may not be visible to naked eyes, there are various NDT's available. These tests and inspection methods are discussed in the next sections.

Features of NDT

(1) To detect and determine the strength and serviceability of the materials and mechanical parts without damaging the materials.

(2) To check or test the materials whether it is fit for the service intended to be performed or not.

(3) It is a preventive measure.

(4) To predict the life of machine or parts of machine by testing its internal structure, also by considering true conditions.

(5) To test different parameters like surface finishing, material hardness, tensile strength, etc. of metals NDT is applied.

The mechanical based sections or machine consists of different parts having spapes like plates, nuts, bolts, rods, clamps etc. All these shapes and different sizes are fabricated by casting, forging, welding, milling, grinding etc. processes. During these processes the defects like porosity, tears, cracks, internal shrinkage, lack of fusion, lack of penetration, under cut etc. takes place. So in meterology it is necessary to test the metals and alloys before using in operating plants, industrial processes. NDT is a very useful technique.

5.14 Non-Destructive Testings

(A) Visual Inspection

In visual inspection nature, geometry and probable type of defect present in the material can be verified. This is the first stage of inspection and suited if appearance of the material is the inspection criterion.

In this method the surface of metal can be inspected using miniature conventional cameras and lens arrangements by taking the photograph. Now-a-days television cameras with screen is used to monitor the inner surface of materials or metallic parts in continuous fashion.

(B) Magnetic Particle Inspection

This is very simple and easy technique of NDT. Magnetic particle inspection is applicable for detecting the flaws and surface defects in the magnetizable material of any size, shape and composition. This method is carried out in following steps :

(1) To magnetize the material which is under test.

(2) To apply magnetic particles (in the form of powder) in dry or liquid form. This powder is finely divided ferromagnetic particles having high permeability and low retentivity.

If the magnetic powder or particles are dry, it is called dry technique.

If the magnetic particles are suspended in liquid it is called wet technique.

Process : Initially the material under test should be magnetized and magnetic field gets induced in. Due to cracks, blowholes (inhomogeneities) the distortion gets produced in the magnetic field resulting to create distortions in the magnetic flux. This distortion in magnetic flux gets developed because the inhomogeneities have different magnetic properties than surrounding material.

Case 1 : Consider the specimen which is a metallic bar as shown in Fig. 5.13(a). It is magnetized with the magnetic field which is perpendicular to the defect.

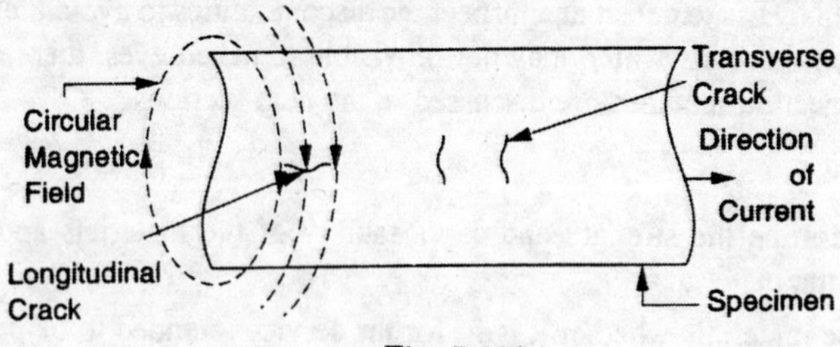

Fig. 5.13(a).

Here, the magnetic field is circular in nature and it will detect only longitudinal cracks or defects *i.e.* those defects which are parallel to longitudinal axis and specimen. This method is not suitable for transverse cracks or defects because the magnetic field is circular in nature. For transverse defects following method is used which is discussed below.

Case 2 : Consider the specimen as shown in Fig. 5.13(b).

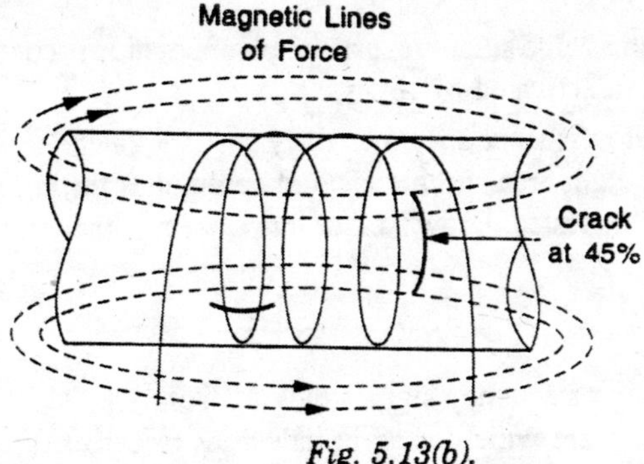

Fig. 5.13(b).

The current carrying conductor is wound over the specimen to produce the magnetic lines of force. If the specimen has some cracks, breaks or inhomogeneities below the surface, it will interrupt the magnetic flux which indicates the flux leakage. This leakage can be detected by applying suitable powder of magnetic particles or materials.

(C) Radiography

Actually, radiography includes utilization of X-rays and gamma rays for NDT's of metals and alloys. In radiography these rays are passed through the specimen (material) under test. These rays impinge on a film or monitor and records the internal structure of specimen.

In this system X-rays or gamma rays are transmitted to specimen and these radiations are recorded so as to produce an image or shadow picture. This image is called a radiograph.

This method is used to test the defects of foundry and welded products. It detects shrink, porosity, lack of fusion, cracks, incomplete penetration and other defects.

Applications of radiography are (1) to test the welds of motor car body, storage tanks, aeroplanes etc. (2) to analyze the defects in the alloy materials like cobalt-nickel steel, duraluminium.

Classification of radiography

(a) **X-ray radiography** : In this process, X-rays are passed through the material under test. As the X-rays can penetrate through the materials they can clearly indicate the cracks and flaws on the film or radiograph. X-rays are electromagnetic radiations.

If the material or specimen under test contains cracks, groove or cavity (inhomogeneities) these inhomogeneities will define different X-ray absorption characteristics. Due to this characteristics variations in the X-ray takes place at these inhomogeneities. These variations or absorption chracteristics should be recorded for analyzing the cracks or cavities but the X-rays are invisible so such recording is carried out on photosensitive film.

If the intensity of X-ray is greater then maximum density gets recorded on photosensitive film. Any cavity or crack if present in the specimen will appear darker on the photosensitive film than other surroundings because maximum transmission of X-rays is at this cavity or crack point.

(b) This process is similar to that of X-ray radiography. Gamma rays are generated by disintegration of radioactive materials. Advantages of this system are penetration power of gamma rays is higher and in small size. In this system the gamma rays are recorded on photosensitive paper so as to analyze the cracks or defects.

(D) Fluoroscopy

In this type of inspection the real-time image gets displayed on fluorescent screen using X-rays. It consists of a screen over which chemicals such as zinc sulphide, barium platinocynide is coated. These materials have property to develop fluorescence when activated by X-rays.

When the X-rays are transmitted towards the specimen, they get penetrated into it and produces an image of cavity or crack on the screen as a light area. In this method continuous testing is possible, as the image being visible the moving objects can be tested. Drawback of this system is that no permanent records of the defects can be obtained.

(E) Ultrasonic Inspection

Ultrasonic signals are the beams of high frequency signals ranging from 1 to 25 MHz, which is beyond the range of human hearing. These signals are widely used in industries to detect the surface and sub surface flaws in the materials. This method is simple, inexpensive and reliable to detect the flaws and cavities in the materials. It is applicable to test steel, aluminium, brass, cast iron, rubber and plastic materials.

Ultrasonic flaw detection : This system requires following necessary equipments :

(1) Ultrasonic frequency signal generator. Mostly it is piezoelectric crystal like Quartz crystal or Rochelle salt.

(2) Transmitter and receiver.

(3) Amplifier.

(4) Cathode ray oscilloscope. (C.R.O.)

Fig. 5.14(a) shows block diagram of ultrasonic flaw detection system which uses pulsed echo phenomenon. The transducer is placed at one end of the material of specimen under test. This transducer will transmit ultrasonic frequency signals towards far end of the material. After transmission, this transducer acts as a receiver so as to receive the ultrasonic echo pulses from the flaw and far end of the material. These received pulses are amplified and displayed on the C.R.O. screen.

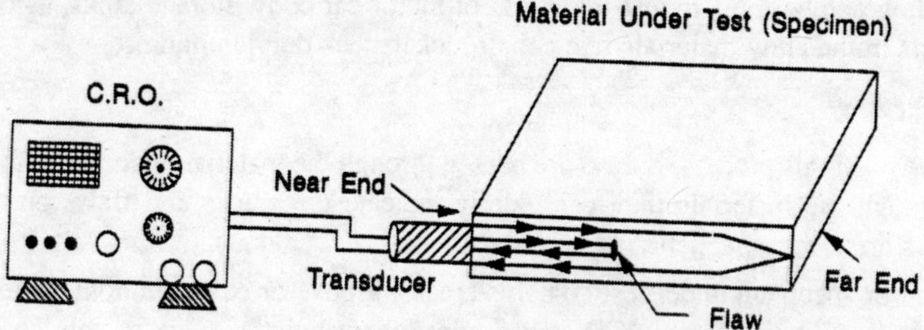

Fig. 5.14(a) Block diagram of ultrasonic flaw detection.

In receiving operation the first pulse relates to the transmitted pulse, then pulse from flaw gets received. Lastly, pulse from far end gets received as shown in Fig. 5.14(b).

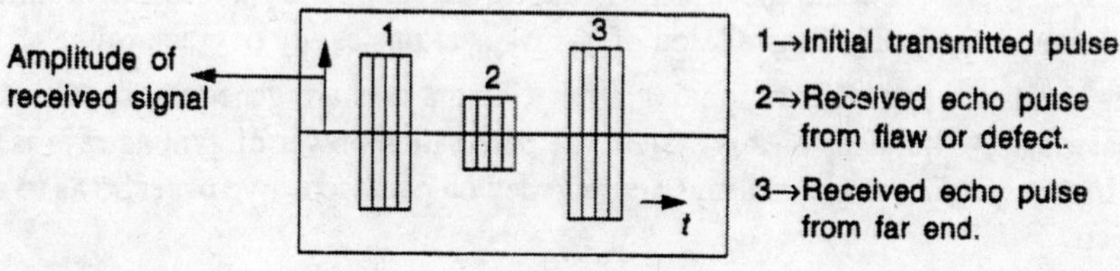

Fig. 5.14(b).

Consider, the pattern displayed on C.R.O. screen. Here the flaw can be detected or traced by analyzing the amplitude and relative position of echo pulse. On C.R.O. the time intervals elapsed between initial pulse and arrival of echoes *i.e.* from flaw and far end are measured. The distance between received echo pulse from flaw and echo pulse from transmitted pulse gives location of flaw from near end. This method uses ultrasonic frequency signals of range 0·8 to 2·5 MHz.

Applications :

Basically this method is useful to detect the flaws and widely used in industries for quality control manufacturing of ships, pressure vessels, airframes, mining equipments, motor vehicles, jet engines, to test rail axes and wheels, rail roads, boiler parts, die blocks etc.

Important features of this system is higher sensitivity, more accuracy, instantaneous flaw detection facility and good penetration power so as to detect the deeper flaws.

Detailed Study of Ultrasonic Testing

Fig. 5.14(c) shows ultrasonic testing method used for flaw detection. It uses ultrasonic probe to transmit ultrasonic signals into the test piece. Probe consists of a piezoelectric crystal with damping body. This assembly is placed in metal or plastic housing. Sometimes the probe may contain double crystals. There are two type of probes.

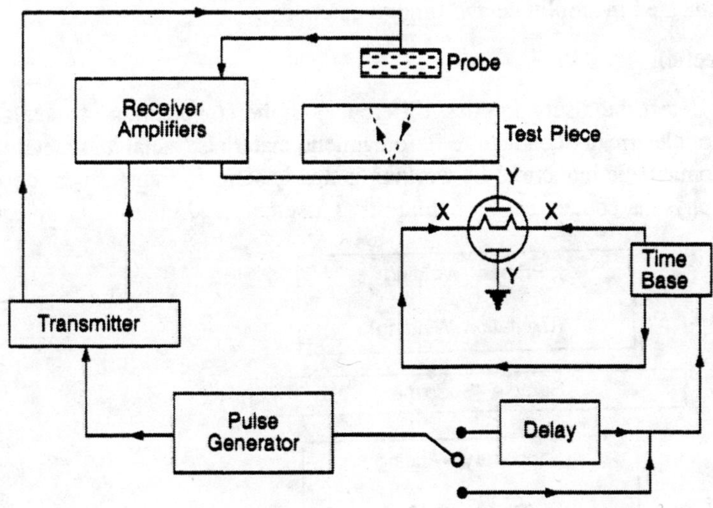

Fig. 5.14(c). Block diagram of Ultrasonic Testing.

(1) **Normal Probes :** In which the ultrasonic signals are transmitted in the direction normal to plane of surface of test piece.

(2) **Angular Probes :** In these type of probes the ultrasonic signals are transmitted into the material

at an angle. The range of angles may be low angle (5° to 22·5°) to high angle (33° to 85°). To transmit maximum ultrasonic energy into test piece couplants are used. Liquid couplant is placed between the transducer and test piece surface. It also provides better acoustic impedance interface. Couplant may be oil, grease, glycerine, amulgums etc.

Method : Initially the pulse generator generates the pulses which trigger the transmitter and energies the transmitter crystal. Similarly, the pulse generator triggers time base and energies right hand X-plates of C.R.T. which moves the luminous spot left to right.

Any echo received by receiving crystal is transmitted to receiving amplifier where it is amplified and energies Y-plates. The receiving amplifier also receives a signal from transmitter at commencement of pulse received from the pulse generator and energies Y-plates. Also receives the echo pulse from flaw and monitored.

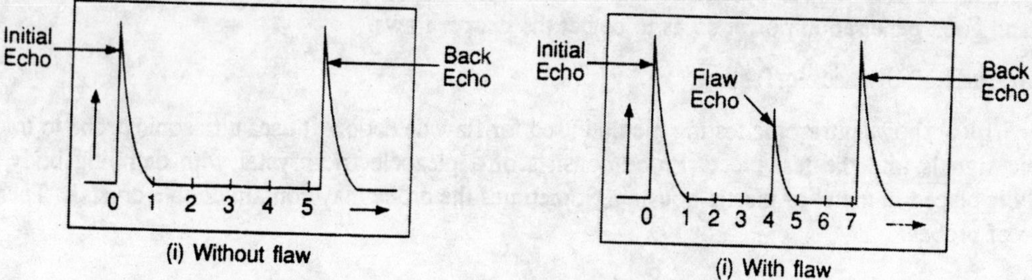

Fig. 5.14 (c): (i) and (ii) Patterns of C.R.T. Screen.

Fig. 5.14(c) (i) and (ii) shows the pattersn of C.R.T. screen indicating without flaw in test piece and with flaw in test piece. It is indicated in amplitude and time manner.

(F) Eddy Current Inspection

This method is useful to detect the discontinuities, differences in the composition of metals. Eddy current inspection is applicable for electrically conductive ferromagnetic materials (metal and metal parts). It works on the principle of electromagnetic induction. Its working is nearly same to transformer action. In this type of inspection there is no physical contact between the testing assembly and material or object under test.

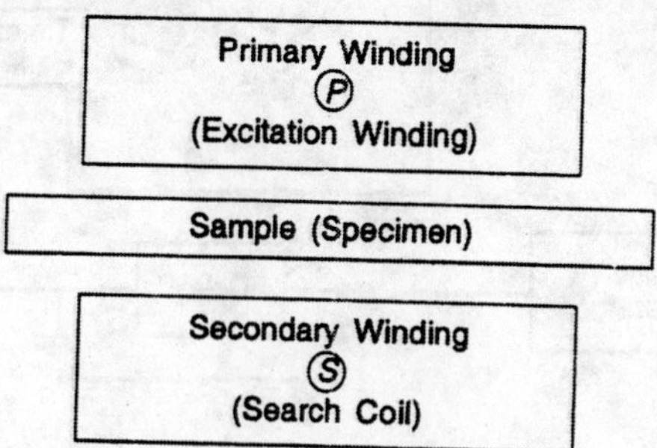

Fig. 5.14 (d): Eddy current inspection method.

Fig. 5.14(d) shows the arrangement of eddy current inspection method.

It consists of primary winding Ⓟ or excitation winding and secondary winding Ⓢ or search coil. An alternating current of fixed frequency is applied to primary winding Ⓟ. Due to this current magnetic flux gets produced which induces an alternating current in the secondary winding Ⓢ with same frequency.

5.15 Level Measurement

Liquid level measurement is an important and oldest parameter in measurements. It is widely used in power plants and petrochemical, paper, sugar industries and also useful for level measurements of exotic and hazardous process matters, fuel handling etc. The liquid level affects pressure and rate of flow in and out of container or vessel due to which it is necessary to measure and control liquid levels.

The service of level measurements is applicable for both solid and liquid materials or process materials. But important factor is to consider the nature and type of material. In following sections various level measurements are discussed. These sections cover both solid and liquid level measurements.

(A) Bubbler Method for Liquid Level Measurement

Fig. 5.15(a) and (b) shows bubbler method for liquid level measurements. It consists of a hollow bubbler tube or dip pipe inserted in the liquid stored in a tank. This tube is fed by air supply from the compressor through pressure regulating valve. Another connection of tube is provided to pressure indicating instruments like pressure gauge or manometer which is calibrated in terms of liquid level. Generally the diameter of bubbler tube is 2 in. to 50 mm. and the tip of pipe should be placed 3 inches above from bottom of tank so as to avoid clogging or blocking due to deposition of particles in liquid.

Pressure of regulator is adjusted so that the air pressure in bubbler tube should be greater than the hydrostatic (pressure) head of fluid or liquid under test. Due to this adjustment air bubble will come out from bottom.

When there is no liquid in tank or liquid level is below the tip of bubbler tube or pipe, compressor air flows out from bottom and pressure indicator will indicate zero reading. As the liquid level starts to increase air pressure in dip pipe or tube changes. This pressure is indicated by pressure indicating instrument. Because of increase in liquid level the air flow gets restricted because of depth of liquid so air pressure acts against liquid head which is back pressure indicated by the instrument. Pressure indicated by pressure measuring instrument is directly proportional to the liquid level.

Precautions :

(1) Materials used for dip pipe or tube should be free from corrosion. The material can be selected according to fluid involved in process.

(2) Air pressure fed to dip pipe or tube should be slightly higher than the maximum pressure of head in tank.

(3) Accuracy depends upon pressure sensing element, so it should be properly calibrated.

(4) Dip pipe or tube should be cleaned properly in regular manner.

Advantages :

(1) Simple method.

(2) Pressure monitoring element or device can be installed at top or bottom of tank. Also it can be placed upto 12·7 meters from tank by using proper piping arrangements.

Applications :

(1) Suitable for measurement of corrosive or abrasive liquid levels.

Fig. 5.15(b) depicts different shapes of the tip of dip pipe or bubbler tube that are available and can be selected according to requirements.

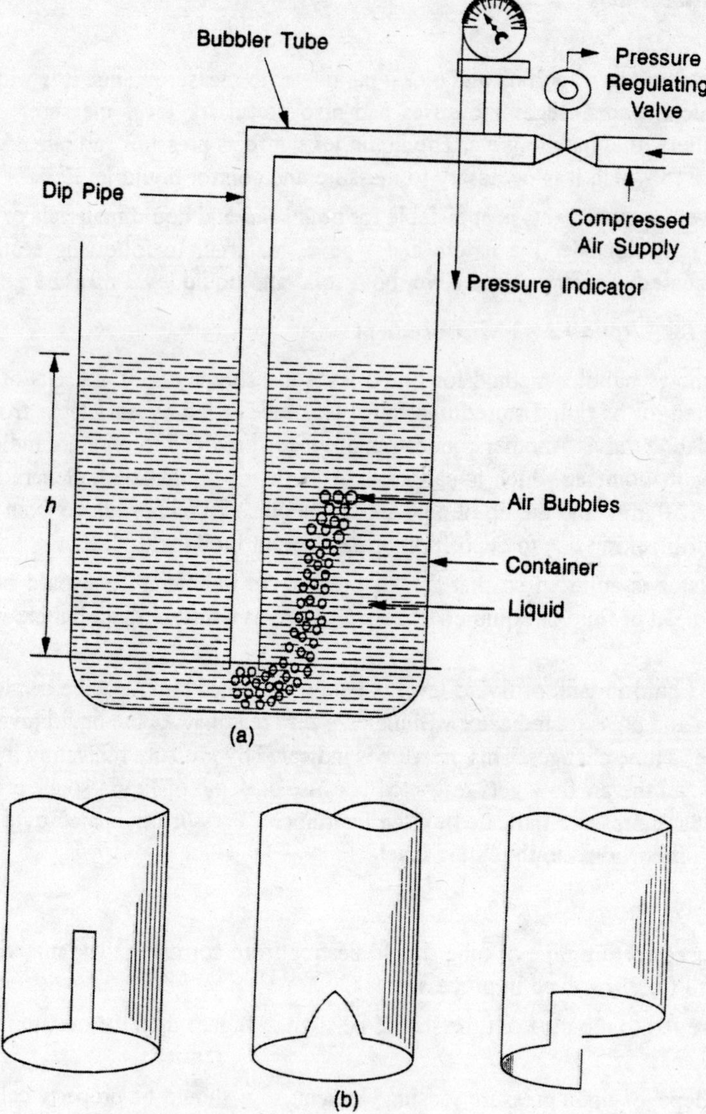

Fig 5.15 : (a) Bubbler method (b) Tips of pipe.

(B) Level Measurement by Thermal Sensors

These type of sensor operates on the principle of temperature difference or difference of thermal conductivity between liquids and vapors. In following sections various thermal sensors are discussed. These device can be used for ON/OFF services.

(a) **Thermal conductivity switch** : Fig. 5.16 shows thermal conductivity switch. It consists of a probe containing a resistive element heater through which current flows using proper circuit arrangements. Temperature sensor is used to sense the temperature of fluid involved in process.

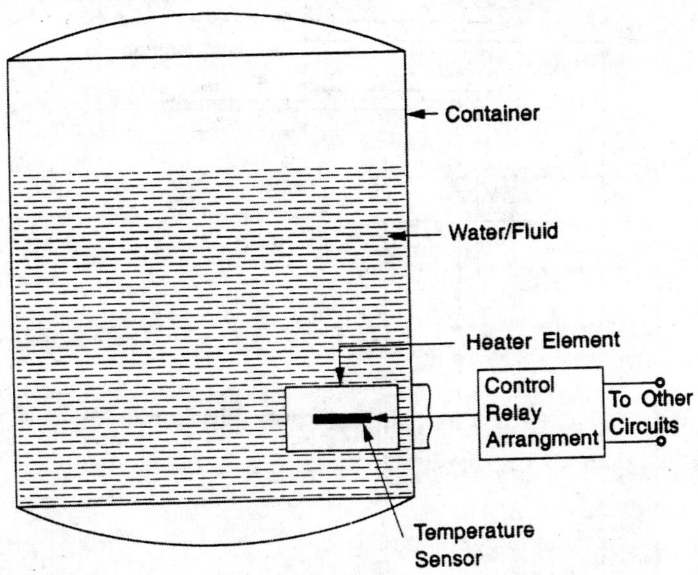

Fig. 5.16 : Thermal conductivity switch.

When this probe is immersed into water or liquid the heater element starts to heat. This heat is transferred to the water, due to this phenomenon the temperature of water becomes very close to temperature of heater. As the water level falls below the probe, temperature of this probe rises because of low thermal conductivity of water vapor or air above the water level.

This means that this system acts as a thermal switch which detects the rise in temperature because of reduction or absence of water or process fluid in container. As this difference gets detected it interrupts the power supply to probe by means of relay arrangements or drives the alarm or shutdown arrangements through relay.

Applications :

(1) In oil delivery trucks using thermistor element probes.

(2) Tank overfill protection arrangements.

(b) **Thermal conductivity switch using two sensors** : Fig. 5.17 shows another arrangement of thermal switch using two sensors. This sensor is immersed in vertical position into liquid or vapor.

It consists of a probe in which two resistive (R_1 and R_2) sensing elements are placed in vertical position. But these elements are kept or displaced at certain distance from each other as shown in Fig. 5.17. Material used for probe is stainless steel. These two resistive elements are supplied by voltage source due to which these elements start to heat. If these elements are immersed in liquid or vapor media (process material) the heat will be transferred to this media and temperature of both (*i.e.* resistive elements and media) becomes closer. The resistance of both elements is same so current flowing through is also same and they will deliver nearly same heat.

If the element R_1 is in liquid media and R_2 in vapor media, more current will flow through R_1 because liquid has more conductivity than vapor so its resistance value is lower. This creates difference in current which is applied to current comparator circuit. Output of comparator circuit is applied to drive the control elements, relay or alarm circuits.

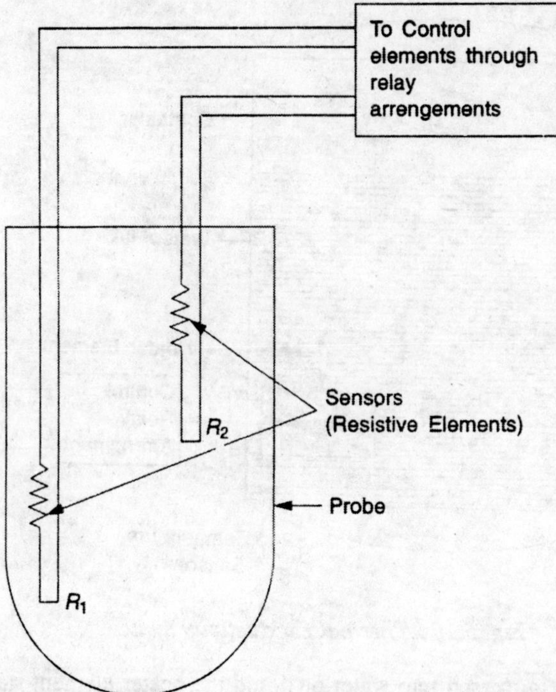

Fig 5.17 : Thermal conductivity switch using two sensors.

Advantages :

(1) Simple construction.

(2) Can be suspended or immersed in a container.

(3) Higher accuracy can be maintained.

Applications :

(1) To detect storage tank levels.

(2) Continuous level detection of liquid/liquid or liquid/vapor media.

(c) **Conductivity level switches** : This method is applicable for point or differential level measurements of conductive liquids or slurries. It requires grounded tank, conductive fluids or liquids, electrodes, relay arrangements and transformer. Fig. 5.18 shows conductivity level switch.

It consists of a metallic grounded tank cointaining conductive liquid in which two electrodes (Electrode 1 and Electrode 2) are installed at different levels *i.e.* high and low. The electrodes may be called probes.

When the liquid level increases conductive path between electrode 1 and grounded tank is formed so the relay 1 gets energized because circuit gets closed. Due to this contact switch S_1 is closed, which operates the alarm circuit, control valves, motors or any other processing equipments.

The high level Electrode 2 is placed so as to indicate maximum or higher level of liquid in tank. In the present case this electrode is above liquid level so no conductive path gets established. This condition is open circuit so no current flows through the entire circuit due to which relay 2 will not be energized and the contact switch S_2 will be opened.

Transformers 1 and 2 are used to provide energizing voltage and current to relays and associated circuitry. Relays 1 and 2 may be solid state or electromechanical relays.

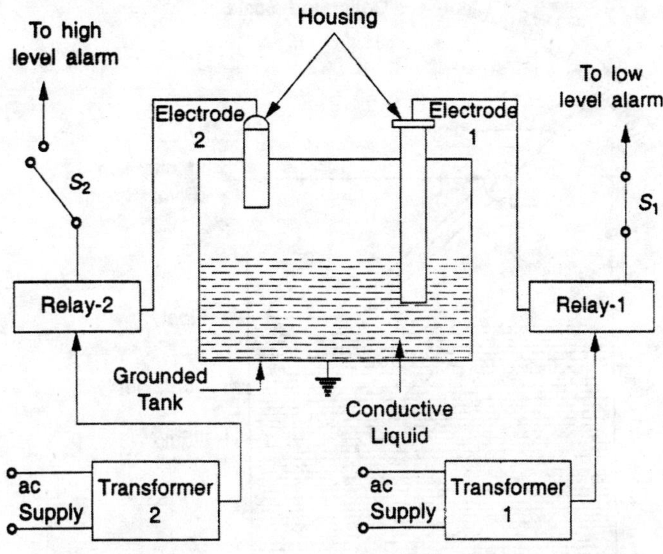

Fig. 5.18 : Conductivity level switch.

Advantages :

(1) Low cost.

(2) No moving part is involved in the system so problem of friction gets avoided.

Limitations :

(1) Spark may be produced when liquid level may touches the electrodes or probes.

(2) Not suitable for chemical industries.

(3) Limited for conductivity upto 10^8 ohm/cm resistivity.

Applications :

(1) In food industries.

(2) Molten glass level measurements which is conductor at higher temperatures.

Metals used for electrodes are stainless steel, Hastelloy, titanium with housing of corrosion free materials like plastic, rubber etc.

(C) Liquid Level Measurements Using Float

Float is a body which has characteristics always to float and follow the changing liquid level. Normally, the floating devices have hollow spheres, cylindrical, and disc shape. Floats can be designed according to requirements *i.e.* whether low or high level measurement is carried out, density of liquid etc. Floats are available in the range of 3 to 7 inches in diameter and materials used for floats are brass, copper, monel, polysulfone, plastics etc. The material used for float should be free from corrosion, oxidation and should be properly sealed. Float rests on the surface of liquids and follows changing levels of liquids. Because of change in liquid level, movement of float takes place which is transmitted to pointer and calibrated scale through mechanism like spring. Following methods are discussed for liquid level measurements using float.

(a) **By using float and shaft arrangements** : Fig. 5.19 shows liquid level measurement using float and shaft arrangements. It consists of a float resting on liquid surface which is connected to pointer using mechanism arrangements.

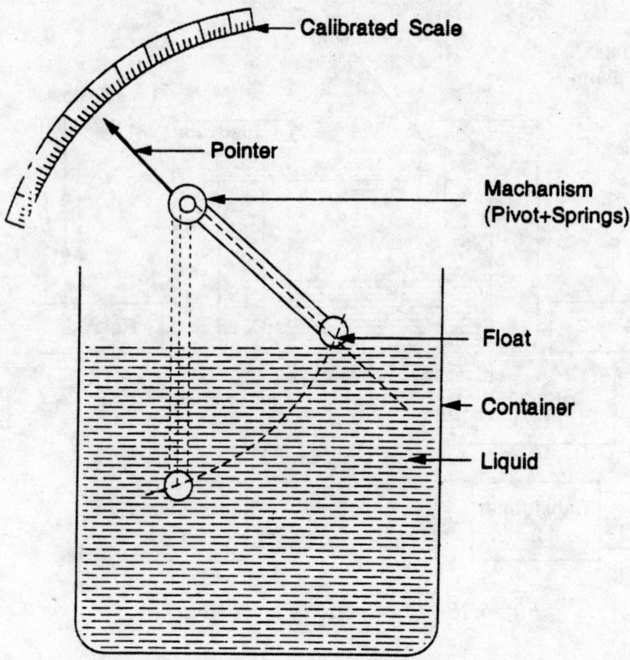

Fig. 5.19 Liquid level measurement using float and shaft.

As liquid level changes, the position of float varies on the surface *i.e.* float moves according to position of liquid level. This movement of float is transmitted to the pointer which slides over calibrated scale through mechanism. The displacement of pointer over calibrated scale is directly proportional to liquid level.

The displacement can be calibrated or converted into electrical voltage by using linear variable differential transformer (LVDT). Displacement of float is applied or connected to the core of LVDT.

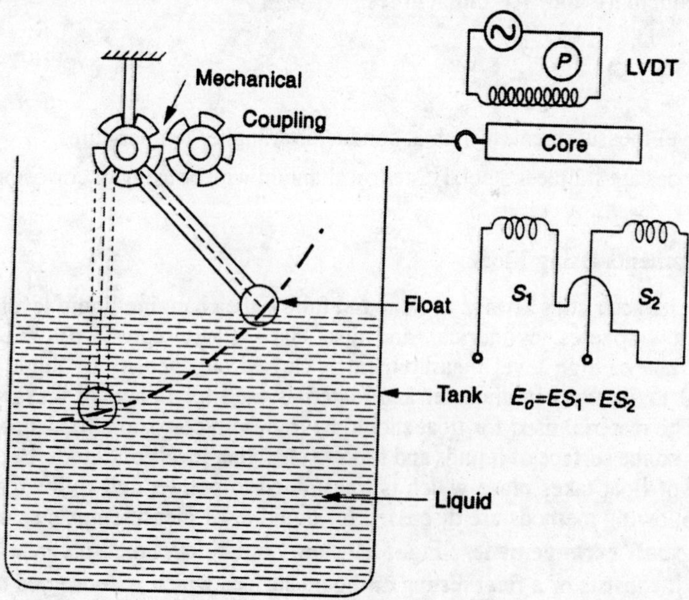

Fig. 5.20 : Using float and LVDT.

(b) **By using float and LVDT** : Fig. 5.20 shows liquid level measurement using float and LVDT. Float resting on the surface of liquid is connected to core of the LVDT through mechanical linkage.

As liquid level varies (increases or decreases) displacement of float takes place. This displacement is applied to the core of LVDT. It produces differential output voltage which is proportional to displacement of core. This voltage can be transmitted to control devices or relays. In this method the float acts as a primary transducer and LVDT functions as a secondary transducer.

(c) **By using float and potentiometer** : This is a simple and direct method to measure the liquid level in terms of electrical voltage. It can be achieved either by using linear or rotary potentiometer. Fig. 5.21

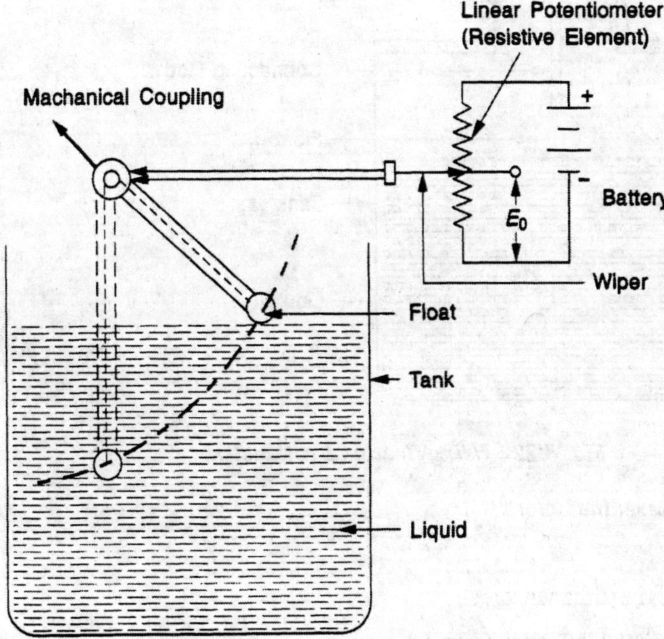

Fig. 5.21 : Using float and potentiometer.

shows implementation of liquid level measurement scheme using linear potentiometer. The float is connected to the wiper of potentiometer through mechanical linkage. As float displaces according to variations in liquid level the wiper slides over resistive element of potentiometer resulting to change in resistance. Change in resistance is directly proportional to displacement of wiper due to float movement.

E_0 is the output voltage drop across potentiometer which is proportional to liquid level. This voltage can be used to operate relay or alarm units or control elements.

(d) **By using float and rotameter** : Fig. 5.22 shows simple arrangement of liquid level measurement. It consists of a float which is connected to a solid shaft. An indicator is placed over this shaft, which slides over a calibrated scale.

When liquid level varies the position of float displaces resulting to slide the pointer over scale. The indication by pointer is proportional to liquid level. Materials used for float and rod arrangement are stainless steel, monel etc. Generally glass indicator is used.

This method is applicable for non-hazardous applications, low level measurements.

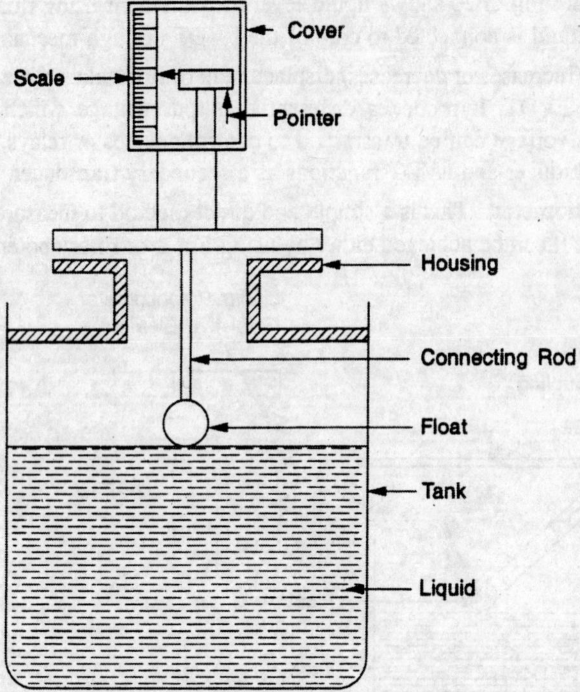

Fig. 5.22 : Using float and rotameter.

Advantages of floating type level indicators

(1) Low cost.

(2) Easy installation and maintanance.

(3) It can operate over large temperature range.

(4) Materials used for fabrication of floats can be selected having corrosion resistance, ruggedness, strong etc.

(5) Available in different shapes such as cylindrical, spherical and disc.

Disadvantages of floating type level indicators :

(1) These elements operate on moderate pressure range.

(2) Not applicable for highly toxic, corrosive and exotic (liquids) fluids.

(3) Accuracy depends upon mechanical **coupling** (assembly) used in construction.

(4) Moving parts may be clogged or blocked due to granules or solid particles.

5.16 Electrical Methods For Level Measurement

(A) Resistive Method : (Contact Point Method)

This is simple method of liquid level measurement. It uses mercury as a conductor. Fig. 5.23 shows resistive method. It consists of a U-shaped mercury column attached to bottom of tank operated by liquid level. Resistances R_1, R_2, ..., R_n of suitable values are placed in the mercury column at various liquid levels through contact rods. As liquid level '*h*' increases the mercury level rises which shorts the successive resistances '*R*' so varies the current '*I*' flowing through network. This current '*I*' is directly proportional to liquid level '*h*' and measured by ammeter.

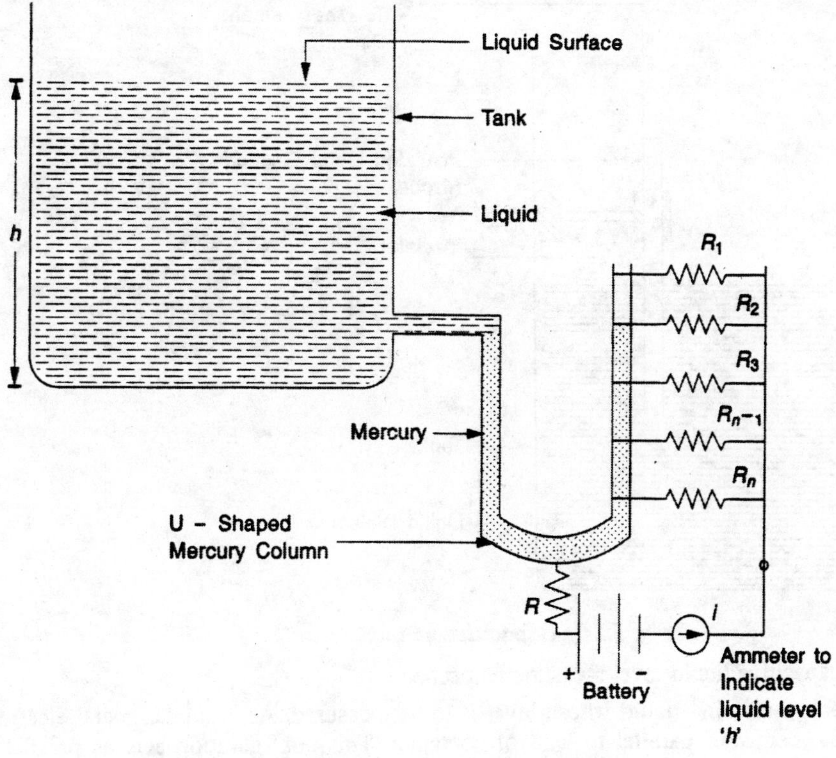

Fig. 5.23 : Resistive method.

Advantages :

(1) Continuous measurement is possible.

(2) Number of resistances can be increased to measure different levels or individual level.

(3) Electrical output can be used to operate relays or other control elements.

(4) Simple construction.

Disadvantages :

(1) Not suitable for explosive and hazardous liquids because possibility of development of arcing at contact rods.

(2) Large number of contact rods are necessary for stepwise level measurements.

(3) Rods may be damaged due to corrosion, moisture, temperature etc.

(B) Capacitive Method

This is also a simple and inexpensive method used for measurement of non-conductive process fluids. In general, the term capacitor is defined as two metal plates separated by some distance and dielectric media like mica, paper, teflon, ceramic, air ,etc. is placed between the metal plates. This means that capacitor is a device having two conductors (metal plates) electrically isolated by a non-conductors *i.e.* dielectric media.

Capacitor stores the charge and measured by unit Farad. In level measurement by capacitive method the liquid, solids, oils, fluids, gasoline (non conductive elements) acts as dielectric media. As the liquid level changes *i.e.* dielectric media variation takes place resulting in change of capacitance.

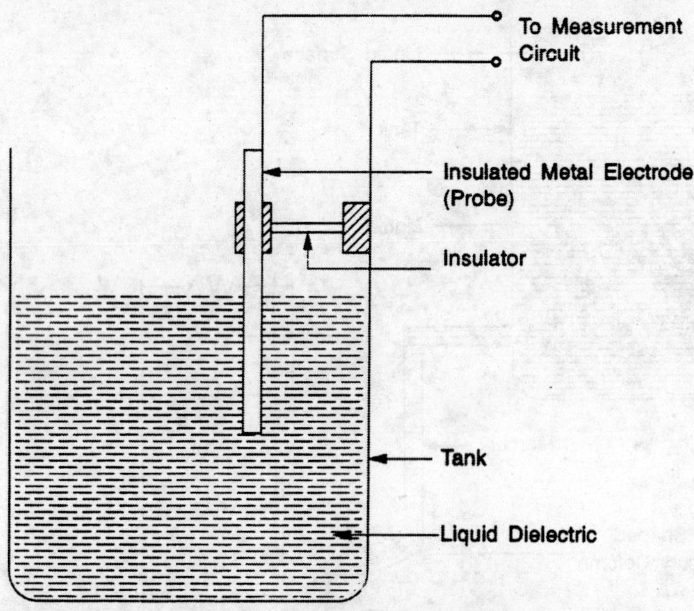

Fig. 5.24 : Capacitive method.

Fig. 5.24 shows capacitive liquid level measurement scheme.

It consists of a tank containing liquid whose level is to be measured. An insulated metal electrode or probe is immersed into container parallel to wall of container. This configuration acts as parallel plate capacitor *i.e.* liquid acts as dielectric media, the probe (electrode) and container (container wall) acts as two parallel plates of the capacitor. Important conditions to be considered are (i) If the liquid in the container is conductive the probe or electrode and liquid forms the plates of capacitor. The dielectric media is insulation of probe. (ii) If the liquid in container is non conductive the walls of container and probe form parallel plate capacitor whereas liquid acts as dielectric media.

Expression for capacitance of parallel plate capacitor is given by :

$$C = \varepsilon_0 \varepsilon_r \frac{A}{d} \text{ Farad ;} \qquad \text{...(5.2)}$$

where ; C = Capacitance in Farad.

ε_0 = Free space permittivity = $8{\cdot}854 \times 10^{-12}$ F/m.

ε_r = Dielectric constant ($\varepsilon_r = 1$ for air).

A = Area of plate (m^2).

d = Distance between two plates (m).

When liquid level increases in tank the capacitance (C) increases (because of dielectric constant of process material) while as liquid level decreasesd the capacitance (C) decreases. This variations in the capacitance is calibrated in terms of liquid level.

(C) Liquid Level Measurement Using Bare Capacitance Probe Method

Fig. 5.25 shows bare capacitance probe which is insulated from vessel and immersed in conductive media which acts as one plate of capacitor while the vessel form another plate. When liquid level increases the total capacitance increases on the other hand when liquid level decreases total capacitance decreases. These variations in the capacitance is calibrated in terms of level (*h*).

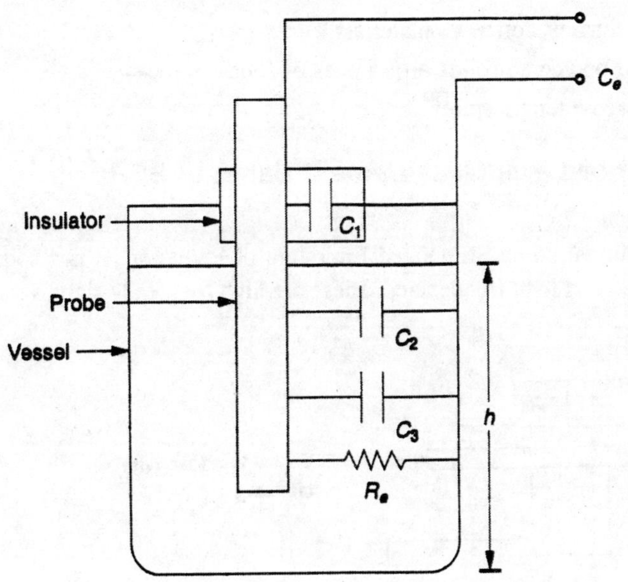

Fig. 5.25 : Bare capacitance method.

From Fig. 5.25 total (equivalent) capacitance of probe is given by :

$$C_e = C_1 + C_2 + C_3 \qquad \text{...(5.3)}$$

where ; C_1 = Dead capacitance of insulator.

C_2 = Capacitance of vapor space.

C_3 = Capacitance of process material.

C_e = Equivalent capacitance.

R_e = Effective resistance between probe and vessel.

The effective resistance R_e is infinite.

This probe can also be used to measure levels of non-conductive materials like fly ash, granular solids, corn oils but it should be insulated.

Advantages :

(1) Simple construction.

(2) Less expensive.

(3) No moving parts, so free from friction and spark.

(4) Suitable for continuous measurements.

(5) Applicable for small measurements.

Drawbacks :

(1) Errors due to change in dielectric constant and conductivity of process material.

(2) Recalibration is necessary if the process material gets changed.

(3) Sensitivity may be affected because of temperature.

Applications :

(1) To measure levels of slurries, corrosive materials.

(2) To determine levels of powdered solids or granules of solids.

(3) Liquified gas levels at low temperatures.

5.17 LASER Level Sensor or Level Measurement Using LASER

The term LASER is Light Amplification by Stimulated Emissioin of Radiation. It is a source of coherent electromagnetic waves at infra red and light frequencies. These are high frequency signals ranging from 430 to 750 tera Hz (THz).

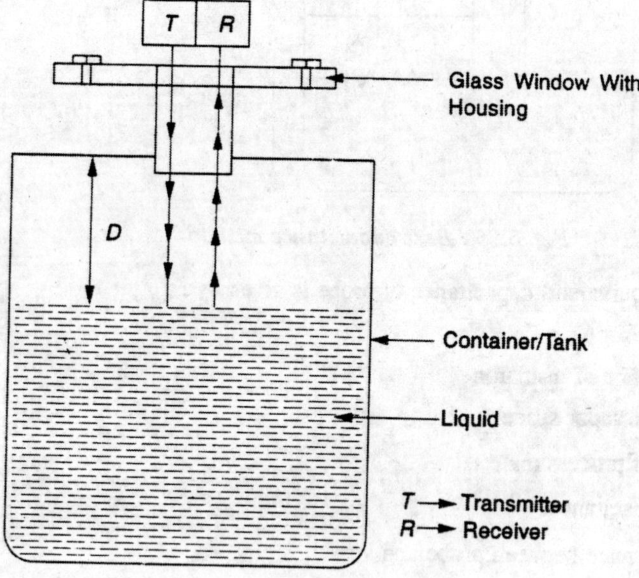

Fig. 5.26 : Laser method.

This method is applicable for solids, liquids with transparent vapors. Fig. 5.26 shows level measurement using laser. It consists of a LASER transmitter and receiver set installed above the container or tank. A glass window is placed on top of the tank using proper housing of rubber or plastic.

From the transmitter end a sharp beam of LASER is focused towards the liquid level by modulation. When this signal reaches at surface of liquid it gets reflected. This reflected laser is received by the receiver. The time period from reflection of signal and to reach at receiver is calculated so as to measure the liquid level. The time period 't' is function of distance 'D' ('D' is distance from liquid level *i.e.* the level from which signal is reflected to the receiver) which is given by :

$$t = f(D) \qquad ...(5.4)$$

$$t = \frac{2D}{C} \quad ...(5.5)$$

where, C = Velocity of light = 3×10^8 m/sec.

The liquid level can be measured by calibrating time period in pulses using counter circuit.

Advantages :

(1) No direct physical contact with process material.

(2) Suitable if vapor can be present with process material.

(3) No moving part is involved in the system so operation is free from friction and spark.

(4) More reliable.

Limitations :

(1) Accuracy depends upon light reflection characteristics of process material.

(2) The process material should be dust free.

(3) The glass window and housing should be clean and dust free.

Applications :

(1) To measure the levels of liquid aluminium, liquid polyethylene, brown tobacco etc.

(2) Levels of wood pulp in paper industries.

(3) To measure levels of hazardous and toxic liquids and acids.

5.18 Ultrasonic Liquid Level Detector

It is conventional method used for measurement of liquid levels. This method uses the principle of reflection of acoustic signals from liquid surface. It consists of an ultrasonic transmitter which generates frequency signals ranging from 35 kHz to 40 kHz. Fig. 5.27 shows ultrasonic liquid level indicator.

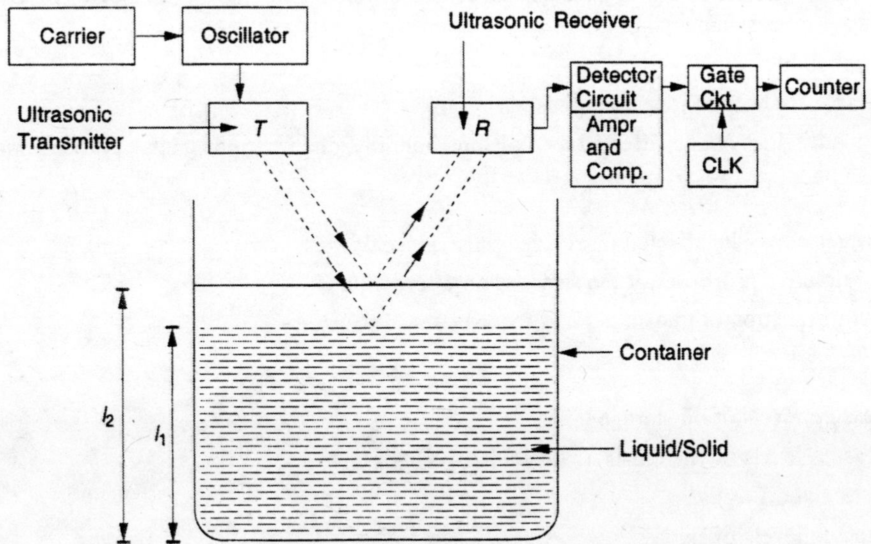

Fig. 5.27 : Ultrasonic method.

It consists of a set of ultrasonic transmitter '*T*' and receiver '*R*' placed on the top of tank. The transmitter transmits ultrasonic wave towards surface of liquid or solid. This beam is reflected back from the surface and received by receiver '*R*'. The measurement is carried out on the basis of time taken by received echo beam to reach the receiver end. As liquid level changes, the transit time required to reach the beam at receiver also changes *e.g.* for level l_1 the transit time will be different than time required for level l_2.

Received echo beam is detected by means of detector circuit. Sometimes amplifier can be used with comparator to detect the echo. By passing detected signal through gate and flip-flop circuit the time period can be measured in the form of pulses. Piezoelectric crystals such as Rochelle salt, barium titanate, quartz can be used as transmitter In some situations of liquid level measurement the ultrasonic transmitter and receiver set may be installed at the bottom of tank as shown in Fig. 5.27(a).

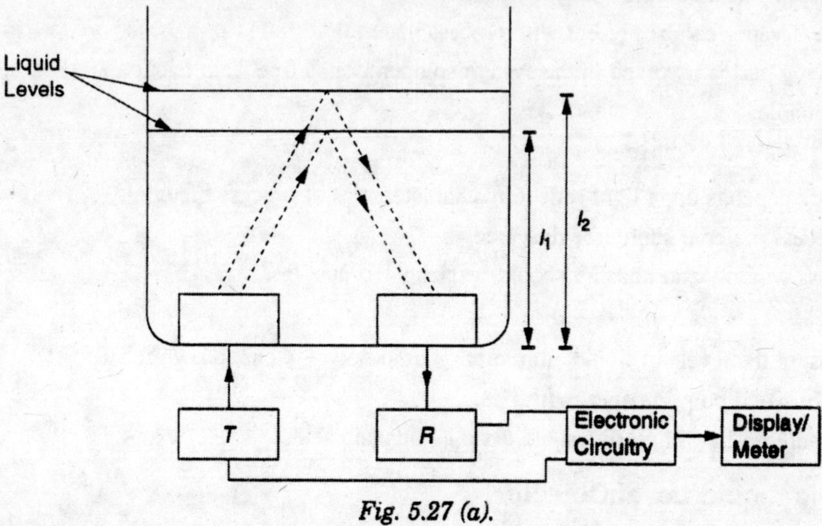

Fig. 5.27 (a).

Advantages :

(1) No moving part is involved in the system hence no problem of development of friction or spark.

(2) No physical contact with process material, so useful for measurements of conductive, corrosive, hazardous process materials.

(3) More reliability.

(4) Applicable for both solids and liquids.

(5) Performance does not get affected by moisture, humidiy, dielectric constants of process materials.

Limitations :

(1) Performance may be affected if received pulse is weak.

(2) Weak refractive properties of the surfaces of process fluids.

(3) Careful installation of transmitter and receiver set is necessary.

Applications :

(1) To measure levels of oil storage tanks.

(2) To measure levels of chemicals, toxic acids, alkalies etc.

(3) In mines and oil wells.

(4) To measure levels of slurries.

(5) To measure levels of petroleum products in oil refineries.

(6) To determine levels of molten metals.

(7) Also suitable for liquid with vapors.

5.19 Microwave Level Detector

These type of detector uses electromagnetic radio waves in microwave *X*-band and *K*-band frequency ranges from 8 to 12·5 GHz and 18 to 26·5 GHz respectively. Maximum available peak power of *X*-band is 10 mW and for *K*-band is 0·6 mW. Microwave signals can be generated by means of microwave tubes like klystron, magnetron etc. and solid state devices such as gunn diode, IMPATT diode, Schottky diode etc. For better performance solid state devices are used to generate and detect microwave signals.

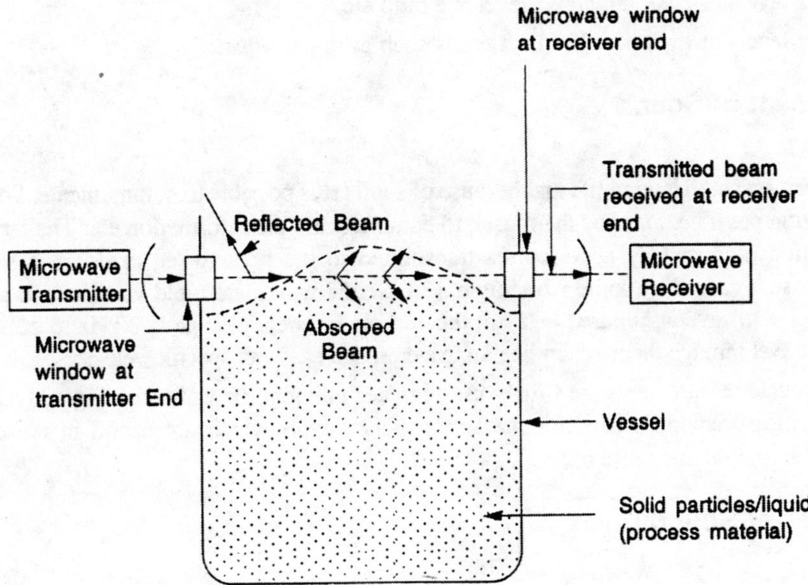

Fig. 5.28 : Microwave method.

Fig. 5.28 shows microwave beam breaker arrangement to detect or measure the level. This system is applicable for solids *i.e.* small size particles, liquids having dielectric range upto 10. Microwave signals do not pass through metals, but can pass through the vessels having walls of fibre glass, plastic and ceramic.

It consists of a microwave transmitter which transmits the signals either in *X*-band or *K*-band through window inside the vessel. Windows are installed at transmitter and receiver ends. Materials used for windows are plastic, teflon, quartz, mica, unleaded glass etc.

When there is air or vapor present inside the vessel, strong microwave signals are received at receiver end and detected by microwave detector. As the level of process material increases it brakes the transmission of microwaves and these signals get reflected resulting in reduction of the signal strength. Due to this weak microwave signal, it gets detected by the detector because of reflection and absorption of signals. In this way the level can be detected.

Advantages :

(1) No moving part is involved in the measurement scheme, so operation is spark and friction free.

(2) Not affected by foam, dust and mist.

(3) Applicable for both conductive and non conductive process materials.

Limitations :

(1) Expensive method.

(2) Complicated construction.

(3) Applicable for limited dielectric constant of process material.

Applications :

(1) Useful to measure levels of toxic, hazardous chemicals or process materials.

(2) To detect levels of coal, minerals, vegetable pulp etc.

(3) To detect levels of light powdered materials such as dry sawdust.

5.20 Measurement of Sound

Sound is an important aspect in human life and because of sound it is possible to communicate. Sound makes speech communication possible, to enjoy the music, to communicate the information etc. The term sound is defined as a sensation produced by nerve-pulses transmitted to the brain when eardrum is subjected to vibrations of air pressure caused by some vibrating object. Sound waves are rapid variations in air pressure in the range from 1 Hz to several hundred kHz but the audible frequency range is 20 Hz to 20 kHz. sound waves propagate or travel through the medium like air, water, metal etc. with specific velocity.

Sound or acoustical measurements are widely useful in designing, testing of recording and reproduction equipments. Sound measurement is used for measuring noise levels which are useful in noise pollution control, to measure aero-dynamic noise etc..

5.21 Parameters of Sound Energy

(A) Sound Intensity

The term sound intensity is defined as average rate of transmission of sound energy in a specified direction through a cross-sectional area of 1 m^2 at right angles to the direction. It is expressed in W/m^2 (watts per

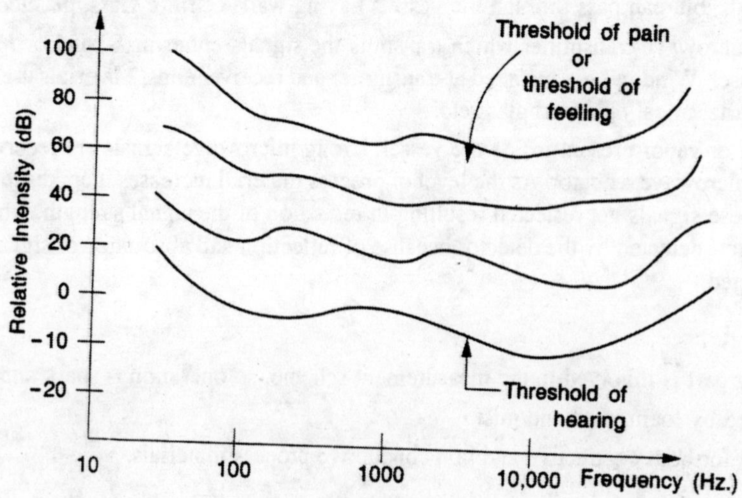

Fig. 5.29 : Fletcher-Munson curves.

square meter). Ear can detect the sound intensities upto 10^{-13} W/m^2. Sound intensity can also be defined as the total power contained in sound waves which hits the eardrum. Sound can be heard only upto a certain level. The minimum sound intensity that can be heard is called as **threshold of hearing**.

When the sound becomes louder *i.e.* when it reaches to upper limit of hearing it is painful and uncomfortable feeling to the ear. This range or upper limit is called **threshold of pain** or threshold of feelings.

Fig. 5.29 shows Fletcher-Munson curves of equal loudness versus frequency. It indicates threshold of hearings and threshold of pain. This graph shows the variations in the intensity with frequency. It uses a logarithmic frequency scale and sound intensity is expressed in decibels above the threshold of hearing because loudness is proportional to logarithm of intensity.

(B) Loudness of Sound

It depends upon the intensity of sound. When the sound strikes or falls on eardrum the intensity of sound will stimulate the ear. Thus, the loudness of sound is the stimulation which is produced by the intensity of sound striking or falling on the eardrum. Loudness depends upon sensitivity of the ear, interest and concentration of listener, psychological conditions etc. Human ear very slowly responds to the increase in sound intensity. When an intensity of particular sound increases by twice to that of original sound value then the original value is not heard twice as loud but it produces sensation of change of loudness. The sound gets heard twice as loud as original sound and its intensity increases by ten times. The response of the ear to change in sound intensity is logarithmic in nature which is indicated in the graph of relative versus frequency.

(C) Decibel

Bel is the unit used to measure sound levels. Bel is named after Alexander Graham Bell, the inventor of telephone. Decibel is logarithmic unit which is one tenth of bel. Decibel means one tenth and decibel is generally written as dB. The quantity measured in decibel expressed as a ratio of relative to reference level of quantity. The quantity may be sound power level or sound pressure level.

Mathematically, decibel can be expressed as :

$$\text{dB} = 10 \log P_2/P_1 \qquad \text{...(5.6)}$$

where, P_1 and P_2 are the power levels.

If $P_2/P_1 = 2$, then $\log 2 = 0{\cdot}301$ from the logarithmic table. (P_2/P_1 is only ratio and its logarithmic value can be determined from logarithmic table).

As $\log 2 = 0{\cdot}301$; so $10 \times 0{\cdot}301 = 3$ dB (approximately).

(D) Sound Pressure Level (SPL)

Mathematically in decibels (dB) the sound pressure level (SPL) is expressed as ;

$$SPL = 10 \log (P_r^2)/(P_{ref}^2)$$

$$SPL = 20 \log (P_r/P_{ref}) \text{ dB} \qquad \text{...(5.7)}$$

where ; (P_{ref}) is r.m.s. reference pressure which is given by

$$P_{ref} = 0{\cdot}0002\ \mu\text{ bar} = 0{\cdot}00002\ \text{N/m}^2.\ (P_r) \text{ is r.m.s. sound pressure (N/m}^2).$$

The range of sound pressure level is from the threshold of hearing at 2×10^{-4} μ bar to 1 m bar at the threshold of pain.

$$\therefore \qquad SPL = 20\log \frac{P_r}{0{\cdot}00002} \text{ dB} \qquad ...(5.8)$$

(E) Sound Power Level

Always the sound pressure energy is transmitted from the sound source. The power of sound energy provides stimulation or sensation to the eardrum in the ear.

Sound power is defined as the total sound energy radiated by a sound source per unit time. The sound power level is also defined in decibels. The sound power level (*PWL*) is the ratio of acoustic power of the source (W_s) to the reference acoustic power (W_{ref}).

$$PWL = 10 \log (W_s)/(W_{ref}) \text{ dB.} \qquad ...(5.9)$$

If the distance between measurement from sound source increases, the sound intensity decreases. The relationship between distance and sound intensity I_s is;

$I_s \propto \frac{1}{d^2}$ which is called as inverse square law.

where d is distance.

5.22 Electroacoustic Transducers (Microphones)

Microphones are the electroacoustic transducers. These transducer converts sound energy or sound pressure variations into analogous electrical signals. These electrical signals lies in the audio frequency (AF) range which are again amplified and fed to recording head or loudspeaker. In communication system, these transducers are placed at the transmitter or sending end so as to convert the sound waves into analogous electrical signals. At the receiving end these electrical signals are converted back into sound waves by means of transducers like loudspeaker.

5.23 Types of Microphone

As discussed above microphones are the transducers which convert the sound signals into analogous electrical signals. These devices have diaphragm which convert sound pressure into displacement, this displacement gets converted into electrical signals. The microphones must satisfy the following requirements :

(1) The output of microphones should not vary much with the rapid variations in frequency.

(2) Its operation should be free from harmonics or distortions.

(3) It should have stable frequency response.

(4) The microphones should be free from stray electric and magnetic fields.

(5) Its sensitivity should be high.

(6) It should have linear relationship between frequency and electrical output (dB).

(7) The construction should be rugged and robust so as to withstand against vibrations, shocks, and rough handling.

(8) It should have large dynamic pressure range.

Sound level meter is used to measure the sound level. Generally, sound level meter consists of a transducer *i.e.* microphone which converts sound pressure into displacement. This displacement is again converted into electrical voltage signal using different methods like capacitive transducer, piezoelectric transducer, moving coil, resistance etc. This is the general discussion regarding microphones. They are classified as :

(a) Piezoelectric Microphone.

(b) Condenser Microphone.

(c) Moving Coil Microphone.

(d) Velocity Ribbon Microphone.

(e) Carbon Microphone.

(A) Piezoelectric Microphone

Piezoelectric microphone works on the principle that when certain crystals or materials are subjected to mechanical stresses *i.e.* vibrations of pressure in sound wave, electric potential gets developed across the faces of crystal. The magnitude of this potential depends upon displacement of crystal due to sound waves. Crystals used are Rochelle salt, ceramic, lead zinconate titanate, ammonium dihydrogen phosphate, barium titanate etc.

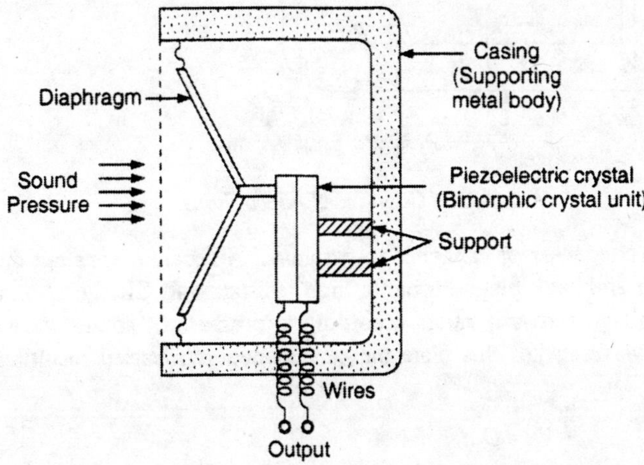

Fig. 5.30 : Piezoelectric crystal microphone.

Fig. 5.30 shows construction of piezoelectric microphone. It consists of a metallic casing in which a piezoelectric crystal is placed on an insulating support. The pressure of sound waves strikes on conical shaped diaphragm. This pressure is transmitted to the crystal directly or through rod arrangement. Thus the crystal is subjected to compression. The crystal element is an assembly of two thin slices called bimorphic crystal unit as shown in Fig. (5.30). Output electric potential is taken across wires which can be connected to amplifier stages for further processing.

This microphone has flat frequency response over the range 30 Hz to 12,000 Hz, very sensitive, light in weight, ruggedness, easy maintenance, output is upto – 65 dB.

Important advantages of this microphone is that it does not require any biasing supply. Sometimes performance may be affected by temperature.

(B) Condenser Microphone (Capacitor Microphone)

The condenser or capacitor microphone consists of a tnin metallic diaphragm suspended or close proximity to a stationary back plate as shown in Fig. 5.31. Metallic diaphragm and back plate forms the plates of

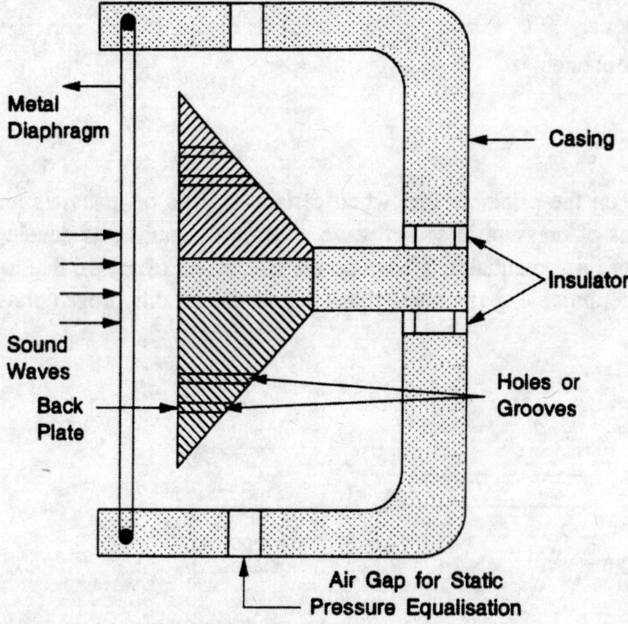

Fig. 5.31 : Condenser microphone.

capacitor, hence known as condenser or capacitor microphone. A separate constant d.c. voltage source is applied between diaphragm and back plate through a high resistance R. Charge Q on the capacitor plates tends to constant so the voltage across it varies according to pressure of sound waves. If one of plate is movable and when sound wave strikes this plate the capacitance gets varied resulting to develop output voltage given by :

$$V_0 = Q\,d \qquad ...(5.10)$$

where ; V_0 = Output voltage of capacitor.

Q = Charge on microphone capacitance (approx. constant).

d = Distance between the two plates.

Generally, the diaphragm is made up of metal sheet like duraluminium of thickness 0·002 to 0·05 mm which constitutes one plate of the capacitor. Distance between diaphragm and back plate may be in between 0·0025 to 0·0005 cm. Fig. 5.32 shows electrical circuit diagram of condenser microphone.

Capillary air-leak is provided to the microphone for equalizing the pressure on both sides of diaphragm so as to avoid bursting of the diaphragm. Holes or grooves are provided to back plate so as to obtain the damping. Because of damping effect control of resonant peaks of diaphragm response can be possible.

Advantages :

(1) Extremely stable performance.

(2) Flat frequency response from 12 Hz to 15 kHz.

(3) Output is upto – 45 dB.

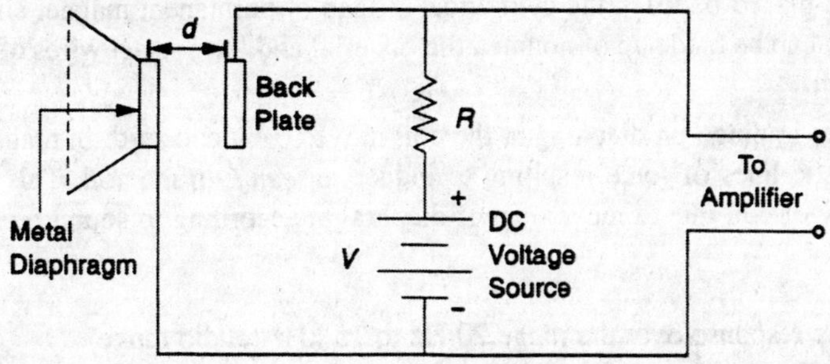

Fig. 5.32 : Electrical circuit diagram for condenser microphone.

Applications :

(1) Sometimes it can be used as a standard microphone for calibrating other microphones.

(2) It is used in sound measuring equipments.

Disadvantages :

(1) Bulky construction.

(2) They require high voltage supply and amplifier unit close to the microphone.

(3) Less ruggedness.

(C) Moving Coil Microphone

Moving coil microphone is also called dynamic microphone. This microphone works on the principle that **when conductor moves in a magnetic field an e.m.f. gets generated *i.e.* Electromagnetic Induction.**

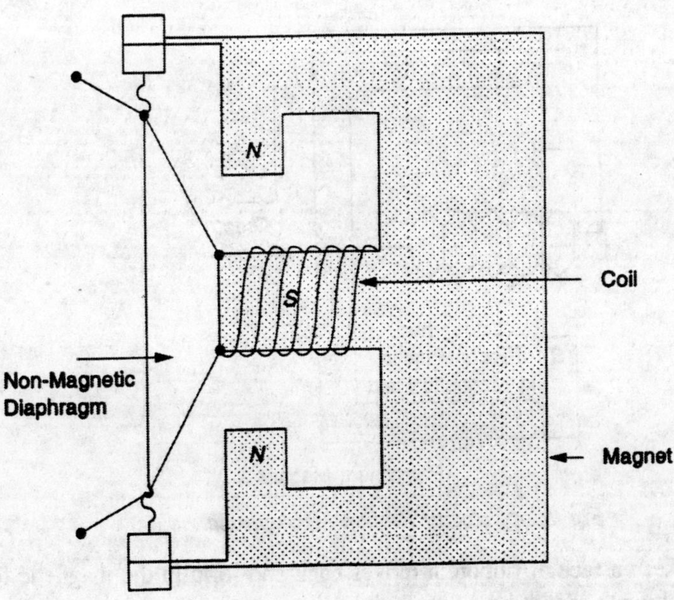

Fig. 5.33 : Moving coil microphone.

Fig. 5.33 shows moving coil microphone. It consists of a coil of conductor wire wound on a non-magnetic material and placed between the cylindrical air gap of permanent magnet and attached to the diaphragm. The diaphragm can be made up of nonmetallic material and the contact wires of coil are attached to the surface of diaphragm.

When the sound waves impinge on diaphragm, the coil moves back and forth in magnetic field due to which the coil cuts magnetic lines of force resulting to induce an *e.m.f.* in the coil. This *e.m.f.* is directly proportional to the velocity of coil due to movement of diaphragm according to sound pressure.

Advantages :

(1) Linear frequency response over the range 20 Hz to 15 kHz (audio range).

(2) No need of any external battery or voltage source for operation.

(3) Good frequency response can be achieved if diaphragm structure used is delicate.

(4) Reliable operation.

Applications :

(1) In music recording and broadcasting.

Limitations :

(1) Because of high inertia of coil frequency response may be poor.

(2) Sometimes it may require impedance matching.

(D) Velocity or Ribbon Microphone

Fig. 5.34 shows construction of velocity microphone. This microphone works on the same principle as of moving coil microphone. The only difference is in the construction. The constructional difference is that the diaphragm is made up of a thin and light ribbon of corrugated aluminium foil. This ribbon is placed between pole pieces of permanent magnet.

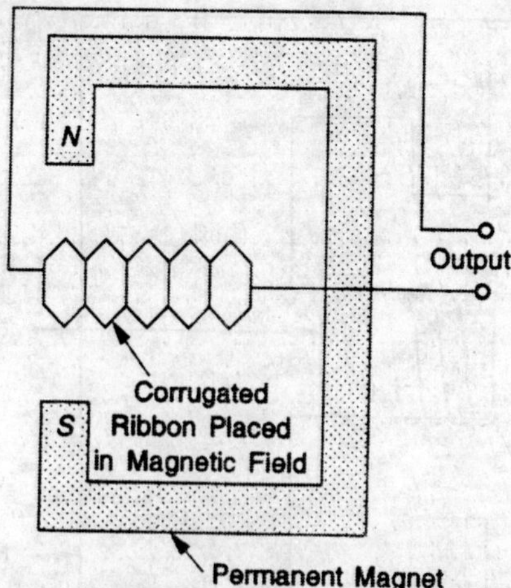

Fig. 5.34 Velocity Ribbon Microphone.

When sound wave strikes on face of ribbon, it moves back and forth in the magnetic field which cuts magnetic lines of force resulting to induce an e.m.f. in the coil.

Advantages :

(1) Good frequency response over the range 20 Hz to 16 kHz.

Applications :

(1) In musical programs like orchestra.

Limitations :

(1) Improper impedance matching.

(2) Operation may be affected by dust or vibrations.

(E) Carbon Microphone

It is a variable resistance microphone. It consists of a metal case and at one end of this case, metal diaphragm is placed. There is a plunger or movable contact through which the diaphragm transmits the varying pressure to carbon granule bed in the microphone. These carbon granules are covered at other end by means of insulated fixed metal contacts. Fig. 5.35 shows construction of a carbon microphone.

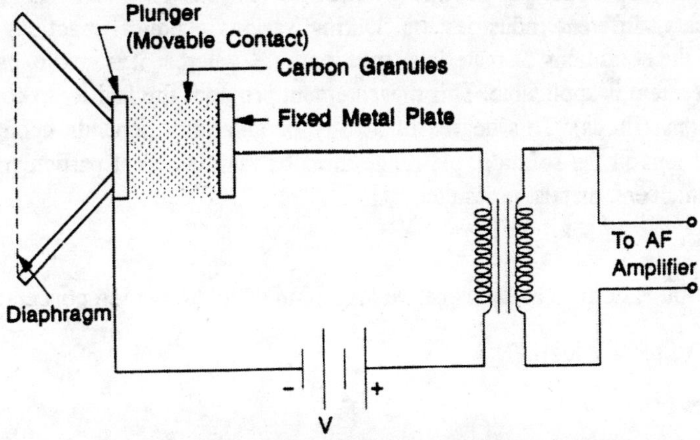

Fig. 5.35 : Carbon microphone.

When sound wave strikes on diaphragm, it pushes the plunger into carbon granules which lowers the contact resistance between them. A steady d.c. bias current is passed through circuit and load resistance from a battery. As the sound pressure decreases resistance increases which varies output current. This current is propotional to sound pressure.

Advantages :

(1) Frequency response ranges from 3 Hz to 4 kHz, so it is useful in speech signal handling circuits.

(2) Rugged in construction.

(3) Inexpensive.

(4) Simple design.

(5) Output is upto – 25 to – 45 dB.

Applications :

(1) Used as a transmitter in telephone handset.

Limitations :

(1) Poor linearity.

(2) Limited frequency response.

(3) Operation may be suffered by unwanted signals or distortions.

In previous section different microphones are discussed. Operation of microphones may be affected or disturbed by means of environmental parameters such as temperature, humidity, vibrations and shocks, noise signals etc. So proper protective arrangements such as casing, housing, isolation methods can be used so as to get good frequency response, more stability and reliable operation of microphones.

For information : For measurement of sound other devices such as hot-wire microphone, Lindval microphone and Reyleigh disc can also be used. These microphones measure the sound by means of their (sound) energy.

5.24 pH Measurement :

pH measurement is widely applicable in chemical industries, various chemical reactions and processes, biomedical measurements, different industries etc. During various chemical reactions or processes it is necessary to determine the conditions of materials (solutions) *i.e.* whether it is acidic, basic or neutral. For this purpose pH measurement is applicable. This measurement provides the facility to control the acidity or alkalinity of the solutions (fluids). This degree of acidity or alkalinity depends upon concentrations of hydrogen and hydroxyl ions in the solution. pH range must be maintained at particular value so as avoid corrosion, to protect containers, metals, pipes etc.

Definition :

pH value of a solution is defined as the negative logarithm of hydrogen ion concentration.

$$pH = -\log_{10}(H^+) \qquad ...(5.11)$$

The pH scale ranges from 0 to 14.

The solution or fluid contains both H^+ and (OH^-) ions. If the (H^+) ions are more in the solution then it is acidic and the solution is alkaline if hydroxyl ion concentration is more.

Case 1 : If the hydrogen ion concentration is 10^{-14} then the hydroxyl ion concentration is 10^0 and the product of these two is 10^{-14}. Therefore the pH value of such solution is

$$pH = -\log_{10}(H^+)$$

$$= -\log_{10}(10^{-14}) = 14$$

This is the dissociation constant which is defined as the product of number of H^+ ions and (OH^-) which is always 10^{-14}.

Case 2 : If hydrogen ion concentration for the solution is 10^0 then pH value for this solution is as follows

$$pH = -\log_{10}(H^+)$$

$$= -\log_{10}(10^0) = 0$$

so the pH value or hydrogen ion concentration can be measured over the range 0 to 14.

Case 3 : For neutral solution the hydrogen and hydroxyl ion concentrations are equal *i.e.* both are 10^{-7}. Therefore for neutral solution the pH value is given by :

$$\begin{aligned} pH &= -\log_{10}(H^+) \\ &= -\log_{10}(10^{-7}) = 7 \end{aligned}$$

e.g. consider the solution having hydrogen ion concentration of 10^{-6} g/liter then the hydroxyl ion concentration is 10^{-8} g/liter so its pH value is calculated as follows

$$\begin{aligned} pH &= -\log_{10}(H^+) \\ &= -\log(10^{-6}) \\ &= 6 \end{aligned}$$

Thus if the solution is acidic the pH value lies between 0 to 7. For alkaline solution pH value lies in the range of 7 to 14. pH value for the neutral solution is 7 *e.g.* pure water.

As the pH scale value decreases below 7 the solution becomes more acidic, if the pH scale value increases above 7 the solution becomes more alkaline. The pH value depends upon following parameters : (1) Chemical nature of acid expressed in degree of dissociation in the solution. (2) relative concentrations of acid and its related base in the solution. (3) temperature of the solution.

The principle of pH value measurement is that the reaction of H^+ ions with special type of glass surface of electrode which is immersed in the chemical solution whose pH value is to be measured. At the surface of this electrode potential difference is developed which can be measured by means of any (analog or digital) measuring system.

The development of potential difference at glass surface relates with the hydrgen ion concentration of the solution which is expressed by Nernst relationship as follows

$$E_g = E_g^{\,0} + \frac{2{\cdot}303\, RT}{F} \log_{10}{}^{(c)} \qquad \text{...(5.12)}$$

where ; E_g = Sum of reference potentials and liquid junction potentials (mV).

$E_g^{\,0}$ = Potential when $(H^+) = 1$ *i.e.* $C = 1$

C = Hydrogen ion concentration.

T = Absolute temperature degrees Kelvin (°C + 273)

R = 1·986 calories per mol degree.

F = Faraday (columbs per mol)

2·303 = Logarithmic conversion factor

The pH value measurement is carried out by immersing a pair of electrodes into the solution whose pH value is to be measured. Then the developed potential is measured (which is across the two electrodes). This complete assembly is called pH cell. The pH assembly consists of (a) reference electrode and (b) measuring electrode.

(a) Reference Electrode

This electrode is kept at constant potential regardless of the pH value of solution under test. Fig. 5.36 shows construction of reference electrode. The function of this electrode is to provide constant potential and to establish electrical conductivity between its internal electrode and measuring electrode. At the bottom end of reference electrode there is a junction where the internal electrolyte makes contact with solution or fluid under test.

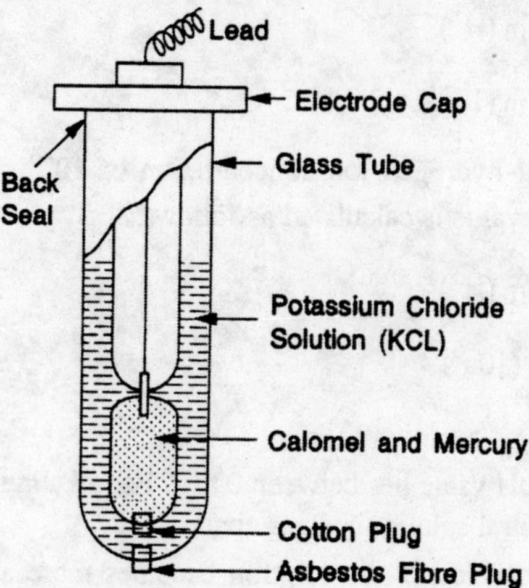

Fig. 5.36 Reference Electrode

Reference electrode consists of a glass tube containing calomel and mercury in a seperate assembly. This calomel and mercury assembly is covered by potassium chloride (KCL) solution. At bottom end small opening is provided so that the potassium chloride gets diffused slowly into the solution under test.

(b) Measuring Electrode

It is also known as glass electrode. Inside this electrode there is a glass bulb which contains (buffer solution) highly acidic buffer solutions. This electrode is closed at bottom end by a tip of glass of special composition. Fig. 5.37 shows the constructional details of measuring electrode.

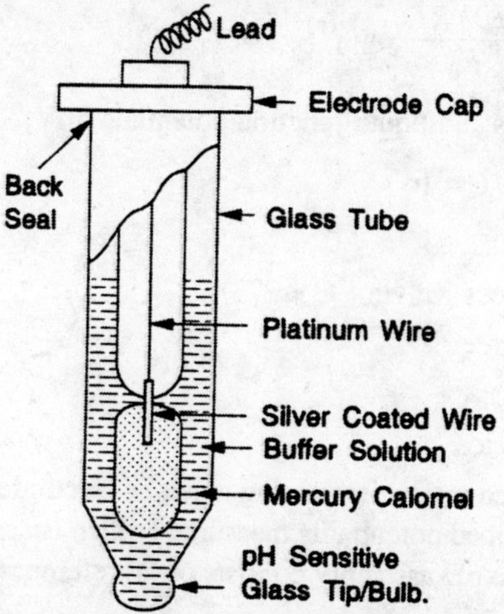

Fig. 5.37 Measuring Electrode

The measuring electrode consists of a glass tube which contains buffer solution of constant pH (= 1). Mercury calomel material is covered by the highly acidic buffer solution. At bottom end this glass tube is closed by means of pH sensitive glass bulb or tip. As the pH values of buffer solution and solution under test are different a potential difference gets developed at the tip or bulb.

Measurement Scheme : Fig. 5.38 shows scheme of pH value measurement.

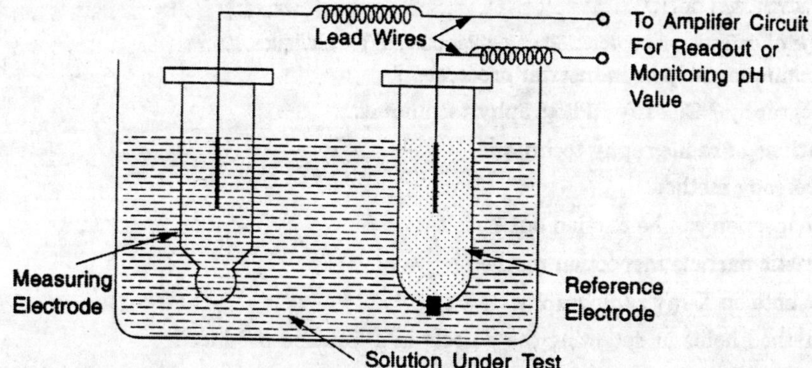

Fig. 5.38 Measurement of pH value

The reference electrode and measuring electrodes are immersed into the solutions whose pH value is to be measured using salt bridge. The algebraic sum of potentials of the two half cells is proportional to the concentration of hydrogen ions in the solution and this is the measure of pH value. The voltage developed by two electrodes is applied to measuring instrument. Sometimes an amplifier stage with proper impedance stage may be introduced with measurement scheme.

EVALUATION COMPONENT

1. Write a short note on biomedical instrumentation.
2. Draw the block diagram of man-instrument system. Explain it.
3. How biopotentials get generated in a human body ?
4. What do you mean by resting potential ?
5. List the transducers used in biomedical applications.
6. Why electrodes are necessary in biomedical applications ?
7. What are microelectrodes ? Discuss various types of microelectrodes.
8. Write short notes on :
   (i) Needle electrodes.
   (ii) Skin surface electrodes.
9. Explain in detail about :
   (i) Ultrasonic blood flowmeters.
   (ii) Electromagnetic flowmeters.
10. Explain direct and indirect method of blood pressure measurement.
11. Write short notes on EEG, ECG, EMG systems.
12. What are the essential requirements that could be fulfilled by the transducers used in biomedical instrumentation ?
13. What are the advantages of sphygmomanometry ?
14. Why jelly or electrolyte is applied at the time of connecting or installation of electrodes on human body ?
15. What is non-destructive testing (NDT) ?

16. Give the significance of NDT.

17. What are the drawbacks of destructive testing ?

18. Compare destructive testing and non-destructive testing.

19. (a) List the advantages of NDT.

(b) What type of defects can be detected using various NDT techniques ?

20. How NDT techniques help in industrial processes ?

21. What is radiography ? Classify radiography techniques.

22. State applications of radiography technique.

23. Explain fluoroscopy method.

24. How visual inspection can be carried out ?

25. Explain magnetic particle inspection method.

26. Write a short note on X-ray radiography.

27. How X-ray method helps in detecting the defects in a finished product ?

28. Explain ultrasonic flaw detection method with neat diagram.

29. How eddy current inspection method can be implemented ? Explain it.

30. Why level measurement is necessary in industries and plants ?

31. Explain bubbler method of liquid level measurement.

32. State the working principle of thermal sensors.

33. How thermal conductivity switch is used for level measurement ?

34. Suggest a suitable method used for level measurement of molten glass.

35. Elaborate how float arrangement is used for liquid level measurement.

36. Whis is the most commonly used level indicator in industrial applications ?

37. List the advantages and disadvantages of float type level indicators.

38. Write a short note on resistive or contact point method used for level measurement.

39. Explain suitable method used for level measurements of non-conductive fluids.

40. How capacitance phenomenon is used for level measurement ?

41. Explain liquid level measurement by capacitance probe method.

42. List the advantages, disadvantages and applications of capacitive level measurement.

43. Explain :

(i) Ultrasonic level measurement.

(ii) Microwave level detector.

44. Define following terms :

(i) Sound

(ii) Sound intensity

(iii) Loudness of sound

(iv) Decibel

45. Elaborate about :

(i) Sound pressure level.

(ii) Sound power level.

46. State the working principle of electroacoustic transducers.

47. Define microphone. State the various types of microphone.

48. Which parameters are to be considered while selecting microphone for particular application

49. **Explain the working of piezoelectric microphone. What are its salient features.**

50. **Write a short note on :**

(i) Capacitor microphone

(ii) Carbon microphone

(iii) Velocity ribbon microphone

(iv) Moving coil microphone

51. **List the parameters due to which performance of microphone gets affected.**

52. Define the term pH.

53. **Why pH value measurement is necessary in industrial applications ?**

54. **Write a short note on pH value measurement using reference and measuring electrodes.**

Chapter 6

SIGNAL CONDITIONING ELEMENTS

Learning Objectives

- To define necessity of signal conditioning elements.
- To study operation of A.C. and D.C. signal conditioning systems.
- To define the term "OP-AMP".
- To discuss the characteristics of ideal op-amp.
- To list the applications of op-amp.
- To analyze inverting amplifier configuration.
- To analyze non-inverting amplifier configuration.
- To design summing amplifier.
- To study transducer bridges.
- To analyze chopper stabilized amplifier.
- To study significance of instrumentation amplifier.
- To discuss operation of sample-and-hold amplifier.
- To design voltage to current and current to voltage converters.
- To design integrator and differentiator circuits.
- To define terms filter, passive filters, active filters.
- To describe significance of filters.
- To elaborate the problems of passive filters.
- To design passive low pass and high pass filters.
- To define advantages of active filters.
- To design first and second order low pass filters.
- To design first and second order high pass filters.
- To design wide band pass filter.
- To design wide band and narrow band reject filter.
- To design all pass filter.
- To discuss the term "Noise" and types of noise.
- To study different sources of noise.
- To discuss protection phenomenon of op-amp.
- To elaborate ground loops.

- To discuss shielding techniques.
- To explain input guarding technique.
- To understand power isolation and filtering method.

6.1 Introduction

Signal conditioning devices are used in instrumentation and measurement systems to modify or improve the output signals (voltage or current) of various transducers. Improvement of signals is necessary so as to make them useful and compatible with next stages such as indicating instruments, recorders, display devices, controlling elements etc.

Signal conditioning devices are used—(1) to provide excitation source to the passive transducers also to amplify their output to suitable higher level. (2) to amplify the output of active transducers. Commonly used signal conditioning devices are amplifiers, adder, integrator, differentiator, filters, bridges, voltage to current and current to voltage converters, analog to digital and digital to analog converters, sample and hold amplifier etc. Special amplifiers like instrumentation amplifier are used to enhance the output signal level of transducers which is less than few millivolts. All these elements are discussed in this chapter.

Operational amplifier (op-amp) is used in various configurations to handle and improve the signals, to convert the signals from one form to another. Op-amp is used in signal conditioning elements because of its number of advantages, which are discussed in detailed manner in this chapter.

In the last section of this chapter noise problems that occur in operations at various stages of circuit, devices, networks etc. is discussed. How these problem of noise effect can be minimized using proper isolation, shielding and grounding techniques is also discussed.

6.2 Necessity of Signal Conditioning Devices

In instrumentation and measurement systems processing and controlling of physical parameters (temperature, pressure, liquid level, light intensity etc.) is carried out by converting these parameters into electrical form by means of suitable transducers.

Depending upon the excitation system or requirement of power supply, transducers are classified as passive and active transducers. For passive transducers like potentiometer, strain gauge thermistors etc. excitation source (either a.c. to d.c.) is necessary because these transducers do not produce their own voltage or current. Similarly amplification stage is also necessary to increase the level of output signal. Active transducers (piezoelectric, thermocouples) produces their own electrical output when physical quantities are applied to them, but amplifier stage is used to bring the output signal upto suitable level. In both cases amplifiers are used because the output of transducers is at low level.

The output signal is amplified because to bring this signal compatible with control elements, indicating and display devices, recorders, controllers etc.

Fig. 6.1 shows generalized block diagram of electronic measurement system.

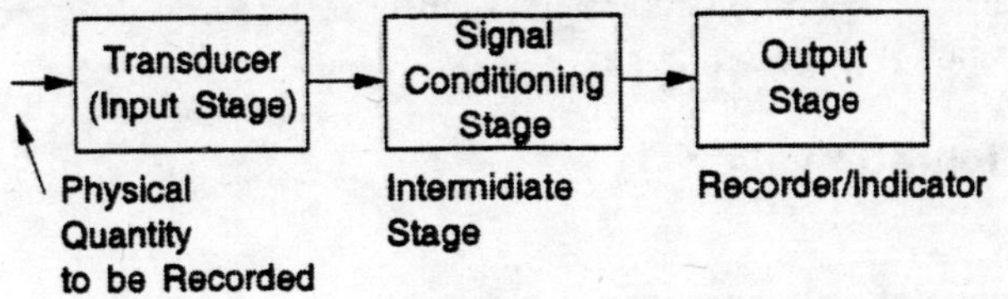

Fig. 6.1.

The input stage is transducer and output stage is recording, or display devices or control elements. These two stages are interfaced by intermediate stage which is signal conditioning stage. Signal conditioning devices are necessary for performing linear operations like amplification, attenuation, integration and addition. These devices also perform non linear operations like clipping, clamping, modulation, demodulation, filteration etc. Another devices are analog to digital converters, digital to analog converters, sample and hold amplifier etc..

Some transducers like strain gauges, thermistor needs bridge amplifiers for calibration which are also signal conditioning devices.

The excitation source used may be a.c. or d.c. voltage signal. Depending upon this signal, conditioning systems are :

(a) D.C. signal conditioning system,

(b) A.C. signal conditioning system.

6.3 DC Signal Conditioning System

Fig. 6.2 shows the block diagram of d.c. signal conditioning system.

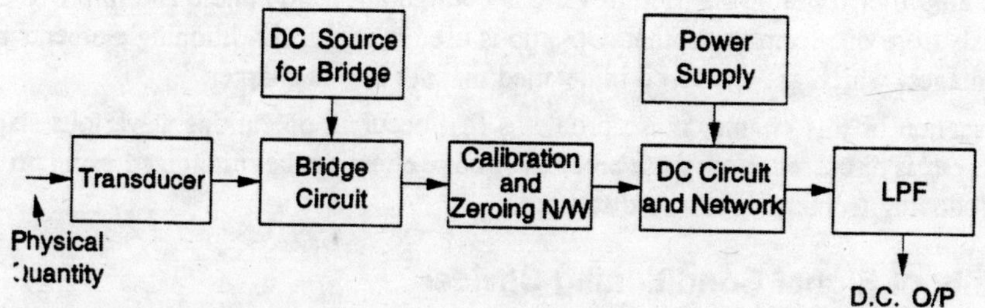

Fig. 6.2. D.C. signal conditioning system.

Transducer is the input stage to which the physical quantity under measurement is applied. Resistive transducers, strain gauges, thermistor needs bridge circuit for measurement of physical quantities in terms of voltage or current. Particular resistive transducer or strain gauge is connected across any one arm of Wheatstone bridge. The bridge is excited by separate d.c. source. Bridge can be balanced by potentiometer arrangement by varying the resistane of an arm.

Calibration and zeroing circuit is provided for indicating the unbalanced condition of bridge.

Output voltage of bridge circuit is amplified by means of d.c. amplifier because it is easy to calibrate at low frequencies and not affected by overloading effect. Drawback of d.c. amplifier is that its performance is affected by drift.

It produces low frequency spurious signals in the output signal. For eliminating this drawback low drift d.c. amplifiers are used. Function of low pass filter is to reject high frequency components from the output signal.

As mentioned earlier that d.c. signal conditioning system is suffered by drift problem so a.c. signal conditioning system is used which is discussed in next section.

6.4 AC Signal Conditioning System

Fig. 6.3 shows block diagram of a.c. signal conditioning system.

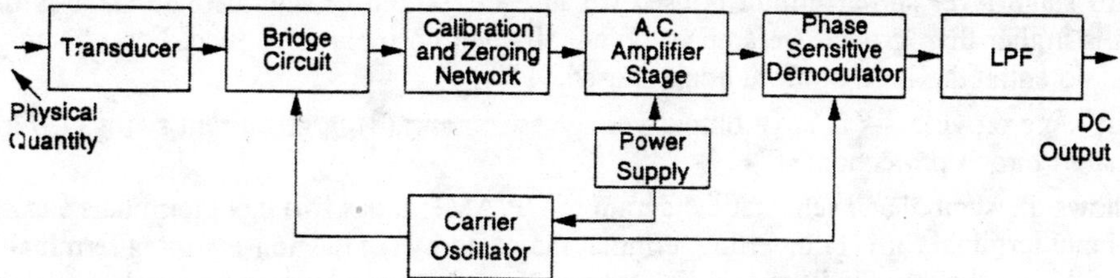

Fig. 6.3 Block diagram of a.c. signal conditioning system.

The transducer such as strain gauge is connected in the arm of bridge which is driven by the carrier oscillator. As the physical input quantity of transducer changes the carrier frequency signal gets amplitude modulated with bridge output. This modulated signal is amplified by means of a.c. amplifier. The amplified signal is demodulated by phase sensitive demodulator. The carrier frequency signal is filtered by phase sensitive detector. Phase sensitive demodulation is used so that polarity of d.c. output can indicate direction of the parameter change in bridge output.

Final stage is low pass filter to reject the high frequency signals and noise signals. Problem in this system is to maintain stable carrier oscillator.

This system is applicable for variable resistance or variable capacitance transducers and can be operated on carrier frequencies of range 50 Hz to 20 kHz. Carrier frequency should be atleast be five times higher than signal frequency.

6.5 Operational Amplifier (OP-AMP)

Operational amplifier is a direct coupled, high gain amplifier with one or more differential amplifiers followed by a level shifter and output stage. It is used for amplifying, controlling and generating the wide range of frequency signals from d.c. level to several mega hertz. Major application area of op-amp is to perform mathematical functions such as addition, subtraction, multiplication, logarithmic, division, antilogarithm, differentiation etc. It is widely used in analog computations, instrumentation and measurement systems. Other applications of op-amp are voltage regulators, filters (active filters), waveform generators, in communication systems etc. In operational amplifiers negative feedback is used.

Definition : Op-amp is basically differential amplifier having high gain, high input impedance (∞) and low output impedance (0), which performs mathematical operations such as addition, subtraction, multiplication etc. over the frequency range of 0 to 1 MHz.

Fig. 6.4 shows the basic block diagram of an OP-AMP.

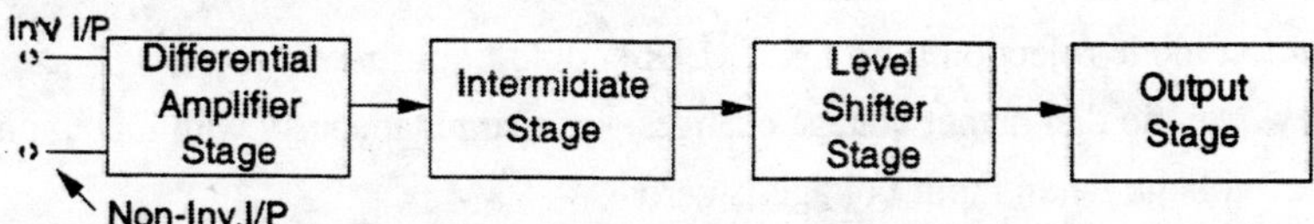

Fig. 6.4. Basic block diagram of an OP-AMP

The input stage has two input terminals (inverting and non-inverting terminals).

The first stage is differential amplifier. It amplifies the difference of two voltage signal applied at input terminals. This amplifier is dual input balanced output differential amplifier. It provides the high CMRR and also determines the input resistance of OP-AMP.

The second stage is intermediate stage which is dual input unbalanced output differential amplifier. The function of this stage is to increase the gain *i.e.* it provides additional gain.

In the third stage level shifter circuit is used because the output of intermediate stage is d.c. output voltage which is higher than ground (reference) potential when no input is applied. For adjusting this d.c. level to ground potential the level shifter circuit is used.

The output stage provides—(1) large output voltage swing and (2) large current swings (3) low output resistance (4) short circuit protection.

Fig. 6.5 shows the symbol and schematic diagram of OP-AMP. It has two input terminals and one output terminal. The input terminals are (1) inverting terminal indicated by (–) (2) non-inverting terminal indicated by (+). If V_1 and V_2 are the input voltages applied to inverting and non-inverting terminals respectively then V_0 is the output voltage at output terminal. This output voltage is controlled by the input voltages. The output voltage (V_0) is directly proportional to the difference of input voltages. Op-amp amplifies this difference voltage as mentioned below,

$$V_0 = A_v V_{id} = A_v (V_1 - V_2) \quad ...(6.1)$$

where ; A_v is the voltage gain of amplifier.

Ideal OP-AMP

The Fig. 6.5 shows basic op-amp representation. This is said to be ideal if it follows the characteristics as mentioned below :

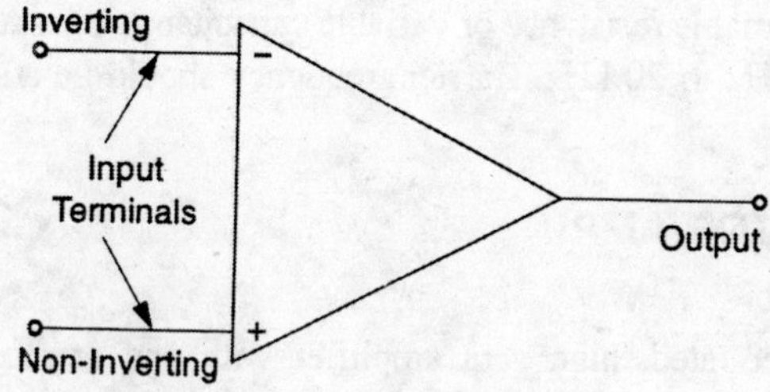

Fig. 6.5

Characteristics of OP-AMP

(1) The voltage gain A_v (open loop voltage gain) should be infinite (∞).

(2) Input resistance R_i should be infinite (∞).

(3) Output resistance R_0 should be zero (0).

(4) Infinite bandwidth.

(5) If $V_1 = V_2$ then $V_0 = 0$ *i.e.* zero offset.

(6) High common mode rejection ratio *i.e.* CMRR should be infinite.

(7) Infinite slew rate. So that output voltage changes occur simultaneously with input voltage changes.

(8) Zero output voltage when input voltage is zero.

OP-AMP Packages

Op-amp is available in following packages :

(1) Metal can (TO) package.

(2) Dual-in-line package (DIP).

(3) Flat package.

These packages contains single, dual or quad op-amps. Most popular and widely used op-amp integrated circuit is IC 741. Fig. 6.6 shows the op-amp IC μA 741 with its pin configuration. It is a single op-amp 8-pin

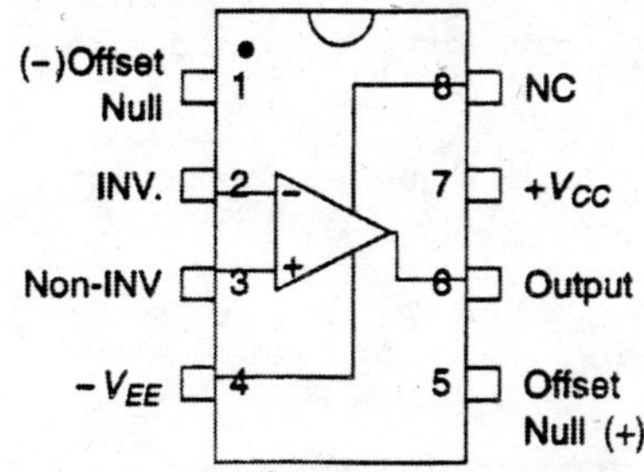

Fig 6.6. Pin connection of OP-AMP IC 741

integrated circuit. From the figure op-amp has five basic terminals such as (1) two input terminals (2) one output terminal (3) two power supply terminals. Detailed pin configuration is discussed below.

Pin 1 (offset null) : Offset null. (–)

Pin 2 (Inverting input terminal) : When input voltage is applied to this pin the output voltage is out of phase with the input. It is indicated by (–) sign.

Pin 3 (Non-inverting input terminal) : When input voltage is applied to this terminal the output voltage is in phase with the input. It is indicated by (+) sign.

Pin 4 and Pin 7 (Power supply terminals) : These terminals are used for power supply connections to the op-amp for biasing. $+V_{CC}$ *i.e.* positive d.c. voltage is applied to pin 7 and $-V_{EE}$ *i.e.* negative d.c. voltage is applied to pin 4. The power supply range is ± 5V to ± 22V.

Pin 5 : Offset null. (+)

Pin 6 (Output terminal) : The output of op-amp is derived at this terminal with respect to ground.

Pin 8 (NC) : Not connected.

6.6 Inverting Amplifier

The Fig. 6.7 shows inverting configuration of an op-amp which is generally used in all op-amp circuits. Here the input signal V_i is applied to the inverting input terminal of the op-amp through input resistance (R_1). The non-inverting input terminal is grounded. The output voltage V_0 is feedback to the inverting input terminal through the resistive network of R_1 and R_F. R_F is the feedback resistance.

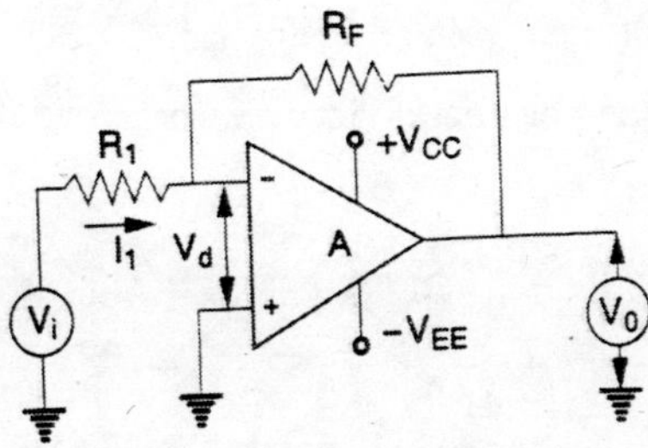

Fig 6.7. Inverting Amplifier

Consider the op-amp is ideal. The current through resistance R_1 is I_1. As $V_d = 0$ node '*a*' is at ground potential. According to virtual ground concept.

$$I_1 = \frac{V_i}{R_1} \qquad \text{...(6.2)}$$

The current flowing through R_1 flows through R_F because op-amp draws no current.

So the output voltage is,

$$V_0 = -I_1 R_F = -V_i \frac{R_F}{R_1} \qquad \text{...(6.3)}$$

Therefore the closed loop gain of inverting amplifier is given by,

$$A_{CL} = \frac{V_0}{V_i} = -\frac{R_F}{R_1} \qquad \text{...(6.4)}$$

The negative sign indicates a phase shift between V_i and V_0. For avoiding the loading effect the value of R_1 should be kept large.

6.7 Non-Inverting Amplifier

Fig. 6.8 shows non-inverting configuration of op-amp. Here the input signal V_i is applied to the non-inverting terminal of op-amp whereas the inverting terminal is grounded. This circuit provides the negative feedback through R_F because output is fed back to the inverting terminal of op-amp. As the input signal is directly applied to the non-inverting terminal the output is in phase with input.

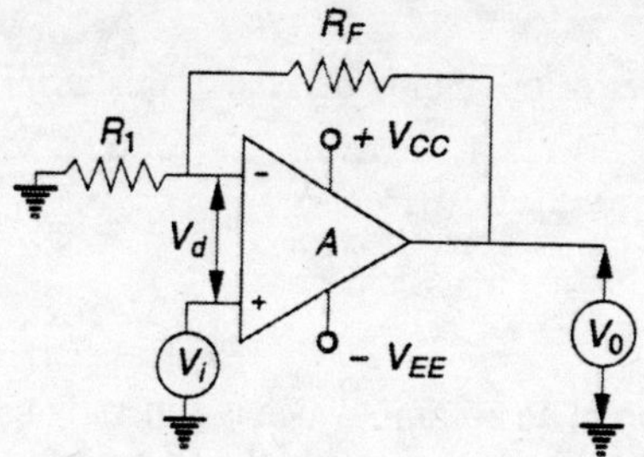

Fig. 6.8. Non-Inverting amplifier

Consider the op-amp is ideal. The differential voltage at the input terminals is $V_d = 0$. So the voltage at inverting input terminal is same as that of input voltage V_i applied.

Here the resistors R_1 and R_F forms the potential divider network.

Therefore,

$$V_i = \frac{V_0}{(R_1 + R_F)} R_1 \qquad \text{...(6.5)}$$

$$\frac{V_0}{V_i} = \frac{(R_1 + R_F)}{R_1} = 1 + \frac{R_F}{R_1} \qquad \text{...(6.6)}$$

The closed loop voltage gain for non-inverting amplifier is :

$$A_{CL} = \frac{V_0}{V_i} = 1 + \frac{R_F}{R_1} \qquad \text{...(6.7)}$$

The gain can be adjusted by selecting proper values of resistance R_1 and R_F.

6.8 Voltage Follower

This circuit is also called as unity gain amplifier or unity gain buffer.

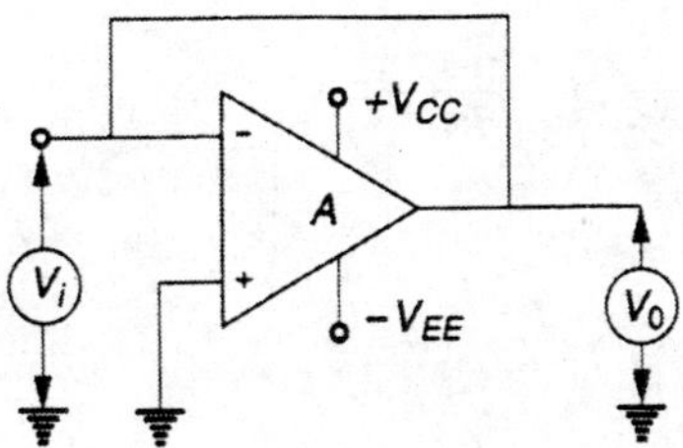

Fig. 6.9. Voltage follower

Fig. 6.9 shows circuit diagram of voltage follower. In this configuration the feedback resistor $R_F = 0$ and R_1 resistance is not used then

$$A_v = 1 \quad ...(6.8)$$

In this configuration output of op-amp follows the input in magnitude and phase *i.e.* output voltage follows the input voltage so it is called as voltage follower. This circuit acts as a voltage source having negligible internal resistance.

From the circuit,

$$V_0 = V_i \quad ...(6.9)$$

This circuit has high input impedance and zero output impedance with unity gain. Due to this reason voltage follower circuit is used as a buffer for impedance matching.

Some applications of the op-amp are discussed in next sections.

6.9 Summing Amplifier

It is also called as adder circuit. This circuit provides the output signal which is a linear sum of number of input signals.

Fig. 6.10 shows the inverting summing amplifier circuit.

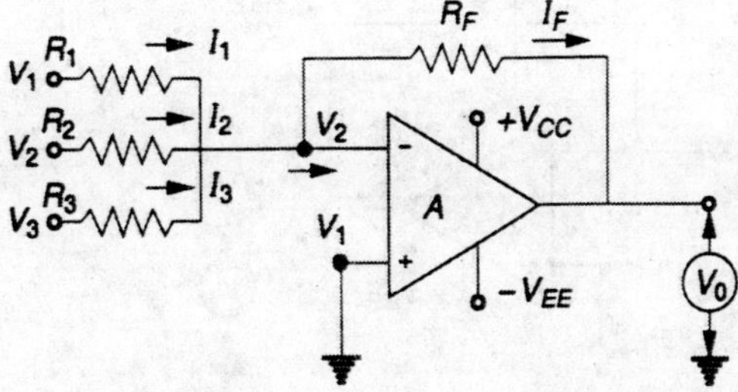

Fig. 6.10. Summing amplifier

Here V_1, V_2 and V_3 are the input voltages applied at inverting input terminal through input resistors R_1, R_2 and R_3 respectively. R_F is the feedback resistor. I_1, I_2 and I_3 are the current flowing through R_1 R_2 and R_3. V_0 is output voltage. Apply Kirchhoff's current equation written at node V_2

$$I_1 + I_2 + I_3 = I_F \quad ...(6.10)$$

For ideal OP-AMP input impedance is infinite and accroding to virtual ground concept $V_1 = V_2 \cong 0V$.

Determine I_1, I_2 and I_3.

$$\therefore \quad I_1 = \frac{V_1}{R_1};\ I_2 = \frac{V_2}{R_2};\ I_3 = \frac{V_3}{R_3}$$

$$\therefore \quad \frac{V_1}{R_1} + \frac{V_2}{R_2} + \frac{V_3}{R_3} = I_F \quad ...(6.11)$$

But

$$I_F = -\frac{V_0}{R_F}$$

$$\therefore \quad \frac{V_1}{R_1} + \frac{V_2}{R_2} + \frac{V_3}{R_3} = -\frac{V_0}{R_F} \quad ...(6.12)$$

$$\therefore \quad V_0 = -\left[\frac{R_F}{R_1}V_1 + \frac{R_F}{R_2}V_2 + \frac{R_F}{R_3}V_3\right] \quad ...(6.13)$$

If $R_1 = R_2 = R_3 = R$ then ;

$$V_0 = -\frac{R_F}{R_1}[V_1 + V_2 + V_3] \quad ...(6.14)$$

The output voltage V_0 is the inverted sum of applied inputs to the inverting terminal of op-amp. This circuit is also called as scaling or weighted amplifier.

Thus the output voltage is proportional to the sum of the input voltages applied to the circuit.

Consider $\frac{R_F}{R_1} = 1/n$, where n is the number of inputs in above equation then the circuit is called as averaging circuit.

6.10 Logarithmic Amplifier

This amplifier provides the output voltage proportional to the logarithm of the input voltage.

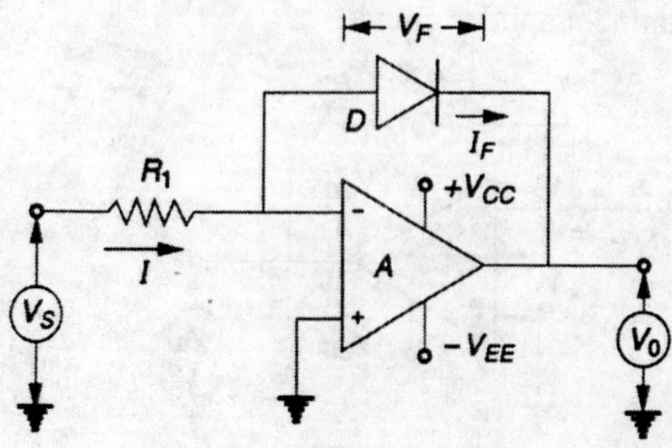

Fig 6.11. Logarithmic amplifier

Logarithmic amplifier uses the non linear volt-ampere relationship of the *p–n* junction diode which is given by ;

$$i_f = I_s\,|\exp(V_f/mV_T) - 1| \quad ...(6.15)$$

where ; i_f = feedback current through diode.

I_s = saturation current of diode.

V_f = forward voltage across the diode.

V_T = thermal voltage $kT/q = T/11{,}000$ volts.

$m = 2$ for silicon devices.

Take logarithm on both sides for above equation

∴ The above equation is expressed as ;

$$l_n\, i_f = l_n\, I_s + \frac{V_f}{mV_T} \quad \text{provided } V_f/mV_T \geq 1 \quad ...(6.16)$$

$$V_f = mV_T\,(l_n\, I_f - l_n\, I_s) \quad ...(6.17)$$

The above Fig. 6.11 shows circuit diagram of logarithmic amplifier. Here a diode is connected in the feedback path of op-amp. Input signal is applied to the inverting input terminal op-amp.

Non-inverting terminal is grounded due to this inverting terminal at virtual ground. So the input current I flowing through R_1 is equal to the current I_F flowing through diode.

$$I = V_s/R_1 \quad ...(6.18)$$

but $\quad I = I_F$

$$V_f = mV_T\,(l_n\, V_s/R_1 - l_n\, I_s) \quad ...(6.19)$$

The output voltage $V_0 = -V_f = -mV_T\,(l_n\, V_s/R_1 - l_n\, I_s)$ ...(6.20)

6.11 Transducer Bridges

In instrumentation systems number of transducers are used for measurement and controlling purpose. As we have studied in Chapter 3 that some transducer (electrical transducers) converts the physical variables into a change in resistance, change in capacitance, change in inductance. It is necessary to calibrate these parameters either in terms of voltage or current. Under such circumstances bridge circuits becomes essential. *e.g.* strain gauges, thermistors.

For temperature measurement thermistors are used. When the temperature which is to be measured is applied to thermistor, its resistance changes according to the resistance co-efficient of temperature. For calibrating or measuring this change in resistance in terms of voltage or current bridge circuit is used. As the input physical parameter changes it imbalances the bridge. This imbalance of bridge is measured in terms of voltage or current which is proportional to the change in physical parameter.

The output of the bridge circuit is interfaced with op-amp for amplification.

Fig. 6.12 shows the basic bridge circuit. This circuit provides the balance voltage at output. Transducer (thermistor or strain gauge) is connected to any one arm of the bridge. V_R is the reference voltage applied to the circuit.

Here, the bridge is used for indicating the imbalance and so it is a measure of fractional change $\alpha = (\Delta R/R)$ in the resistance of transducer. R_X is the transducer resistance which is changing with respect to change in physical quantity affecting the transducer.

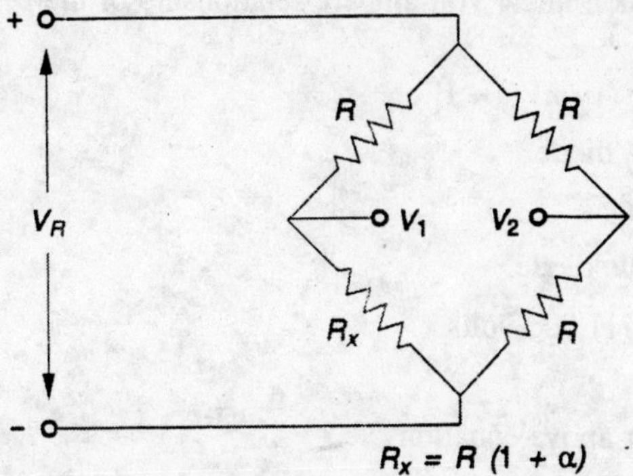

Fig. 6.12 Bridge circuit

$\therefore$ $R_X = R\,(1 + \alpha)$; R = Resistance of reference condition.

Circuit Analysis (a) For Output Voltage

$$\therefore \quad V_1 = \frac{[R\,(1+\alpha)]\cdot V_R}{R + R\,(1+\alpha)} = \frac{V_R\,(1+\alpha)}{2+\alpha} \qquad ...(6.21)$$

$$\therefore \quad V_2 = \frac{V_R}{2R}\cdot R = \frac{V_R}{2} \qquad ...(6.22)$$

$$\therefore \quad V_0 = V_1 - V_2 = \frac{V_R\,(1+\alpha)}{2+\alpha} - \frac{V_R}{2} = \frac{V_R\,\alpha}{4\,(1+\alpha/2)} \qquad ...(6.23)$$

But $\alpha \leq 1$

$$\therefore \quad V_0 = \frac{V_R\cdot\alpha}{4}$$

(b) For Output Current

$$I_0 = \frac{V_R}{R}\,\frac{\alpha}{4\,(1+\alpha/2)} \qquad ...(6.24)$$

Due to change in physical quantity the bridge gets unbalanced and this unbalanced voltage is amplifed by instrumentation amplifier as shown in Fig 6.13.

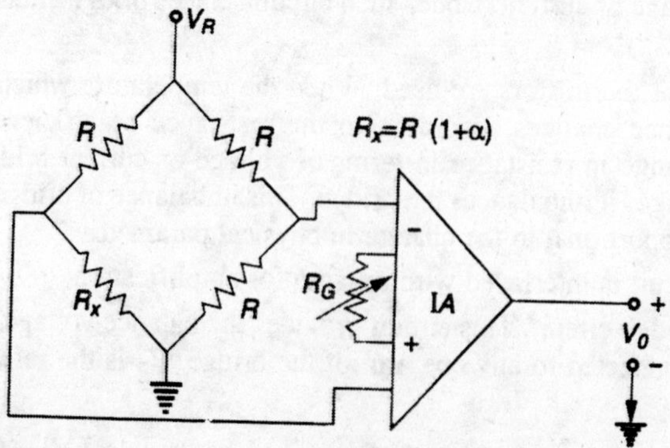

Fig. 6.13.

Equation 6.23 indicates that the output signal is a linear function of variation in the transducer parameters.

The output signal (voltage or current) can be amplified by using various amplifier configurations but it is necessary to consider the factors as mentioned – supply of the bridge, linearity of the circuit, the transducer element must be grounded or floated, distance between transducer element and amplifier, if distance is more than use transmission leads.

6.12 Chopper Stabilized Amplifier

Chopper stabilized amplifier is used because of following advantages : (1) low drift (2) higher d.c. gain (3) wide bandwidth achievement.

This amplifier avoids the problem of drift occurring in d.c. amplifier. This topic is discussed in previous section of d.c. signal conditioning. This amplifier increases the level of small d.c. voltage signals.

Firstly, we shall discuss about chopper amplifier. Fig. 6.14 shows chopper amplifier.

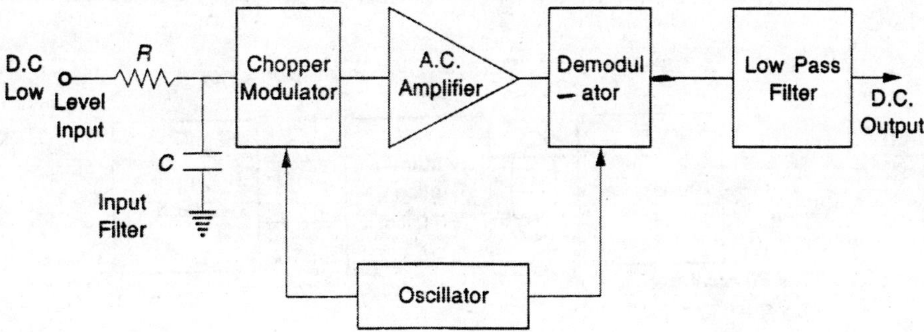

Fig. 6.14. Chopper amplifier

In this circuit the input d.c. signal is converted into a.c. signal by means of chopper or modulator. Peak to peak amplitude level of converted a.c. signal is same to that of d.c. input signal. Converted a.c. signal is amplitude modulated using the carrier frequency of oscillator. This amplitude modulated signal is amplified by a.c. amplifier having low drift and higher gain. Amplified signal is demodulated for converting a.c. signal into higher level d.c. signal. Chopper frequency is 2 kHz having bandwidth 200 Hz.

Fig. 6.15 shows chopper stabilized amplifier.

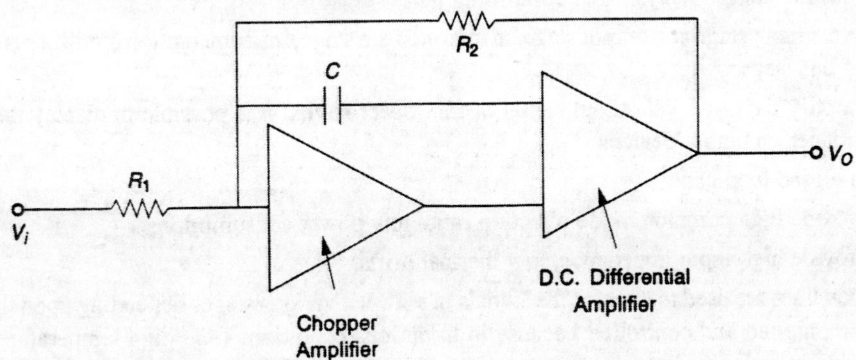

Fig. 6.15. Chopper stabilized amplifier

This amplifier provides stabilized gain with higher bandwidth. In this circuit the low frequency signal gets amplified by chopper amplifier where high frequency signals are bypassed. At the input of d.c. amplifier higher and lower frequency signals are applied and reconstructed. Function of chopper amplifier is to increase the gain at lower frequency signals resulting in reduction of drift effect of d.c. amplifier.

6.13 Instrumentation Amplifier

This is a special class of amplifier widely used in instrumentation area, measurement and control of various parameters like temperature, pressure humidity etc. Generally instrumentation amplifier is used with transducers. Transducer is a device which converts one form of energy into another and used with the measuring system for monitoring and analyzing the process.

Basic necessity of instrumentation amplifier is amplify the low output signals of electrical transducer for controlling the operations in preceding stages. Sometimes the transducers are located at far away from measurement system also their output may be low with high source impedance in such cases instrumentation amplifier is used.

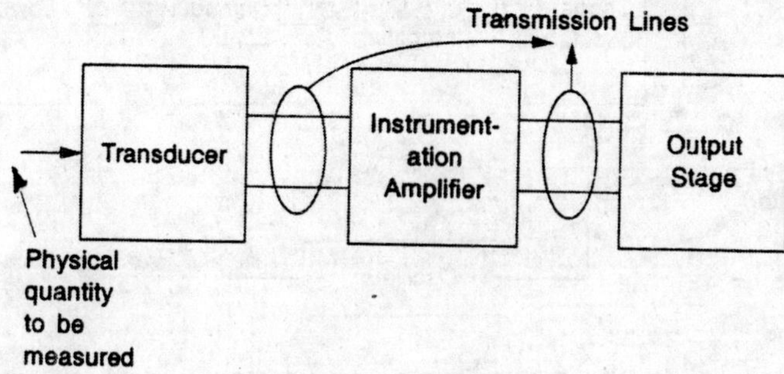

Fig 6.16. Block diagram of an instrumentation system.

Fig. 6.16 shows the block diagram of an instrumentation system. The physical quantity which is to be measured or monitored is applied to the input stage which is a transducer and preamplifier. At this stage the physical quantity gets converted into electrical signal by means of suitable transducer and amplifies the output of transducer to certain level.

The function of output stage is to monitor the output of transducers on display devices, oscilloscopes, recorders, various meters for analysis and controlling purposes.

In between the input stage and output stage an intermediate stage, instrumentation amplifier is interfaced. The function of this stage is

(1) to amplify the low level output signal of transducer so that it is possible to display the output of transducer on output devices.

(2) high closed loop gain

(3) to provide high common mode rejection ratio, low power consumption.

(4) to provide high input impedance, low thermal drifts.

Transmission lines are used to transfer the signals in between various stages depending upon the quantity which is to be monitored and controlled because in hazardous conditions (*e.g.* high temperature measurement) the transducers are placed at some distance from the measuring devices.

Important features of instrumentation amplifier are as follows :

(1) Simple gain adjustment.

(2) It provides high input impedance and low output impedance.

(3) High common mode rejection ratio.

(4) Possible to handle differential inputs.

(5) Gain is high with maximum accuracy.

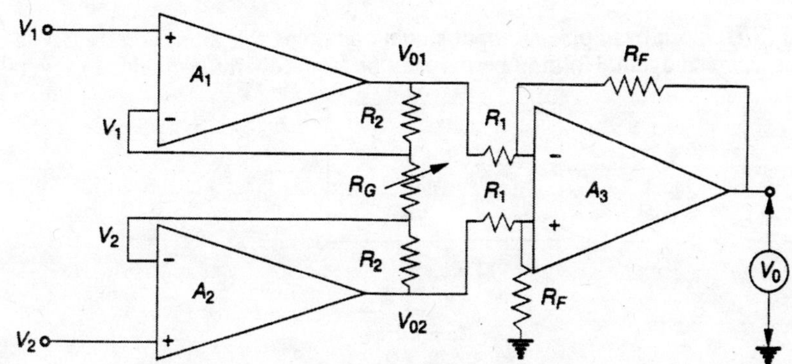

Fig. 6.17. Instrumentation amplifier.

Fig. 6.17 shows the basic instrumentation amplifier using three op-amps. Op-amps A_1 and A_2 are connected in non-inverting configuration. Inverting terminals of A_1 and A_2 are connected to resistance R_G which provides gain adjustment. Op-amp A_3 is connected in differential mode.

Circuit Analysis

Let the output of op-amp A_1 is V_{01} and output of A_2 is V_{02}.

When $V_2 = 0$; op-amp A_1 works as a non-inverting amplifier and A_2 acts as an inverting amplifier.

$$V_{01} = (1 + R_2/R_G)\, V_1 - R_2/R_G\, V_2 \qquad ...(6.25)$$

When $V_1 = 0$; op-amp A_2 works as a non-inverting amplifier and A_1 acts as an inverting amplifier.

$$V_{02} = (1 + R_2/R_G)\, V_2 - R_2/R_G\, V_1 \qquad ...(6.26)$$

These equations are derived by applying the input voltage and by making other input as zero (according to the superposition theorem).

Now the differential gain of A_1 and A_2 is,

$$\frac{\Delta V_0}{\Delta V_s} = 1 + \frac{2R_2}{R_G} \qquad ...(6.27)$$

where ;

$$\Delta V_0 = V_{02} - V_{01} \qquad ...(6.28)$$

$$\Delta V_s = V_2 - V_1 \qquad ...(6.29)$$

The op-amp A_3 is connected in differential amplifier configuration.

Gain of A_3 is R_F/R_1

R_F is the feedback resistor.

The overall gain of instrumentation amplifier is as follows :

$$\frac{V_0}{V_2 - V_1} = \left[1 + \frac{2R_2}{R_G}\right]\frac{R_f}{R_1} \qquad ...(6.30)$$

The resistance R_G is provided for adjusting the gain of instrumentation amplifier without disturbing the complete circuit arrangement.

Instrumentation amplifier using three op-amps are available in monolithic integrated circuits (IC). Most commonly used are analog devices AD521, INA101, INA102, INA104 (Burr-Brown).

6.14 Sample and Hold Amplifier

Sample and hold (S/H) circuit samples an input signal and holds this signal on its last sampled value until the input signal is sampled again. Holding period may be from few milliseconds to several seconds.

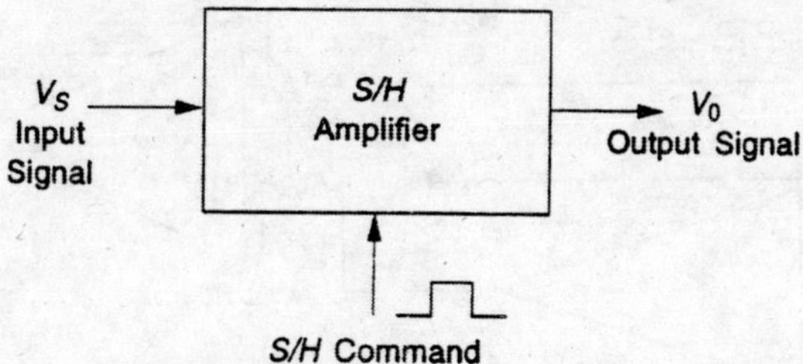

Fig. 6.18. Block diagram of S/H amplifier.

Fig. 6.18 shows the block diagram of S/H circuit. It has three input terminals as V_s is the input signal, V_0 is the output signal and the third is S/H command. This command signal is used to provide the pulse either to sample or to hold the input signal.

When the S/H command is applied to the circuit the output voltage V_0 increases towards V_s and the output voltage becomes equal to V_s for remainder of the pulse, this mode is called sampling (tracking) mode. As the S/H command is removed the circuit holds V_0 at the value of V_s that is hold mode.

If sometime delay takes place in changing the position of switch from sample mode to hold mode that time delay is called as aperture time (generally 50 ns).

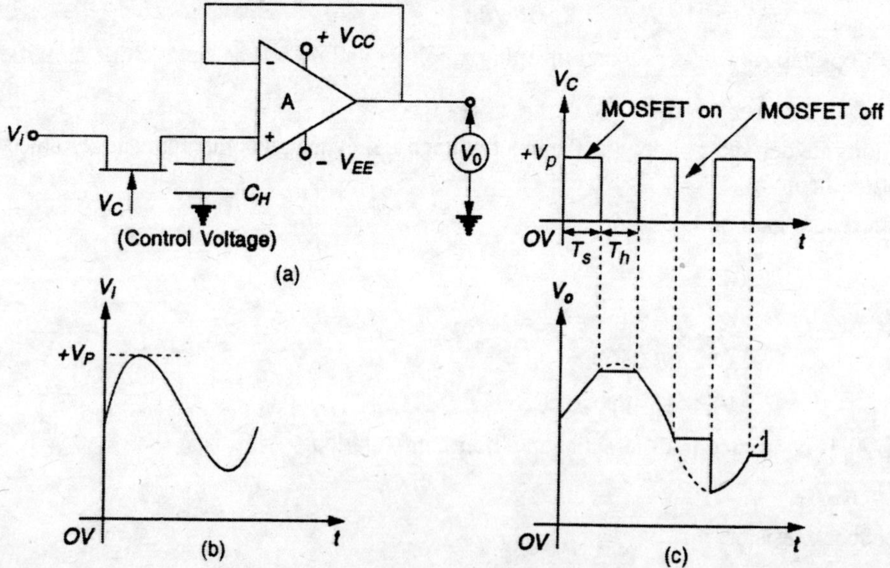

Fig. 6.19 (a) shows the sample and hold circuit. It uses N-channel E-MOSFET which acts as a switch and controlled by the control voltage V_c. The function of capacitor C_H is to store the charge.

Analog input signal V_i which is to be sampled is applied to the drain of E-MOSFET and the control voltage is applied to its gate as shown in Fig. 6.19 (b).

When the positive control voltage V_c is applied to the E-MOSFET, it turns ON and capacitor C_H starts to charge towards the input voltage V_i. In this way the input voltage V_i gets stored across capacitor C_H. This voltage appears at the output through op-amp A which is voltage follower. Because the discharge path for C_H is amplifier A.

If the control voltage V_c is zero the E-MOSFET is OFF. As the input impedance of A is high capacitor C_H cannot discharge. Thus capacitor holds the voltage across it.

The time period T_s during which the voltage across capacitor is equal to input voltage that time period is called SAMPLE PERIOD. The time period T_h of control voltage V_c during which the capacitor held constant voltage is called HOLD PERIOD T_h as shown in Fig. 6.19 (c). Generally, the capacitor used for S/H circuit are teflon, polystyrene and mylar because of low leakage.

Sample and hold circuit is used in analog-to-digital converters and pulse code modulation systems.

6.15 Converters

(A) Voltage to Current Converter (Transconductance Amplifier)

In some applications such as signal transmission over long distances, relay operations, zener diode testers etc. voltage to current converters are used. Sometimes the output of transducer is voltage and the next stage is current operated, in such cases this type of converters are used.

The function of VIC is to convert a input voltage signal into proportional output current signal. VIC circuit can be arranged by two methods depending upon the load arrangement :

(1) Voltage to current converter using floating load.

(2) Voltage to current converter using grounded load.

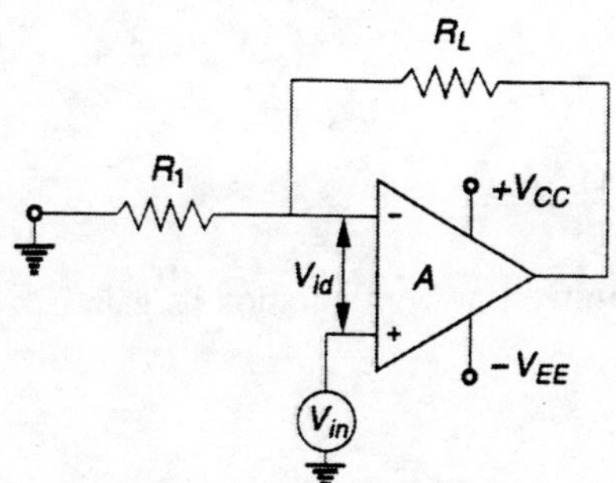

Fig. 6.20. Voltage to current converter using floating load.

(1) **Voltage to current converter using floating load** : Fig. 6.20 shows a voltage to current converter using floating load in which the load resistance R_L is not grounded. Here the input voltage signal is applied to non-inverting input terminal of op-amp.

According to Kirchhoff's voltage law the equation for input section is,

$$V_{in} = V_{id} + V_f \qquad ...(6.31)$$

As the gain is large $V_{id} = 0$ (according to virtual ground concept)

$$V_{in} = V_f \qquad ...(6.32)$$

V_{in} is given by, $\quad V_{in} = I_0 R_1$

Therefore $I_0 = V_{in}/R_1$ ...(6.33)

Thus the input voltage signal is converted into current signal.

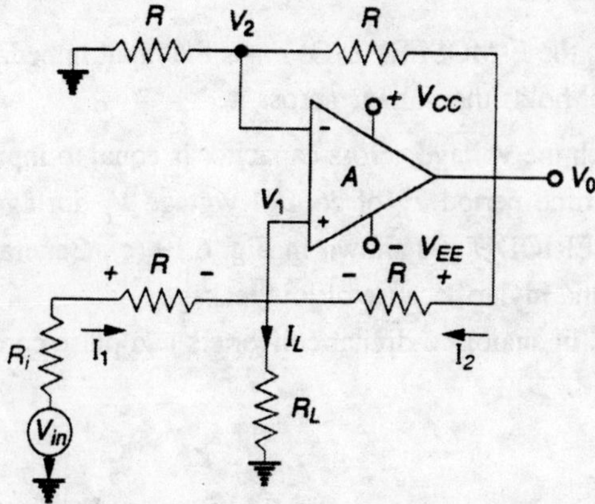

Fig. 6.21. Voltage to current converter using grounded load.

(2) **Voltage to current converter using grounded load** : In this circuit (Fig. 6.21) arrangement one terminal of load resistor is grounded and load current is conrolled by an input voltage.

Op-amp is connected in non-inverting configuration.

Current equations for the circuit are,

$$I_1 + I_2 = I_L \quad \text{...(6.34)}$$

$$\frac{V_{in} - V_1}{R} + \frac{V_0 - V_1}{R} = I_L \quad \text{...(6.35)}$$

$$V_{in} + V_0 - 2V_1 = I_L R \quad \text{...(6.36)}$$

$$V_1 = \frac{V_{in} + V_0 - I_L R}{2} \quad \text{...(6.37)}$$

As op-amp is connected in non-inverting configuration its gain is,

$$1 + R/R = 2$$

The output voltage is,

$$V_0 = 2\,V_1$$

$$= V_{in} + V_0 - I_L R$$

Therefore $V_{in} = I_L R$

$$I_L = V_{in}/R \quad \text{...(6.38)}$$

In this way the voltage is converted into current signal. Output current depends upon input voltage and resistor R.

(B) Current to Voltage Converter (Transresistance Amplifier)

Some transducers or devices produce current signal at the output proportional to the input parameters. The produced current can be converted into voltage signal to make compatible with next stage otherwise for measurement purpose. Photosensitive devices (Photodiode, photocell) produces current signal at the output proportional to incident light energy. This energy can be measured by converting the current signal into proportional voltage signal. It is also used in digital-to-analog converters.

Consider the Fig. 6.22 of current to voltage converter using inverting configuration of op-amp.

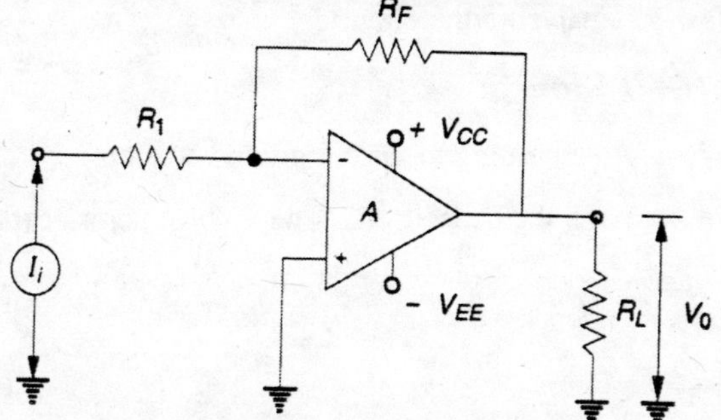

Fig. 6.22. Current to voltage converter.

Gain of this amplifier is,

$$\frac{V_0}{V_i} = -\frac{R_f}{R_1} \quad \text{...(6.39)}$$

$$V_0 = -\frac{V_i}{R_1} R_f \quad \text{...(6.40)}$$

but, $V_i/R_1 = I_i$

Therefore, $V_o = -I_i \,.\, R_f$ ...(6.41)

Thus the current signal is converted into proportional voltage signal.

6.16 Integrator and Differentiator Circuits

Both integrator and differentiator are very useful circuits. They are used in analog computers, processing of different signals, waveform generation circuitry etc.

(A) Integrator

Integrator circuit provides the output voltage which is integral of the input voltage waveform. The output voltage is proportional to the integral of input voltage. Fig. 6.23 shows the circuit diagram of integrator in inverting configuration.

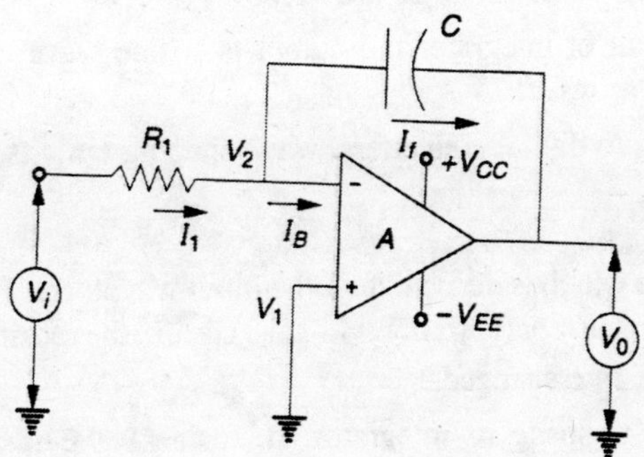

Fig. 6.23. Integrator circuit

In integrator circuit the feedback resistor R_F is replaced by capacitor C. The input signal is applied to the inverting input terminal through input resistor R_1. V_i is the constant voltage applied to the inverting input terminal of op-amp.

Applying Kirchhoff's current law for analyzing the circuit.

$$I_1 = I_B + I_f \quad \text{...(6.42)}$$

But I_B is negligible or = 0 and $I_1 = I_f$ (According to virtual ground concept)

I_f is the feedback current flowing through the capacitor C. The voltage across capacitor C is V_c.

$$I_f = C\, dV_c/dt \quad \text{...(6.43)}$$

$$I_1 = \frac{V_i - V_2}{R_1}$$

but

$$I_1 = I_f$$

$$\frac{V_i - V_2}{R_1} = C\frac{d\,V_c}{dt} = C\frac{d\,(V_2 - V_0)}{dt} \quad \text{...(6.44)}$$

As $V_1 = 0$, according to virtual ground concept $V_2 = 0$

$$\frac{V_i}{R_1} = -\,C\frac{d\,V_0}{dt} \quad \text{...(6.45)}$$

Integrating the above equation on both sides with respect to time for calculating the output voltage.

$$\int_0^t \frac{V_i}{R_1}\,dt = -\,C\int_0^t \frac{d\,V_0}{dt}\,dt \quad \text{...(6.46)}$$

Therefore the output voltage is,

$$V_0 = \frac{-1}{R_1\,C}\int_0^t V_i\,dt \quad \text{...(6.47)}$$

where R_1C is the time constant of integrator circuit.

Above equation shows that the output voltage is directly proportional to the inverted integral of input voltage and inversely proportional to the time constant R_1C.

For accurate integration the time period T of input signal must be more than the time constant R_1C.

The capacitor used in integrator circuit should have minimum leakage. Generally teflon, polystyrene and mica capacitors having the range of 0·001 to 10 μF are used.

If sine wave is applied to the input of integrator the output is cosine wave, simultaneously if square input is applied the output is rectangular wave.

Integrator circuit is used in analog to digital converters, wave shaping circuits, analog computers etc.

(B) Differentiator

This circuit provides the output voltage which is derivative of the input waveform. Fig. 6.24(a) shows circuit arrangement of differentiator in inverting configuration. In this circuit the positions of resistor R_1 and capacitor C of the integrator circuit are interchanged.

The function of differentiator is opposite to integrator. If cosine wave is applied at the input of differentiator output is sine wave. Similarly if the input is triangular wave output is square wave.

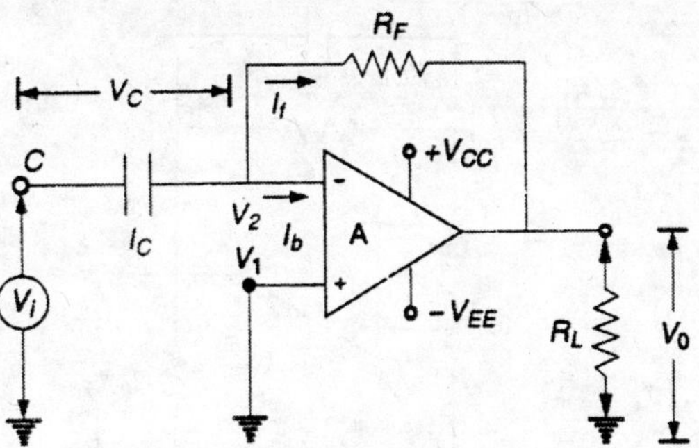

Fig. 6.24. (a) Differentiator circuit

From figure,

According to Kirchhoff's current law :

$$I_c = I_b + I_f \qquad ...(6.48)$$

Since $I_b = 0$; $I_c = I_f$ (According to virtual ground conept and input impedance is infinite)

$$I_c = C\frac{d\,V_c}{dt} = C\frac{d\,(V_i - V_2)}{dt} \qquad ...(6.49)$$

and,

$$I_f = \frac{V_2 - V_0}{R_F} \qquad ...(6.50)$$

as

$$I_c = I_f \qquad ...(6.51)$$

$$C\frac{d\,(V_i - V_2)}{dt} = \frac{V_2 - V_0}{R_F} \qquad ...(6.52)$$

According to virtual ground concept $V_1 = V_2 = 0$ as gain is large.

$$C\frac{d\,V_i}{dt} = -\frac{V_0}{R_F} \qquad ...(6.53)$$

Therefore the output voltage is,

$$V_0 = -R_F\,C\frac{d\,V_i}{dt} \qquad ...(6.54)$$

$R_F\,C$ is the time constant of differentiator circuit. Drawback of this circuit are (1) the gain inputs of this circuit $\frac{R_F}{X_c}$ increases with the increase in frequency at a rate of 20 dB/decade which makes the circuit unstable. (2) the input impedance X_c decreases with the increase in frequency; it introduces noise signals at high frequency signals. These problems gets removed by using practical differentiator circuit as shown in Fig. 6.24(b) which uses two additional components R_1 and C_F. This circuit is also called practical differentiator.

Differentiator circuit is used in wave shaping circuits, communication systems, analog computers etc.

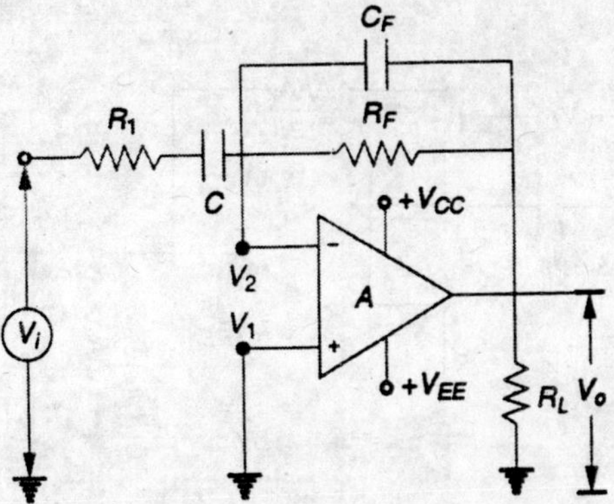

Fig. 6.24. (b) Practical differentiator.

6.17 Subtractor

Basic differential amplifier mode of OP-AMP can be used as a subtractor. It consists of inverting and non-inverting amplifiers.

Consider the Fig. 6.25, it has two input voltages V_a and V_b. Its operation is explained by using superposition theorem.

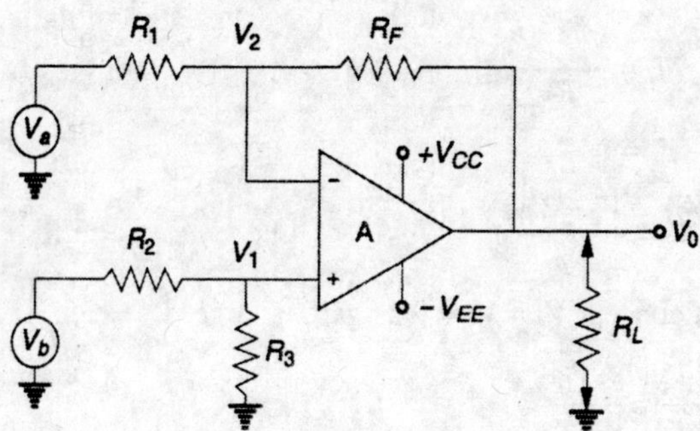

Fig. 6.25. Subtractor circuit.

Case I : Consider $V_b = 0$; in this case the OP-AMP is in inverting mode. So the output voltage V_{0a} is;

$$\therefore \quad V_{0a} = -\frac{R_F}{R_1} \cdot V_a \quad \text{...(6.55)}$$

Case II : Consider $V_a = 0$; in this case OP-AMP acts as a non-inverting amplifier and its output voltage V_{0b} is ;

$$\therefore \quad V_{0b} = \left(1 + \frac{R_F}{R_1}\right) V_1 \quad \text{...(6.56)}$$

The voltage V_1 is; (According to voltage divider method).

$$V_1 = \left(\frac{R_3}{R_2 + R_3}\right) . V_b$$

(Here R_2 and R_3 acts as a voltage divider network).

$$\therefore \quad V_{0b} = \left(\frac{R_1 + R_F}{R_1}\right)\left(\frac{R_3}{R_2 + R_3}\right) . V_b \quad \text{...(6.57)}$$

But $R_1 = R_2$ and $R_F = R_3$

$$\therefore \quad V_{0b} = \frac{R_F (V_b)}{R_1} \quad \text{...(6.58)}$$

∴ From eqn. (6.55) and (6.58)

$$V_0 = V_{0a} + V_{0b} \quad \text{...(6.59)}$$

$$\therefore \quad V_0 = -\frac{R_F}{R_1}(V_a - V_b) \quad \text{...(6.60)}$$

The term $-\frac{R_F}{R_1}$ indicates gain of differential amplifier and $(V_a - V_b)$ is the difference of voltages applied at input terminals.

6.18 Filters

The output of transducers are electrical signals in accordance with the input physical quantities. These outputs are applied to the signal conditioning elements so as to measure the output of transducers or used for controlling the processes. But the output of signal conditioning element contains unwanted signals (distortions or noise) or spurious signals. Due to these signals precise and accurate measurement is not possible so it becomes necessary to eliminate or remove such type of unwanted signals. Filters are the circuits or elements which remove or reject such type of signals.

e.g. : In industries or plants there are number of transformers, electrical motors, relays, power supplies etc. are used. These devices or sometimes the components of circuitry involved itself produes the noise signals, unwanted voltage signals gets induced because of stray electric and magnetic fields. These are unwanted or spurious signals. For suppressing these unwanted or spurious signals, filter circuits are necessary.

The term filter is defined as the network or circuit elements which passes the wanted frequency signals and simultaneously rejects or attenuates the unwanted signals.

These unwanted or spurious signals originate from transduction stage or from signal conditioning elements. Filters are frequency selective networks.

(A) Significance of Filters

(1) They provide accurate and precise measurements of electrical parameters (voltage and currents).

(2) It is possible to measure peak-to-peak or RMS (Root Mean Square) or average values of the parameters as stated above.

(3) Filter helps to analyze the harmonics or distortions present in the output signals or waveforms.

(4) Eliminates the effects of harmonics (unwanted signals) on preceding stages.

(5) Provides better resolution for measurements.

(6) The output of the transducers can be reproduced faithfully.

(7) Due to elimination of unwanted signals higher bandwidth can be achieved or maintained during measurements.

(8) Better controlling is possible because of accurate measurements of wanted signals, which is desirable in feedback control system.

(9) Accuracy depends upon Signal to Noise (S/N) ratio *i.e.* if the noise signals gets minimized the strength of wanted signals is increased which provides higher accuracy and resolution in measurements.

In other words we can say that filtering is the process of removing the unwanted interference and noise signals so as to recover the required signals or data. Filter helps to suppress the drift, high frequency noise signals, transients and distortions.

(B) Classification of Filters

Basically, depending upon applications the filters are classified as mechanical, electrical, hydraulic, acoustical, pneumatic etc. But generally in instrumentation systems electrical filters are used.

Electrical filters : These are the frequency selective networks which passes desired band of frequencies and attenuate the frequency signals outside the desired frequency band.

Electrical filters are classified as :

(A) Analog filters and digital filters.

(B) Active filters and passive filters.

(C) Audio Frequency (AF) filters and Radio Frequency (RF) filters.

Active and passive filters : These filters are discussed in following sections. Initially, we will study Passive Filters.

6.19 Passive Filters and their Analysis

Designing of these filters is carried out by means of passive components like resistance, capacitance, inductance. These filters are nothing but the networks of resistance, capacitance and inductance *i.e.* RC filters and LC filters.

The passive filters are sub divided as :

(a) Low pass filter
(b) High pass filter
(c) Band pass filter
(d) Band reject filter
(e) All pass filter

Passive filters are suffered by some problems such as :

(1) These filters require source and termination (load) impedances.

(2) Bulky and non-linear inductances are necessary for low and very low frequency filters.

(3) They require proper isolating stages.

(4) For the cascading of passive filters isolation stages are necessary.

(5) No sharp cut off may be possible.

Now we will study the types of passive filters.

(A) Low Pass Filter (LPF)

As the name indicates this filter passes the low frequency signals and attenuates the high frequency signals. The frequency signals are passed upto the cut off frequency (f_C) as shown in the ideal frequency response. The Fig. 6.26(a) shows ideal frequency response of low pass filter. Upto cut off frequency (f_C) the lower frequency signals are passed with constant gain upto f_C. If input signal frequency is increased (higher than f_C)

the gain starts to decrease and higher frequency signals are rejected. Practically this response cannot be obtained.

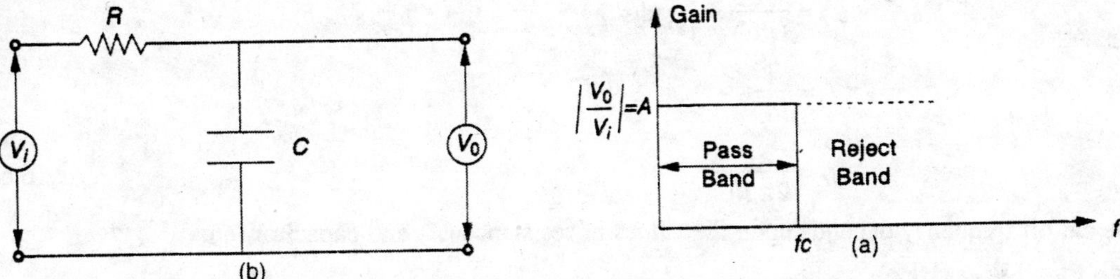

Fig. 6.26 (a) Ideal Frequency of LPF. (b) Circuit diagram of LPF.

Circuit analysis of low pass filter : Fig. 6.26(b) shows circuit diagram of low pass filter using RC network (passive element). V_i is input voltage and V_0 is the output voltage signal. If lower frequency signal are applied to the input of circuit, capacitive reactance of capacitor C increases and it acts as open circuit due to which the output voltage (V_0) equals to input voltage (V_i) *i.e.* ($V_0 = V_i$) voltage gain is equal to unity. If frequency of input signals is increased *i.e.* higher frequency signals are applied to the circuit, capacitor C offers low reactance. Due to this output voltage decreases than input voltage (V_i). Thus voltage gain decreases if the frequency of input signal is increased.

Fig. 6.26(c) shows the response of low pass filter. From this response it is observed that if the input frequency is increased more than cut off frequency (f_c) gain starts to decrease gradually. Upto cut off frequency the gain is constant.

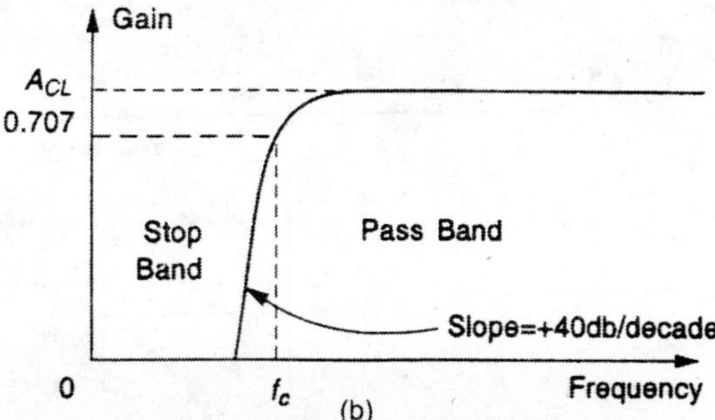

Fig. 6.26. (c) Response of low pass filter.

Circuit analysis : Consider transfer function of the circuit.

$$\frac{V_0\,(j\omega)}{V_i\,(j\omega)} = \frac{1/sC}{R + 1/sC} = \frac{1}{1 + sRC} = \frac{1}{1 + s\tau} \quad ...(6.61)$$

as

$$\tau = RC$$

For sinusoidal transfer function substitute $s = j\omega$.

$$\frac{V_0\,(s)}{V_i\,(s)} = \frac{1}{1 + (j\omega)\,RC} = \frac{1}{1 + (j\omega)\,\tau}$$

$$\therefore \text{ Gain } A = \left|\frac{V_0\,(j\omega)}{V_i\,(j\omega)}\right| = \frac{1}{\sqrt{1 + (\omega RC)^2}} = \frac{1}{\sqrt{1 + (\omega\tau)^2}} \quad ...(6.61a)$$

Gain drops at 0·707 at cut off frequency f_c.

$$0{\cdot}707 = \frac{1}{1 + (\omega_c\,RC)} \quad ...(6.62)$$

or cut off frequency (f_C) is,

$$\omega_C = \frac{1}{(RC)} = \frac{1}{\tau} \text{ rad/s.}$$

$$2\pi f_C = \frac{1}{RC}$$

$$f_C = \frac{1}{2\pi RC} \qquad ...(6.63)$$

The cut off frequency depends upon the values of resistance (R) and capacitor (C).

(B) High Pass Filter (HPF)

High pass filter passes high frequency signals to the output and rejects the low frequency signals.

Fig. 6.27(a) shows ideal frequency response of high pass filter. The gain is zero upto cut off frequency (f_c). Beyond this frequency the gain is constant (A). Above cut off frequency (f_c) the input signals gets reproduced at the output with constant gain (A). The range of frequency signals from 0 to f_c gets rejected. Practically this type of response cannot be possible.

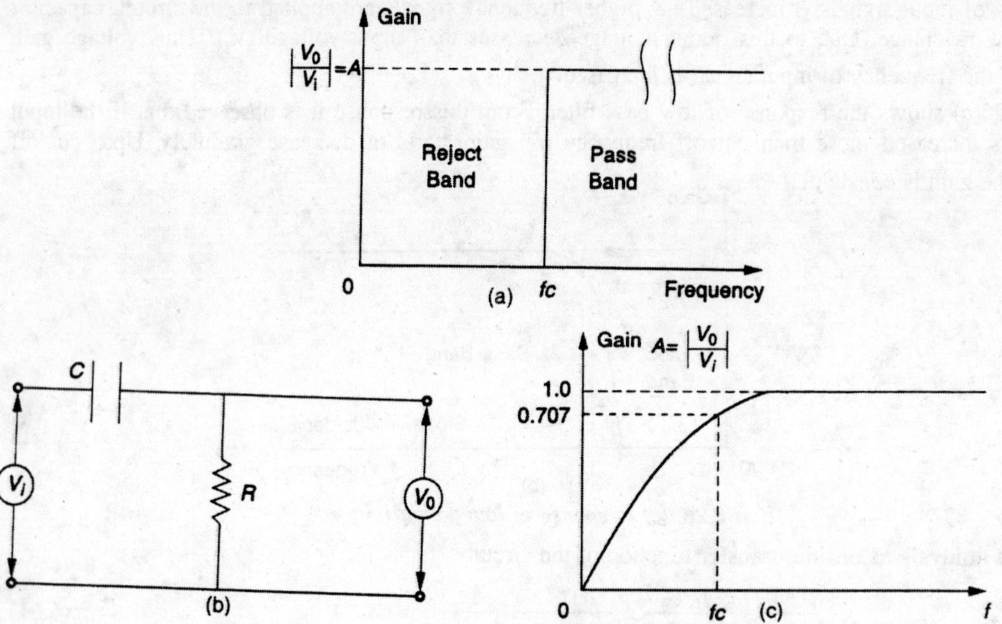

Fig. 6.27 (a) Ideal frequency response of HPF. (b) Circuit diagram of HPF (c) Response of high pass filter.

In above Fig. 6.27(b) shows the circuit diagram of high pass filter using RC network. V_i is the input voltage signal and V_0 is the output voltage signal. If the frequency of applied input signal is low, capacitive reactance is high due to which small output voltage is obtained resulting minimum gain. As the frequency of applied input signal is higher, capacitive reactance is low so the output voltage (V_0) equals to input voltage (V_0) *i.e.* voltage gain is equal to unity.

Fig. 6.27(c) shows response of high pass filter, it shows that the circuit passes high frequency signals and rejects the low frequency signals.

Analysis of high pass filter : Consider transfer function of the circuit.

$$\frac{V_0(s)}{V_i(s)} = \frac{R}{R + 1/sC} = \frac{sCR}{sCR + 1} = \frac{s\tau}{1 + \tau s} \quad \text{...(6.64)}$$

where ; $\tau = RC$

for sinusoidal transfer function of the circuit substitute $s = j\omega$.

$$\frac{V_0(j\omega)}{V_i(j\omega)} = \frac{RC(j\omega)}{1 + RC(j\omega)} = \frac{j\omega\tau}{1 + j\omega\tau} = \frac{1}{1 + 1/j\omega\tau}$$

Gain $$A = \left|\frac{V_0(j\omega)}{V_i(j\omega)}\right| = \frac{\omega\tau}{\sqrt{1 + (\omega\tau)^2}} \quad \text{...(6.65)}$$

Thus high pass filter passes high frequency at constant gain and rejects low frequency signals.

The cut off frequency is $f_c = \dfrac{1}{2\pi RC}$ ...(6.66)

(C) Band Pass Filter

Fig. 6.28 shows ideal response of band pass filter. f_{c_1} and f_{c_2} are the cut off frequencies. This filter passes frequency signals between the range fc_1 and fc_2. While frequency signals from 0 to f_{c_1} and higher than f_{c_2} gets rejected. This means the signals from f_{c_1} to f_{c_2} gets reproduced at the output faithfully. This is the ideal frequency response, practically which cannot be possible.

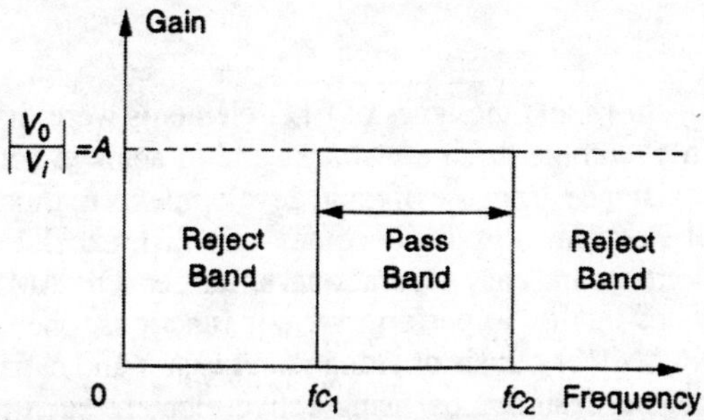

Fig. 6.28. Ideal frequency response of band pass filter.

(D) Band Stop Filter

This filter is also called as notch filter. Ideal frequency response of band stop filter is shown in Fig. 6.29. This filter passes frequency signals from the range of 0 to f_{c_1} and f_{c_2} onwards at constant gain by rejecting the frequency band from f_{c_1} to f_{c_2}.

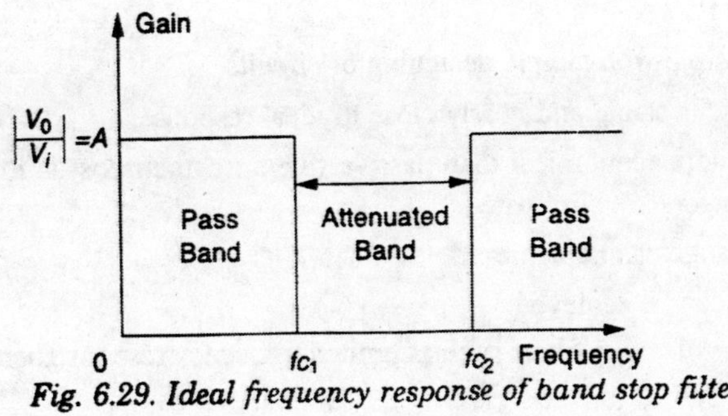

Fig. 6.29. Ideal frequency response of band stop filter.

(E) All Pass Filter

As the name implies this filter passes all frequencies without rejection. This filter is used when the phase shift of frequency signals at the output is required, but the magnitude remains constant.

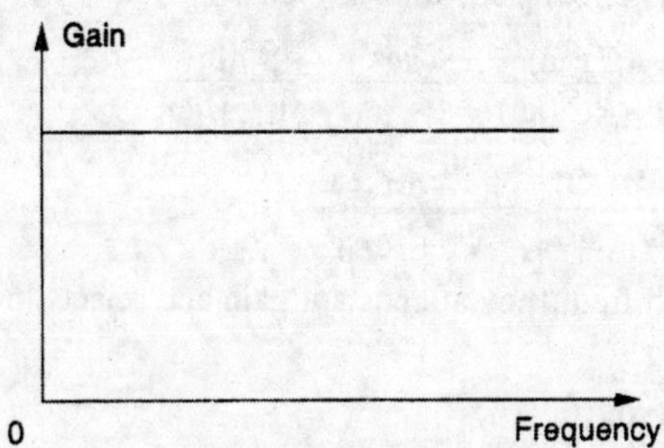

Fig. 6.30. Ideal frequency response of all pass filter.

Fig. 6.30 shows ideal response of all pass filter, practically it cannot be achieved. Phase shift can be possible over the range of 0° to 360° over the frequency range of 0 to ∞.

This type of filters are used in MODEM's (Modulator and Demodulator), data transmission networks, equalizer circuits etc.

6.20 Active Filters

In previous section the concepts, functions and the types of filter elements were discussed. In passive filters some drawbacks were observed. To overcome these drawbacks and to achieve better filteration quality the concept of active filters comes into existence. Because of rapid developments in monolithic integrated circuits the functioning and performance of electronic circuits becomes very fast, excellent and reliable. Also their designing and testing becomes economical and easy. Operational amplifier (OP-AMP) is one of the integrated circuit (IC) *e.g.* IC 741 which provides facility to perform various functions, one of them being **Filtering**.

The filter circuit which uses OP-AMP or transistor along with resistor and capacitor networks are called as **Active Filters**. Active filters can be designed by using active element such as OP-AMP and passive elements *i.e.* resistors and capacitors. Because of introducing OP-AMP, the filter circuit offers number of advantages which are listed below.

(1) As OP-AMP provides high input and low output impedance the active filter circuit is not suffered by the problem of loading effects at source or load terminals or ends. *i.e.* good isolation is possible.

(2) OP-AMP has constant and high gain characteristics so without attenuation the input signals get reproduced at the output.

(3) Use of inductors is completely eliminated so the problems associated with inductors are completely avoided.

(4) Easy tuning is possible with simple designing of circuits.

(5) Frequency response is stable and nearly close to ideal response.

(6) Active filters are more economical than passive filters *i.e.* their cost is low because of elements such as bulky inductors are eliminated.

(7) Good "Q" factor range can be achieved.

(8) Higher bandwidth can be achieved.

There are some limitations of active filters such as limited frequency response below 100 kHz and power requirements for operation is more.

(A) Low Pass Filter

Fig. 6.31(a) shows basic low pass filter using op-amp and RC network. OP-AMP is in unity gain configuration

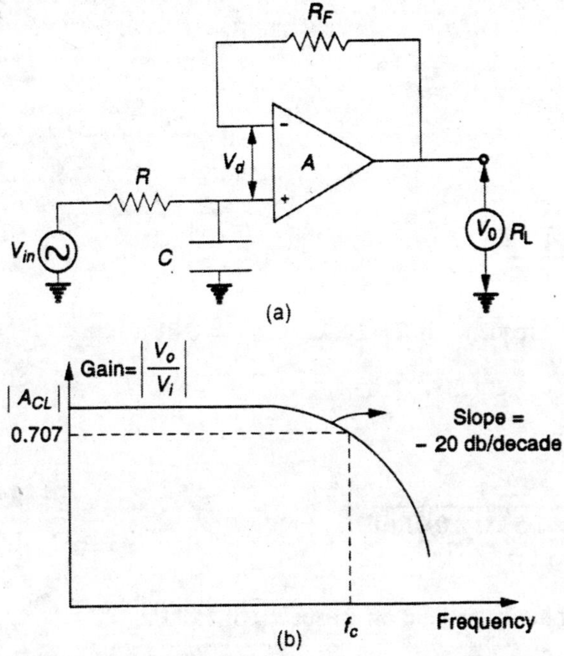

Fig. 6.31 (a) Circuit diagram of basic active LPF. (b) Frequency response of LPF.

i.e. $R = R_F$. At d.c. level input the capacitive reactance of capacitor C is infinite and the d.c. resistance path to ground for both input terminals must be equal. According to virtual ground concept the difference voltage V_d across inverting and non-inverting terminals of op-amp is zero. Hence the voltage across capacitor C is equal to output voltage V_0. The circuit is unity gain amplifier *i.e.* voltage follower. The input voltage V_{in} gets distributed in R and C. As stated above the capacitor voltage V_0 is given by :

$$V_0 = \frac{1/j\omega C}{R + 1/j\omega C} V_{in} \qquad \text{...(6.67)}$$

where ; ω is frequency of input voltage signal (rad/sec)

$$\omega = 2\pi f$$

So the closed loop gain A_{CL} is given by :

$$A_{CL} = \frac{V_0}{V_{in}} = \frac{1}{1 + j\omega RC} \qquad \text{...(6.68)}$$

From above equation it is observed that :

(1) If input signal frequency is very low *i.e.* $\omega = 0$; the closed loop gain $|A_{CL}| = 1$.

(2) When input signal frequency is very high *i.e.* $\omega = \infty$, closed loop gain $|A_{CL}| = 0$. It indicates that this filter passes low frequency input signals and rejects high frequency input signals hence it is a low pass filter.

Above Fig. 6.31(b) shows frequency response of low pass filter with roll of – 20 db/decade. When input frequency signal increases beyond f_c the gain $|A_{CL}|$ decreases at the rate of 20 db/decade. Cut off frequency is defined as the frequency of input voltage at which gain $|A_{CL}|$ is decreased to 0·707 times to that of its lower frequency value.

Cut off frequency is $\omega_c = 1/RC = 2\pi f_c$

$$f_c = \frac{1}{2\pi RC} \quad \text{...(6.69)}$$

where ; f_c = Cut off frequency in Hz

R = Resistance in Ohm

C = Capacitance in Farad

The value of resistance $R = \dfrac{1}{2\pi f_c C}$

Example 6.1 : Calculate the cut off frequency if $R = 1{\cdot}5\ \text{k}\Omega$ and $C = 0{\cdot}01\ \mu\text{F}$

Solution :

We know ;
$$f_c = \frac{1}{2\pi RC}$$
$$= \frac{1}{2\pi \times 1{\cdot}5\ \text{k}\Omega \times 0{\cdot}01\ \mu\text{F}}$$
$$\therefore\ f_c = 10{\cdot}61\ \text{kHz}$$

Example 6.2 : Design a low pass filter having out off frequency of 15 kHz.

Solution :

In designing, the parameter which is to be calculated is R by assuming suitable value of capacitor C.

Step 1. Given data f_c = 15 KHz

Assume $C = 0{\cdot}01\ \mu\text{F}$

Step 2. Calculate value of R

$$R = \frac{1}{2\pi f_c C}$$
$$R = \frac{1}{2 \times \pi \times 15\ \text{kHz} \times 0{\cdot}01\ \mu\text{F}}$$
$$\therefore\ R = 1{\cdot}06\ \text{k}\Omega\ \text{(Practical value is } 1{\cdot}5\ \text{k}\Omega)$$

Example 6.3 : Design a low pass filter having cut off frequency of 1·5 kHz and pass band gain of 3.

Solution :

Step 1. Given data f_c = 1·5 kHz, A_{CL} = 3.

Step 2. Calculate the value of resistance R

Assume $C = 0{\cdot}01\ \mu\text{F}$

$$R = \frac{1}{2\pi f_c C}$$
$$R = \frac{1}{2\pi \times 1{\cdot}5\ \text{kHz} \times 0{\cdot}01\ \mu\text{F}}$$
$$R = 10{\cdot}61\ \text{k}\Omega$$

Step 3. Gain is $A_{CL} = 1 + R_F/R_1$ (For Non-Inverting mode)

$$3 = 1 + R_F/R_1$$

$$R_F/R_1 = 2$$

$$R_F = 2\,R_1$$

Let $R_1 = 1$ kΩ; Therefore $R_F = 2$ kΩ is selected.

(a) First order low pass Butterworth filter

Fig. 6.32(a) shows first order low pass Butterworth filter circuit. It consists of an op-amp in non-inverting configuration with RC network. R_1 and R_F are gain deciding elements of op-amp.

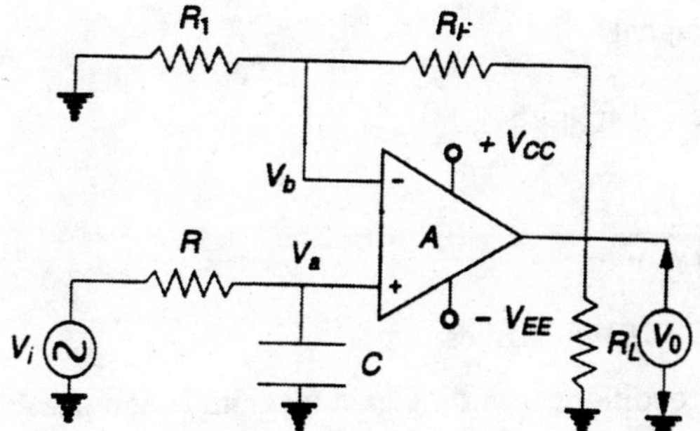

Fig. 6.32. (a) First order low pass Butterworth filter circuit.

Step 1. Calculate V_a *i.e.* voltage across capacitor C by using potential divider method.

$$V_a = \frac{-j\,X_c}{R - j\,X_c}\,V_i \qquad \text{...(6.70)}$$

but
$$-j\,X_c = \frac{1}{2\pi\,jf_c}$$

$$V_a = \frac{1/2\pi\,jf_c}{R + 1/2\pi\,jf_c}\,V_i$$

$$V_a = \frac{V_i}{2\pi\,jf_cR + 1} \qquad \text{...(6.71)}$$

As the op-amp is in non-inverting configuration its gain is $A_{CL} = 1 + R_F/R_1$ (for pass band gain of the filter).

Output voltage
$$V_0 = (1 + R_F/R_1)\,V_a \qquad \text{...(6.72)}$$

$$V_0 = (1 + R_F/R_1)\,\frac{V_i}{(2\pi\,jf_cR + 1)}$$

$$\frac{V_0}{V_i} = \frac{A_{CL}}{2\pi\,jf_cR + 1} \qquad \text{...(6.73)}$$

But $f_h = 1/2\pi\,RC$ *i.e.* high cut off frequency

$$\frac{V_0}{V_i} = \frac{A_{CL}}{1 + (f/f_h)\,j} \qquad ...(6.74)$$

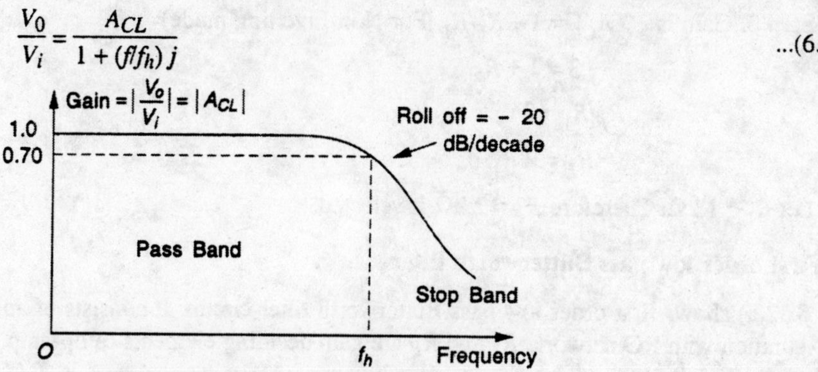

Fig. 6.32 (b) Frequency Response

Here ; f = Frequency of input signal.

$\frac{V_0}{V_i}$ = Gain of filter as a function of frequency.

$$\text{Gain magnitude } \left|\frac{V_0}{V_i}\right| = \frac{A_{CL}}{\sqrt{1 + (f/f_h)^2}}$$

Phase angle $\phi = -\tan^{-1}(f/f_h)$ in degrees.

Now we will study following conditions regarding gain magnitude and phase angle.

Case I : Let $f < f_h$ *i.e.* input signal frequency is less than higher cut off frequency.

Then,
$$\frac{V_0}{V_i} \cong A_{CL}$$

It indicates that the gain of low pass filter is constant in the input frequency range of 0 to higher cut off frequency (f_h).

Case II : Let $f = f_h$ *i.e.* both input frequency signal and higher cut off frequency signals are equal.

$$\frac{V_0}{V_i} = 0{\cdot}707\, A_{CL}$$

Means the gain is $0{\cdot}707\, A_{CL}$ at higher cut off frequency.

Case III : Let $f > f_h$ *i.e.* input frequency signal to greater than high cut off frequency.

$$\frac{V_0}{V_i} < A_{CL}$$

In this case beyond f_h the gain starts to decrease as frequency of input signal increases.

When the frequency of input signal is increased ten times *i.e.* one decade, the voltage gain is divided by 10 or if the frequency increases by 10 the gain decreases by 20 db *i.e.* 20 log 10 = 20.

(b) Second order low pass Butterworth filter

As the first order low pass Butterworth filter uses one RC network stage. If one more stage *i.e.* two stages of RC networks are cascaded the result would be second order low pass Butterworth filter. It provides combined response of two RC stages to produce a – 40 db/decade roll off. This response is twice steep to that of first order low pass filter.

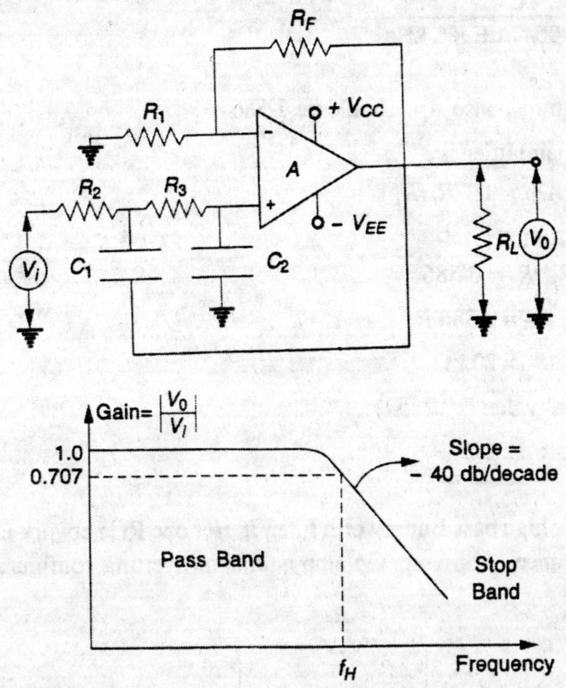

Fig. 6.33 (a) Circuit diagram of second order L.P. Butterworth filter (b) Frequency response

Fig. 6.33(a) shows second order low pass Butterworth filter. It uses two RC networks *i.e.* R_2C_1 and R_3C_2 with op-amp in non-inverting configuration. Capacitor C_1 is connected to output of op-amp which provides positive feedback path.

Pass band gain of filter is :

$$A_{CL} = 1 + R_F/R_1 \qquad ...(6.75)$$

The high cut off frequency $f_H = \dfrac{1}{2\pi\sqrt{R_2R_3C_2C_1}}$

From circuit if $R_2 = R_3$ and $C_2 = C_1$ the Butterworth response can be obtained by making $A_{CL} = 1{\cdot}586$.

Fig. 6.33(b) shows the frequency response of second order low pass Butterworth filter. Gain is decreasing beyond f_H with a slope of – 40 db/decade.

Example 6.4 : Design a second order low pass Butterworth filter for cut off frequency of 5 kHz.

Solution :

Step 1. Given data $f_c = 5$ kHz.

Assume $R_2 = R_3 = R$ and $C_2 = C_1 = C$

Assume op-amp is in non-inverting configuration.

Step 2. Calculate R.

Assume $C = 0{\cdot}0047$ μF

$$R = \frac{1}{2\pi f_c C}$$

$$R = \frac{1}{2 \times \pi \times 0{\cdot}0047\ \mu\text{F} \times 5\ \text{kHz}}$$

$$R = 6{\cdot}77\ \text{k}\Omega$$

Step 3. For Butterworth response A_{CL} should be 1·586

Gain for non-inverting amplifier is :

$$A_{CL} = 1 + R_F/R_1$$

$$1 + R_F/R_1 = 1{\cdot}586$$

$$R_F/R_1 = 0{\cdot}586$$

$$R_F = 0{\cdot}586\, R_1$$

Choose $R_1 = 20\ \text{k}\Omega$

$R_F = 11{\cdot}72\ \text{k}\Omega$ (Practical value is 12 kΩ)

So ; $R_1 = 20\ \text{k}\Omega$ and $R_F = 12\ \text{k}\Omega$

(B) High Pass Filter

Fig. 6.34(a) shows first order high pass Butterworth filter. It uses one RC network *i.e.* R and C. R_F is feedback resistance, R_1 is input resistance of op-amp. Op-amp is in non-inverting configuration.

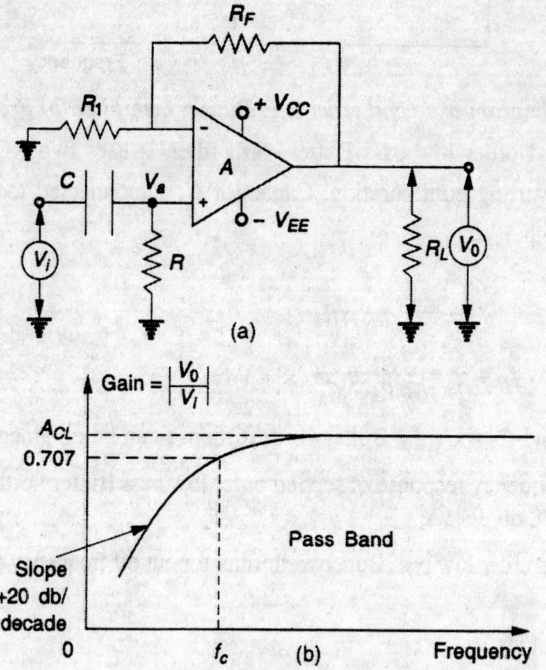

Fig. 6.34 (a) Circuit diagram of first order high pass Butterworth filter (b) Frequency response.

Fig. 6.34(b) shows response of this filter. The frequencies greater than cut off frequency f_c are passed at output upto maximum frequency of closed loop bandwidth of op-amp.

Consider the circuit diagram of high pass Butterworth filter :

Step 1. Calculate V_a *i.e.* voltage at node *a*.

By using potential divider method this voltage is calculated.

$$V_a = \frac{R}{(R - jX_C)} \cdot V_i \quad \text{...(6.76)}$$

but $$-jX_C = 1/2\pi jf_c$$

$$V_a = \frac{R}{(R + 1/(2\pi jf_c))} V_i$$

$$V_a = \frac{2\pi jf_c CR}{(1 + 2\pi j R f_c)} V_i \quad \text{...(6.77)}$$

Step 2. As the op-amp is connected in non-inverting configuration its gain is :

$$A_{CL} = (1 + R_F/R_1) \quad \text{...(6.78)}$$

Step 3. Calculation of output voltage.

Output voltage $V_0 = (1 + R_F/R_1)\, V_a$

$$V_0 = (1 + R_F/R_1) \frac{V_i (2\pi jfCR)}{(1 + 2\pi jfCR)}$$

$$\frac{V_0}{V_i} = A_{CL} \left| \frac{j(f/f_c)}{1 + j(f/f_c)} \right| \quad \text{...(6.79)}$$

where ; $f_c = 1/2\pi RC$ cut off frequency of filter

f = Frequency of input signal

Gain magnitude $$\left| \frac{V_0}{V_i} \right| = \frac{A_{CL}(f/f_c)}{\sqrt{1 + (f/f_c)^2}} \quad \text{...(6.80)}$$

Phase angle $\phi = -\tan^{-1}(f/f_c)$ in degrees. ...(6.81)

Like first order low pass filter, we will study following conditions.

Case I : Let $f < f_c$ *i.e.* input signal frequency is less than cut off frequency f_c.

$$\frac{V_0}{V_i} < A_{CL}$$

In this condition the gain is decreasing.

Case 2 : Let $f = f_c$ *i.e.* input signal frequency f is equal to cut off frequency f_c.

$$\frac{V_0}{V_i} = 0{\cdot}707\, A_{CL}$$

It indicates that the gain is 0·707 A_{CL} at cut off frequency f_c as shown in Fig. 6.34(b)

Case III : Let $f > f_c$ *i.e.* input signal frequency f is greater than cut off frequency f_c.

$$\therefore \quad \frac{V_0}{V_i} \cong A_{CL}$$

Means the gain A_{CL} is constant in the input frequency range from fc to maximum frequency of closed loop bandwidth of op-amp.

Second order high pass Butterworth filter

This circuit can be obtained by simply interchanging the resistors and capacitors of second order low pass filter as shown in Fig. 6.35(a).

Fig. 6.35(b) shows frequency response of this filter having slope of + 40 db/decade.

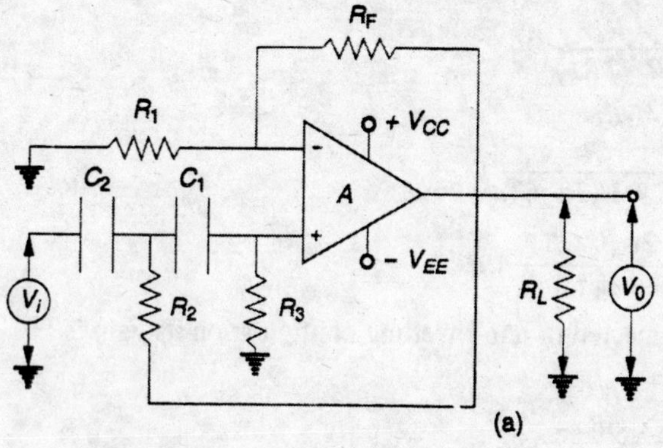

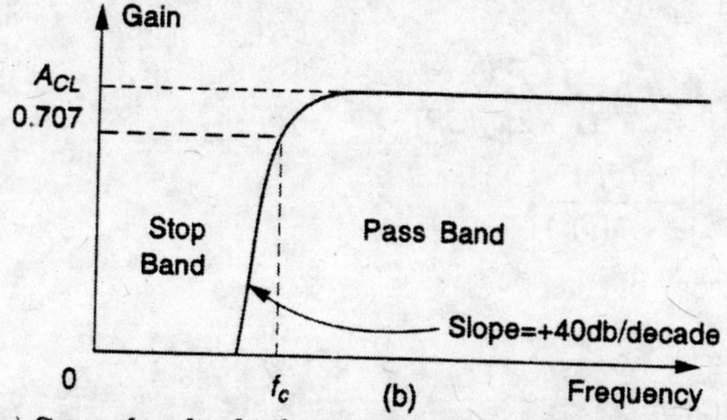

Fig. 6.35 (a) Second order high pass Butterworth filter. (b) Frequency response

Cut off frequency is given by :

$$f_c = \frac{1}{2\pi\sqrt{R_2R_3C_1C_2}} \quad ...(6.82)$$

If $R_2 = R_3 = R$ and $C_1 = C_2 = C$; the cut off frequency is given by :

$$f_c = \frac{1}{2\pi RC} \quad ...(6.83)$$

Example 6.5 : Design a second order low pass filter at cut off frequency of 5 kHz.

Solution :

Step 1. Given data $f_c = 5$ kHz.

Step 2. Calculate R. Assume $C_1 = C_2 = C = 0{\cdot}0047\ \mu F$

$$R = \frac{1}{2\pi f_c C}$$

$$R = \frac{1}{2\pi \times 5 \text{ kHz} \times 0{\cdot}0047\ \mu F}$$

$$R = 6{\cdot}7 \text{ k}\Omega$$

Therefore, $R_1 = R_2 = R = 6{\cdot}5$ kΩ (Practical value)

$$C_1 = C_2 = C = 0{\cdot}0047\ \mu F$$

Step 3. For Butterworth filter assume non-inverting amplifier gain to be as :

$$A_{CL} = (1 + R_F/R_1) = 1{\cdot}586$$

$$R_F/R_1 = 0{\cdot}586$$

Assume $R_1 = 20\ \text{k}\Omega$

$$R_F = 0{\cdot}586 \times 20\ \text{k}\Omega = 11{\cdot}72\ \text{k}\Omega$$

So the designed values are : $R = R_2 = R_3 = 6{\cdot}5\ \text{k}\Omega$

$$C = C_1 = C_2 = 0{\cdot}0047\ \mu\text{F}$$

$R_1 = 20\ \text{k}\Omega$; $R_F = 11{\cdot}72\ \text{k}\Omega$ if $A_{CL} = 1{\cdot}586$ for second order Butterworth filter.

(C) Band pass Filter

This filter passes a certain range or band of frequencies and rejects the frequencies outside this band. Band pass filters can be designed by cascading both high pass and low pass filter. Band pass filters are classified as :

(a) Wide band pass filter.

(b) Narrow band pass filter.

This classification is on the basis of Q-factor or figure of merit which can be expressed as :

$$Q = \frac{f_c}{BW} = \frac{f_c}{f_H - f_L} \quad \text{...(6.84)}$$

where ; BW is the bandwidth.

f_H = Higher cut off frequency.

f_L = Lower cut off frequency.

f_c = Center frequency = $\sqrt{f_H f_L}$

If the Q-factor is less than 10 ($Q < 10$) then it is wide band pass filter.

If Q-factor is greater than 10 ($Q > 10$) it is narrow band pass filter.

(a) Wide Band Pass Filter

It can be designed by combining or cascading high pass and low pass filter circuits. There are number of methods to design the wide band pass filter circuit. Fig. 6.36 shows circuit diagram for wide band pass filter.

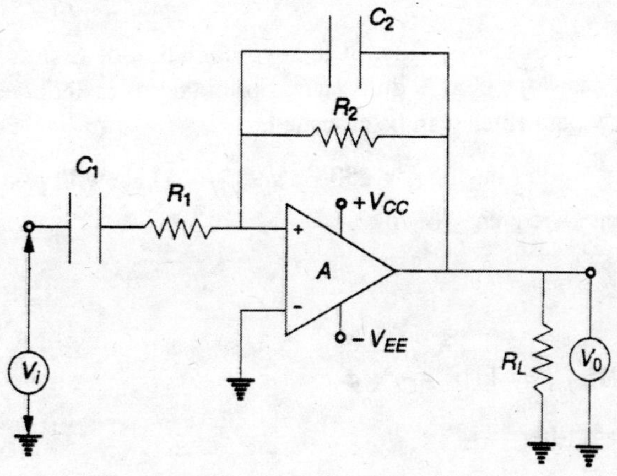

Fig. 6.36. Circuit diagram of wide band pass filter.

Here the op-amp is in non-inverting configuration. Resistance R_1 and capacitor C_1 forms high pass filter circuit having input impedance Z_H.

$$Z_H = R_1 + j\omega C_1 \quad (6.85)$$

This circuit has cut off frequency $f_H = \frac{1}{2\pi R_1 C_1}$ ...(6.86)

Similarly the resistance R_2 and capacitor C_2 forms low pass filter with an impedance of Z_L.

$$Z_L = R_2 \parallel (1/j\omega C_2) = \frac{R_2}{1 + j\omega\, C_2 R_2} \quad ...(6.87)$$

This circuit has cut off frequency $f_L = \frac{1}{2\pi R_2 C_2}$ ...(6.88)

The frequency band which is passed or transmitted to the output is between f_H and f_L.

where ; f_H = Higher cut off frequency

f_L = Lower cut off frequency

Thus the pass band is $f_H \le f \le f_L$ and frequencies outside this range gets rejected.

Fig. 6.37 shows frequency response of wide band pass filter.

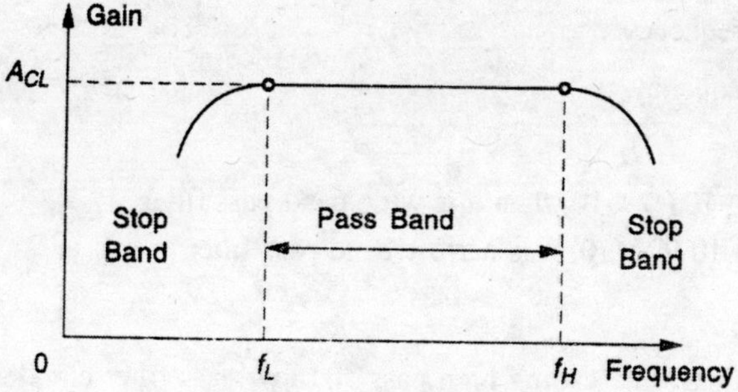

Fig. 6.37 Frequency response

Another circuit for wide band pass filter

Fig. 6.38 shows circuit diagram of wide band pass filter. It uses combination of first order high pass filter and first order low pass filter which has slope of ± 20 db/decade. For obtaining ±40 db/decade slope second order high pass and second order low pass filters can be cascaded.

Example 6.6 : Design a wide band pass filter having $f_L = 250$ Hz and $f_H = 4$ kHz with pass band gain 4. Also find out Q-factor and center frequency of filter.

Solution :

Step 1. Given data :

$$f_L = 250 \text{ Hz}, f_H = 4 \text{ kHz}, A_{CL} = 4$$

Step 2. Calculate R_1 for high pass filter.

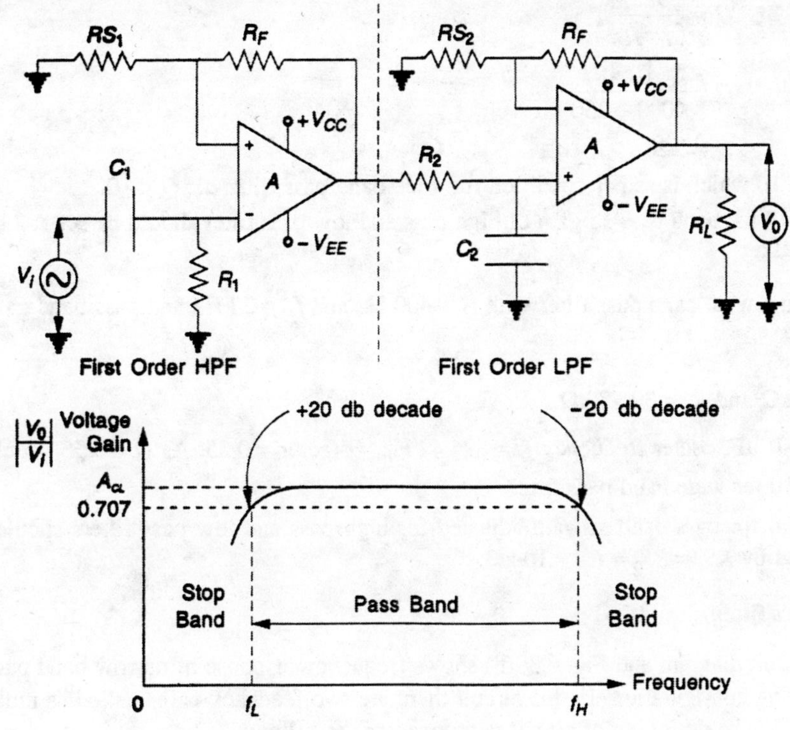

Fig. 6.38 Another circuit of wide band pass filter.

$$R_1 = \frac{1}{2\pi f_H C} \quad \text{assume } C = 0{\cdot}01\ \mu\text{F}$$

$$R_1 = \frac{1}{2\pi \times 4\ \text{kHz} \times 0{\cdot}01\ \mu\text{F}}$$

$$R_1 = 3{\cdot}97\ \text{k}\Omega$$

Step 3. Calculate R_2 for low pass filter ($f_L = 250$ Hz)

$$R_2 = \frac{1}{2\pi f_L C} \quad \text{assume } C = 0{\cdot}01\ \mu\text{F}$$

$$R_2 = \frac{1}{2\pi \times 250 \times 0{\cdot}01\ \mu\text{F}}$$

$$R_2 = 63{\cdot}66\ \text{k}\Omega$$

Step 4. Calculate center frequency f_c

$$f_c = \sqrt{f_L f_H}$$

$$f_c = \sqrt{250 \times 4\ \text{kHz}}$$

$$f_c = 1\ \text{kHz}$$

Calculate Q-factor

$$Q = \frac{f_c}{f_H - f_L}$$

$$Q = \frac{1000}{4000 - 250}$$

$$Q = 0{\cdot}26$$

So Q is less than 10 which is as per condition for wide band pass filter *i.e.* $Q < 10$.

Step 5. As gain = 4 for band pass so gain of high pass and low pass filter should be set at 2 by making $RS_1 = RS_2 = R_F = 10\ \text{k}\Omega$.

Example 6.7 : Design a wide band pass filter with $f_L = 400$ Hz and $f_H = 2$ kHz and pass band gain of 4.

Solution :

$R_1 = 7{\cdot}95\ \text{k}\Omega$ and $R_2 = 39{\cdot}78\ \text{k}\Omega$,

Assuming $C = 0{\cdot}01\ \mu\text{F}$, center frequency $f_c = 894{\cdot}42$ Hz, Q-factor $= 0{\cdot}55$. As $Q = 0{\cdot}55$ which satisfies the condition of $Q < 10$ for wide band pass filter.

Gain specified is 4 for pass band so gain adjusted for high pass and low pass filters should be same *i.e.* 2 which is adjusted by $RS_1 = RS_2 = R_F = 10\ \text{k}\Omega$.

(b) Narrow band pass filter

Fig. 6.39(a) shows circuit diagram and Fig. 6.39(b) shows frequency response of narrow band pass filter. It uses op-amp in inverting configuration. In this circuit there are two feedback paths called as multiple path feedback via R_3 and C_2. The designing of circuit components is as follows :

$$R_1 = \frac{Q}{2\pi f_c\, C A_{CL}}\ ;\ R_2 = \frac{Q}{2\pi f_c\, C\,(2Q^2 - A_{CL})} \qquad \text{...(6.89)}$$

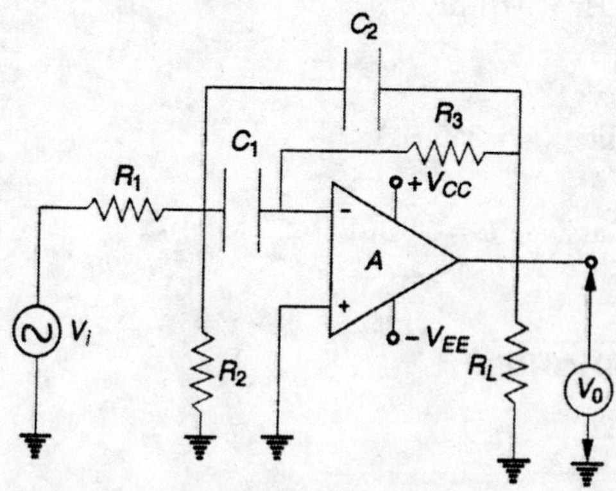

Fig. 6.39(a) Circuit diagram of narrow band pass filter.

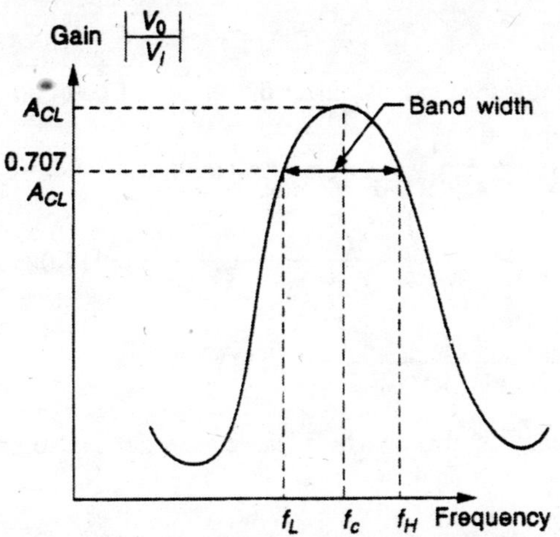

Fig. 6.39(b) : Frequency response.

$$R_3 = \frac{Q}{\pi f_c C}\text{ ; and gain } A_{CL} = \frac{R_3}{2R_1} \text{ but } A_{CL} < 2Q^2 \quad ...(6.90)$$

Because of multiple path feedback it is possible to change the center frequency f_c to new center frequency f_c' without changing the gain and bandwidth. This design can be obtained by changing R_2 to R_2' as follows :

$$R_2' = R_2 (f_c/f_c')^2 \quad ...(6.91)$$

Example 6.8 : Design a narrow band pass filter so that $f_c = 2$ kHz, $Q = 20$ and $A_{CL} = 10$.

Assume $C_1 = 0{\cdot}01\ \mu F$

Solution :

$$R_1 = \frac{Q}{2\pi f_c C A_{CL}} = \frac{20}{2\pi \times 2 \text{ kHz} \times 0{\cdot}01\ \mu F \times 10} = 15{\cdot}91 \text{ k}\Omega$$

$$R_2 = \frac{Q}{2\pi f_c C (2Q^2 - A_{CL})}$$

$$= \frac{20}{2\pi \times 2 \text{ kHz} \times 0{\cdot}01\ \mu F\ (2 \times 20^2 - 10)}$$

$$R_2 = 201{\cdot}46\ \Omega = \text{Select } 200\ \Omega$$

$$R_3 = \frac{Q}{\pi f_c C} = \frac{20}{\pi \times 2 \text{ kHz} \times 0{\cdot}01\ \mu F} = 318{\cdot}30 \text{ k}\Omega$$

Example 6.9 : Design narrow band pass filter having $f_c = 2{\cdot}5$ kHz, $Q = 15$ and $A_{CL} = 10$. If the center frequency is changed to 3·5 kHz by keeping A_{CL} and bandwidth constant design necessary components.

Solution :

Step 1. Given data :

$$f_c = 2{\cdot}5 \text{ kHz}, Q = 15, A_{CL} = 10\ f_c' = 3{\cdot}5 \text{ kHz}$$

Step 2. Calculate the values of components.

Assume $C = 0{\cdot}01\ \mu F$

By using above mentioned formulas we can calculate the values of components.

$$R_1 = \frac{15}{2\pi \times 2{\cdot}5\ \text{kHz} \times 0{\cdot}01\ \mu\text{F} \times 10} = 9{\cdot}54\ \text{k}\Omega$$

$$R_2 = \frac{15}{2\pi \times 2{\cdot}5\ \text{kHz} \times 0{\cdot}01\ \mu\text{F}\ (2 \times 15^2 - 10)} = 217{\cdot}02\ \Omega$$

$$R_3 = \frac{15}{\pi \times 2{\cdot}5\ \text{kHz} \times 0{\cdot}01 \mu\text{F}} = 190\ \text{k}\Omega$$

Step 3. Now the center frequency is changed to 3·5 kHz *i.e.* $f_c' = 3{\cdot}5$ so only the component newly designed is R_2'

$$R_2' = R_2\ (f_c/f_c')^2$$

$$R_2' = 217\ (2{\cdot}5\ \text{kHz}/3{\cdot}5\ \text{kHz})^2$$

$$R_2' = 110\ \Omega$$

(D) Band-Reject Filter (Band-Stop Filter or Band-Elimination Filter)

In this filter the rejection of frequencies in particular band takes place while frequency signals outside this range or band gets reproduced at the output. Band-reject filters are classified into two types on the basis of Q-factor.

(a) Wide-band reject filter ($Q < 10$).

(b) Narrow-band reject filter ($Q > 10$).

(a) Wide-band reject filter

This filter can be designed from low pass and high pass filter in parallel combination. The output of this

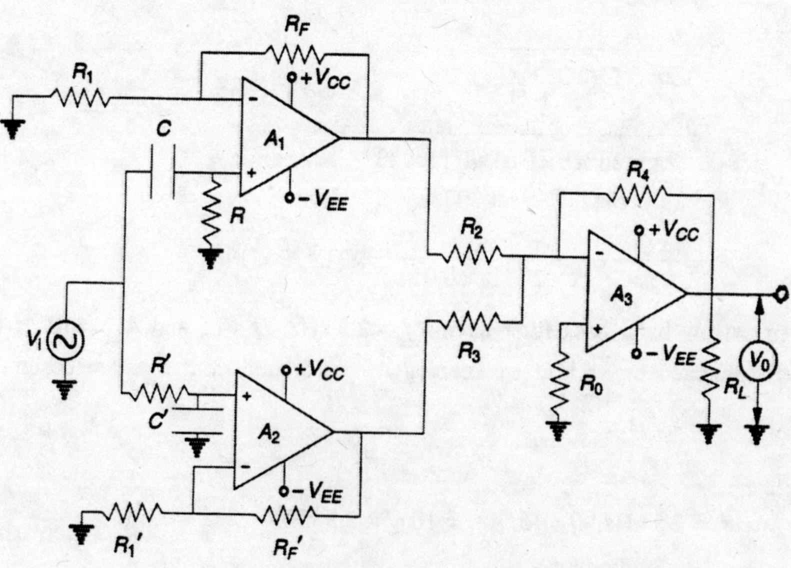

Fig. 6.40(a) : Circuit diagram of wide-band reject filter.

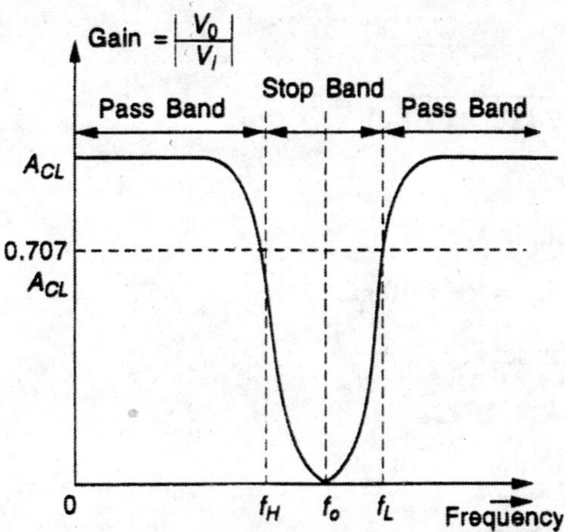

Fig. 6.40(b) Frequency response.

filter is applied to summing amplifier. These two filters attenuate the frequencies in stop band having range between f_H and f_L.

Fig. 6.40(a) shows circuit diagram of wide-band reject filter. Fig. 6.40(b) shows frequency response of this filter. Let f be the input frequency so $f < f_H$ transmitted through low pass filter and $f > f_H$ is passed by high pass filter. This filter should follow the following conditions :

(1) The pass-band gain of low pass and high pass filters (A_1 and A_2) must be equal.

(2) Low cut off frequency f_L of high pass filter must be greater than f_H *i.e.* high cut off frequency of low pass filter.

Op-amp A_3 is summing amplifier and whose gain can be 1. So $R_2 = R_3 = R_4$ and $R_0 = R_2 \| R_3 \| R_4$.

Example 6.9 : Design wide-band elimination filter having $f_H = 400$ Hz and $f_L = 2$ kHz.

Solution :

Step 1. Given data :

$$f_H = 400 \text{ Hz}, f_L = 2 \text{ kHz}.$$

Step 2. The designing of components which are used in wide band-elimination filter are as follows :

For high pass filter calculate R,

$$R = \frac{1}{2\pi f_H C}$$

Assume $C = 0{\cdot}01$ μF

$$R = \frac{1}{2\pi \times 400 \times 0{\cdot}01 \ \mu\text{F}}$$

$$\therefore \ R = 39 \text{ k}\Omega$$

For low pass filter calculate R'

$$R' = \frac{1}{2\pi f_L C}$$

Assume $C = 0{\cdot}01\ \mu F$

$$R' = \frac{1}{2\pi \times 2\ \text{kHz} \times 0{\cdot}01\ \mu\text{F}}$$

$$\therefore \qquad R' = 7{\cdot}95\ \text{k}\Omega$$

Step 3. For LPF and HPF the gain is 2

$$\therefore \qquad R_1 = R_F = R_1' = R_F' = 10\ \text{k}\Omega$$

For summing amplifier the gain is set at 1

$$\therefore \qquad R_2 = R_3 = R_4 = 10\ \text{k}\Omega$$

$$\therefore \qquad R_0 = R_2 \parallel R_3 \parallel R_4 = 3{\cdot}3\ \text{k}\Omega$$

(b) Narrow band-reject filter (Notch filter)

This filter uses twin-*T* networks *i.e.* one *T*-network consists of two resistors and a capacitor while other *T*-network uses two capacitors with one resistor. Fig. 6.41(a) shows the twin-*T* network which is a passive filter.

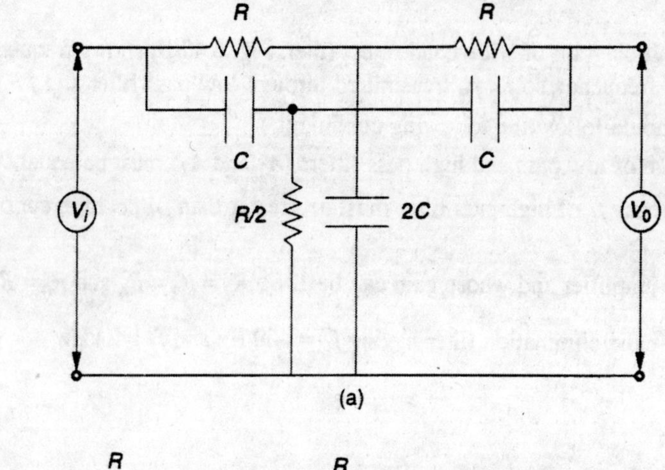

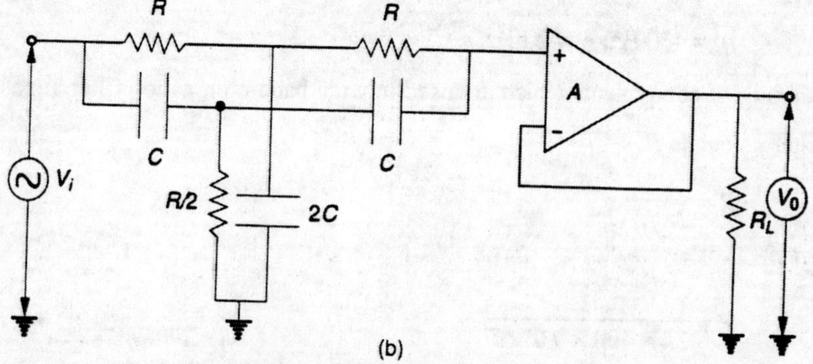

Fig. 6.41. (a) Twin-T network, (b) Circuit diagram of narrow band reject filter.

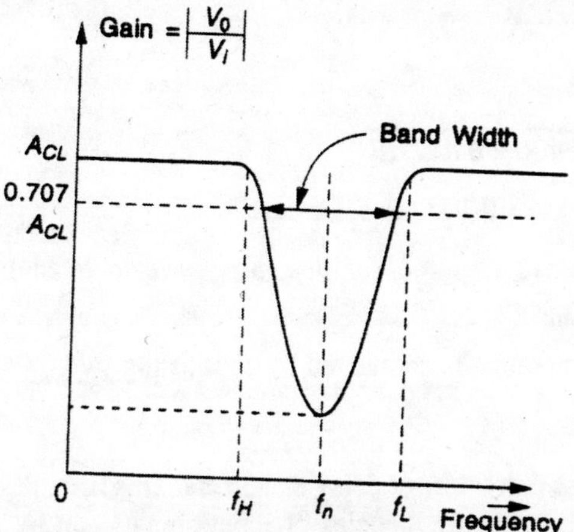

Fig. 6.41 : (c) Frequency response.

The frequency at which maximum attenuation occurs is known as notch-out frequency which is given by :

$$f_n = \frac{1}{2\pi RC} \qquad \text{...(6.92)}$$

This passive *twin-T* network has disadvantages of relatively low Q factor which can be increased by using voltage follower stage of OP-AMP as shown in Fig. 6.41(b).

Fig. 6.41(c) shows frequency response of narrow band reject filter.

Precautions For Notch Filter :

(1) This filter is suffered by noise drift produced due to temperature. To avoid this drift 1% silver mica or polycarbonate capacitors and 1% to 0·1% metal film resistors are preferable for designing of this filter.

(2) For proper tuning of *twin-T* precise components should used to design the filter.

Mostly for the designing of active notch filter at specific notch frequency (f_n) $C \leq 1$ μF is selected and required value of R is calculated by the equation :

$$f_n = \frac{1}{2\pi RC}; R = \frac{1}{2\pi f_n C} \qquad \text{...(6.93)}$$

Notch filters are widely used in biomedical instrumentation, communication networks etc.

Example 6.10 : Design active notch filter for 400 Hz.

Solution :

Step 1. Given data $f = 400$ Hz.

Step 2. Calculating R

$$f_n = \frac{1}{2\pi RC}$$

choose; $C = 0{\cdot}047\ \mu F$

$$R = \frac{1}{2\pi f_n C}$$

$$R = \frac{1}{2\pi \times 400 \times 0{\cdot}047\ \mu F}$$

$$R = 8{\cdot}46\ k\Omega = 9{\cdot}1\ k\Omega \text{ is selected.}$$

Step 3. $R/2$ resistance component required for designing is calculated by connecting two 9·1 kΩ resistances in parallel fashion.

Similarly, the 2 C capacitor component is calculated by connecting two 0·047 μF in parallel fashion.

(E) All Pass Filter

All pass filters are also called as phase correctors or delay equalizers. This filter passes all frequencies without attenuation but the phase shift of different frequencies of input signals can be adjusted over wide range. However the amplitude of signal remains constant.

In some applications the signals are transmitted over long distances through transmission lines during this operation some times phase changes, this phase change can be compensated by all pass filter *e.g.* telephone lines.

Fig. 6.42(a) shows all pass filter circuit.

Circuit Analysis

Calculate voltage at node '*a*' *i.e.* V_a by potential divider method.

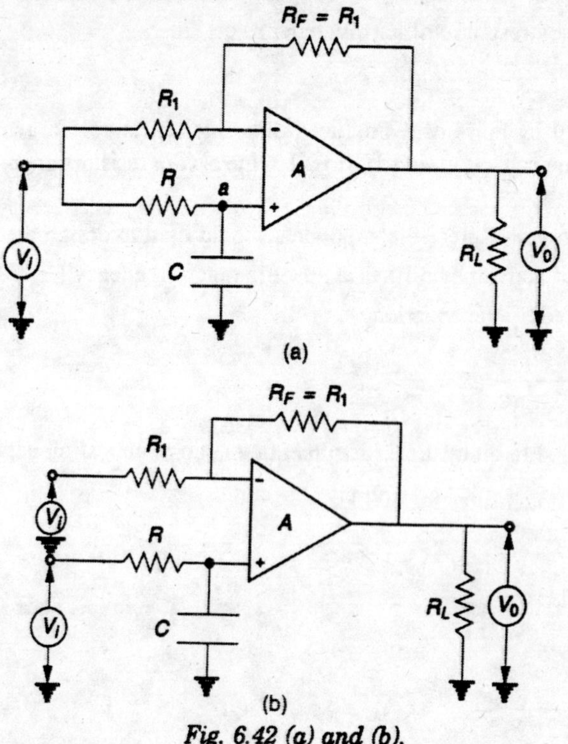

Fig. 6.42 (a) and (b).

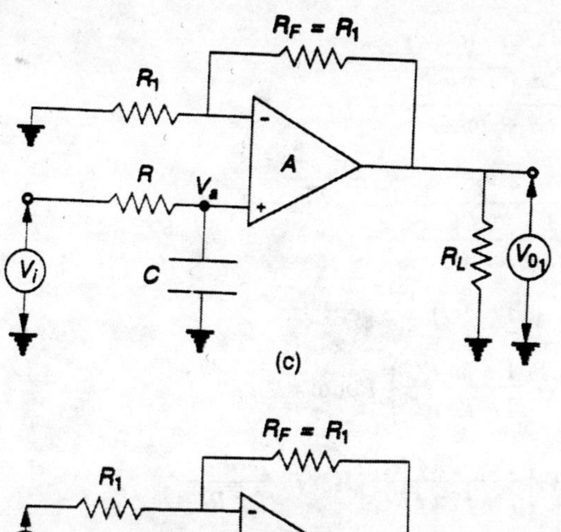

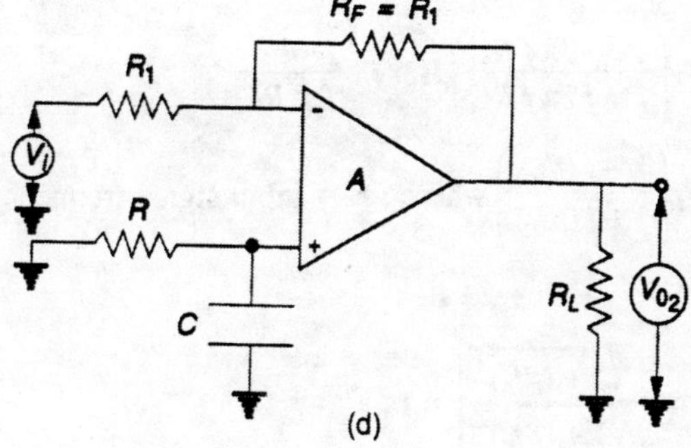

Fig. 6.42 (c) and (d) : All pass filter.

$$V_a = \frac{1/j\omega C}{R + 1/j\omega C} V_i$$

$$V_a = \frac{V_i}{j\omega CR + 1} \qquad ...(6.94)$$

The circuit can be realized as the lower and upper sources are short circuited. This circuit is shown in Fig. 6.42 (b)

Case I : Consider the op-amp is in non-inverting configuration. Assume the output is V_{01} in this mode as shown in Fig. 6.42 (c)

$$V_{01} = (1 + R_F/R_1)\ V_a = 2\ V_a \qquad ...(6.95)$$

$$V_{01} = \frac{2\ V_a}{1 + j\omega\ RC}$$ as $R_F = R_1$, by putting equation (6.94) into (6.95).

Case II : Consider op-amp is in inverting configuration. Assume V_{02} is the output. As shown in Fig. 6.42(d)

$$V_{02} = -\ R_F/R_1\ V_i = -\ V_i \text{ (as } R_F = R_1) \qquad ...(6.96)$$

∴ The total output is $V_0 = V_{01} + V_{02}$ (according to superposition theorem).

$$V_0 = \frac{2\ V_a}{1 + j\omega\ RC} - V_i \qquad ...(6.97)$$

Substitute the value of V_a in equation (6.97)

$$\therefore \qquad V_0 = \frac{2 \cdot \dfrac{V_i}{1 + Rj\omega C}}{1 + j\omega RC} - V_i \qquad \text{...(6.98)}$$

$$\therefore \qquad V_0 = \frac{\dfrac{2\,V_i}{1 + j\omega\,RC}}{1 + j\omega\,RC} - V_i \qquad \text{...(6.99)}$$

By solving equations (6.98) and (6.99), we get;

$$\therefore \qquad V_0 = V_i \left(\frac{1 - j\omega\,RC}{1 + j\omega\,RC} \right) \quad \text{Put } \omega = 2\pi f$$

$$\therefore \qquad V_0 = V_i \left(\frac{1 - j\,2\pi f RC}{1 + j\,2\pi f RC} \right) \quad \text{But } f = \frac{1}{2\pi\,RC} \qquad \text{...(6.100)}$$

$$\therefore \qquad V_0 = V_i \left(\frac{1 - j\,(f/f_c)}{1 + j\,(f/f_c)} \right) \quad \text{where ; } f \rightarrow \text{Input signal frequency.}$$

Now,

$$\text{The amplitude or Gain } \left| \frac{V_0}{V_i} \right| = \left| \frac{\sqrt{1 + (f/f_c)^2}}{\sqrt{1 + (f/f_c)^2}} \right| = 1 \qquad \text{...(6.101)}$$

$$\therefore \qquad \text{Phase angle } \phi = \tan^{-1} [-\,(f/f_c)] - \tan^{-1} [(f/f_c)]$$

$$\therefore \qquad \phi = -\,2\tan^{-1}\,(f/f_c)$$

$$\therefore \qquad \phi = -\,2\tan^{-1}\,\omega RC. \qquad \text{...(6.102)}$$

From this equation phase angle can be determined if the values of f, R, C are known.

The negative sign of phase angle indicates that the output voltage is lagging V_i by phase angle ϕ but if the connections of R and C are interchanged the V_0 leads V_i by ϕ.

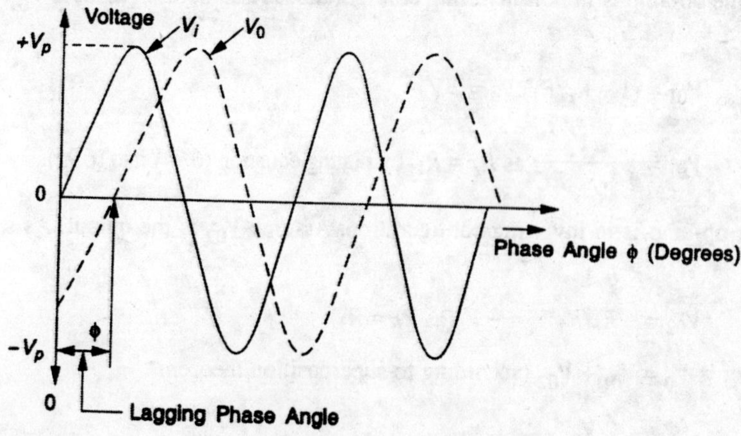

Fig. 6.43. Input and output waveforms of all pass filter.

Fig. 6.43 shows input and output waveforms of all pass filter. The output voltage (V_0) is lagging input voltage (V_i) by phase angle (ϕ).

Example 6.11 : For the all pass filter determine phase angle (ϕ) between input and output at $f = 2$ kHz.

Solution :

Step 1. Given data $f = 2$ kHz

Step 2. Calculate phase angle (ϕ)

$$\therefore \quad \phi = -2 \tan^{-1} (\omega RC)$$

Assume $R = 10$ kΩ; $C = 0{\cdot}01$ μF

$$\therefore \quad \phi = -2 \tan^{-1} (2\pi \times 2000 \times 10 \times 1000 \times 0{\cdot}01 \times 10^{-6})$$

$$\therefore \quad \phi = -102{\cdot}97^\circ$$

Example 6.12 : For a all pass filter find the phase angle ϕ if the frequency of input voltage is 4·5 kHz. Suggest what modifications are necessary to obtain positive phase shift (ϕ).

Solution :

Step 1. Given data $f = 4{\cdot}5$ kHz

Step 2. Calculate the phase angle (ϕ).

$$\therefore \quad \phi = -2 \tan^{-1} (\omega RC)$$

Assume $R = 15$ kΩ; $C = 0{\cdot}01$ μF

$$\therefore \quad \phi = -2 \tan^{-1} (2\pi \times 4{\cdot}5 \times 1000 \times 15 \times 1000 \times 0{\cdot}01 \times 10^{-6})$$

$$\therefore \quad \phi = -153{\cdot}46^\circ$$

Step 3. For obtaining positive phase shift (ϕ) interchange the connections of R and C in the circuit of all pass filter.

6.21 Noise Problem, Grounding and Shielding

The term noise is defined as any unwanted signal with desired or information signal. Noise is random in nature and it is difficult to predict and analyze. Output of amplifier contains the signals which are not related with input signals. These unwanted or not related signals in the output are called noise signals. Noise signal results in superimposition of signals, phase change, variations in waveform shape, and amplitude variations. The noise signals introduced in op-amp or electronic circuits or electrical networks are of two types :

(A) External noise

In electronics the external or interference noise gets induced from external noise sources such as switching of machinery, powerlines, motors, transformers, electromagnetic interferences, radio broadcasting stations, ignition systems, electrostatic and inductive pick up's, voltage spikes due to switching of circuits, stray signals between adjacent conductors, electrical arcing etc..

(B) Internal Noise

The internal noise or self induced noise may be caused by the circuit itself. Examples of internal noise signals are stray electric and magnetic circuits, drift, temperature variations in circuit components, switching of circuits.

Other noise signals may be produced or present in electronic ciircuits like transistors, OP-AMPs etc. are thermal noise, Schottkey noise, $1/f$ noise.

Thermal or Johnson noise gets produced because of thermal agitation of current carriers in conductors or semiconductors. It is due to random motions of molecules, atoms and electrons. If resistor component is considered as circuit element the noise voltage developed across it is :

$$V_{rms} = \sqrt{(4\ kTRf)} \qquad \text{...(6.103)}$$

where ; k = Boltzman's constant ($1{\cdot}38 \times 10^{-23}$)

V_{rms} = rms noise voltage

T = Temperature (°K)

R = Resistance of noise source

f = Frequency (Hz)

In active devices the noise is generated by thermal agitation.

The $1/f$ noise is also called as a Pink noise. It is defined as a noise source having noise density which varies inversely with the square root of frequency. It is given by :

$$n = k\sqrt{1/f} \qquad \text{...(6.104)}$$

where ; n = Pink noise

k = Value of n at f = 1 Hz

6.22 Sources of Noise Signals

As discussed in previous section there are various types of noise like internal noise, external noise, white noise etc. The noise signals gets impressed or mixed with the output signals of transducers. There are various sources of noise out of which most common noise sources are mentioned.

(1) Inductive pickup from power sources

This covers the noise from fluorescent lighting, electrical arcing, transmitting devices, opening and closing of inductive circuits or devices.

(2) Ground Loops

If in a circuit more than one ground is provided it forms the ground loops due to which noise signals get induced in the loop and directly gets coupled with signal transmission line.

(3) Improper temperature compensation

If proper and effective temperature compensation is not provided to the transducers such as strain gauges, resistance-temperature sensors the system sensitivity gets changed resulting to introduce noise or drift in the system. It also affects the performance of amplifiers, measuring instruments and circuit components.

(4) Transients in AC Line

Sometimes transients and fluctuations in the line voltage may be present which also induces noise signals in electronic circuitry. *e.g.* due to lightening effect fluctuations may take place.

(5) Improper insulation

Sometimes due to improper insulation of cables and wires, leakage paths, moisture, etc. interference signal may be introduced in the output signal of transducer or amplifier.

(6) Loading effect

If the instrument or amplifier with low input impedance is connected in parallel at the output of transducer it creates the loading effect by reducing the sensitivity of system.

(7) Drift due to temperature

Sometimes due to increase in temperature (*e.g.* transistors, power transistors) of components or circuit according to thermoelectric effect, the unwanted voltage (DC) in microvolt gets induced which is "drift".

(8) Due to improper soldering

If improper *i.e.* loose and open contact soldering is used, or dry solder connections take place, it also induces noise signals.

(9) Improper common mode rejection may introduce noise signals

Generally, in common use the noise signals are present in different process industries like paper, sugar, oil refineries, power plants etc. where high power consumption and high voltage complex networks are used.

It is very necessary to minimize noise interference so as to obtain error free, faithful and reliable functioning of op-amp circuits. For eliminating or minimizing noise suitable shielding, isolation and grounding techniques are used. It helps to protect the op-amp from noise interference. In next sections these methods are discussed.

6.23 Protection Circuits for OP-AMP

As discussed in previous sections that operational amplifiers are used in different signal conditioning devices for improving, processing and measuring the output of transducers. Supply voltages (+ V_{CC} and − V_{EE}) are applied for biasing the op-amp. Important aspect is that these voltages should not exceed during operation than the specified ratings. Sometimes due to transients, variations, distortions or switching (turn on and turn off) of power supply voltage level may be increased which is greater than supply voltage. Due to this the op-amp may be destroyed or damaged.

Op-amp may be damaged if input signals are applied but supply voltages are not applied. So to protect op-amp from damage, protection circuits such as overvoltage protection, overcurrent protection circuits can be used.

Protection circuit for OP-AMP using semiconductor diodes and zener diodes : Fig. 6.44 shows op-amp protection circuit using resistances and diodes. Here resistance *R* is connected across input terminals which limits the transients produced due to high common mode voltage signals. The supply voltages are applied through high voltage diodes *D* to protect op-amp failure or damage from transients and power supply reversals.

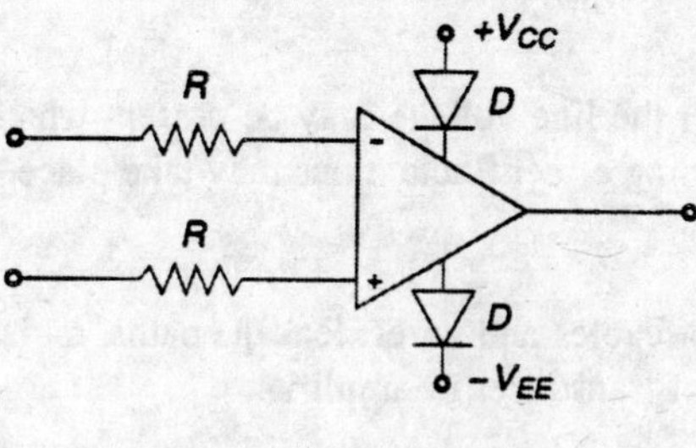

Fig. 6.44.

Zener diodes D_z may be connected in power supply line so as to protect op-amp from high voltages which is shown in Fig. 6.45.

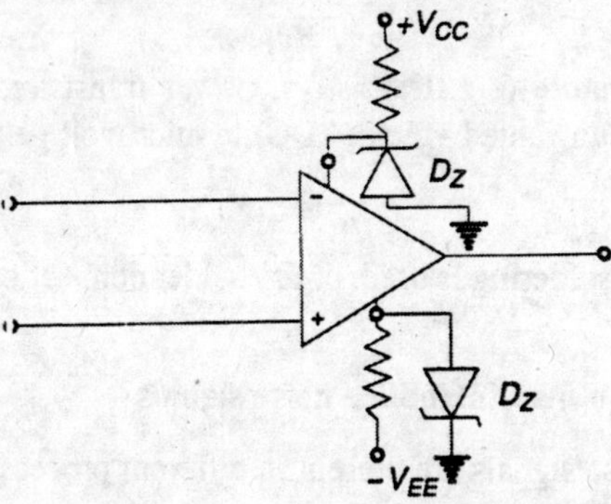

Fig. 6.45.

6.24 Ground Loops or Earth Loops

In measurement systems and electronic circuits voltages are measured with respect to zero or ground. If separate or individual ground or earth is provided (*i.e.* more than one ground point) then ground loop gets formed or developed. A ground or earth loop is defined as it is a closed loop (path) connected between two or more ground points. Another cause for ground loops is that in many circuits different ground points are connected to metal chassis as a reference. But due to (a) improper ground connections to chassis, (b) poor (dry and loose) soldering points, (c) a resistance between ground and circuit point the earth loop gets formed.

Fig. 6.46 shows how the ground or earth loop gets formed in op-amp circuit. In this circuit the earth loop is formed due to separate ground provided to voltage source and amplifier. Disadvantage of this type of loops is that the stray electric and magnetic fields induces unwanted voltages and flow of current in the circuit or system. This affects the circuit components as well as measurement system by introducing error signals.

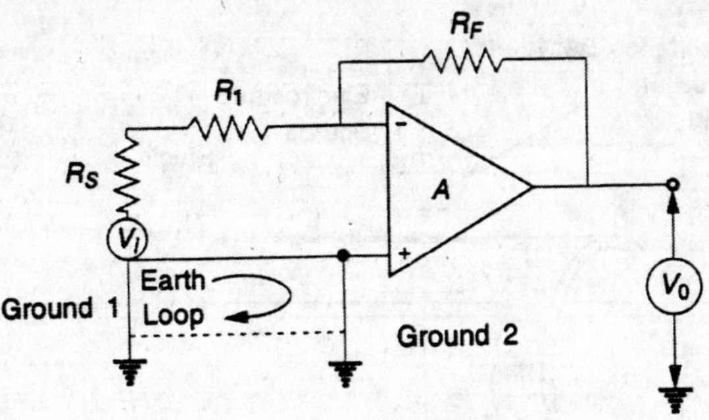

Fig. 6.46.

To avoid this problem the remedy is to provide single ground point or only one ground point to the circuit, generally at the input of circuit. Fig. 6.47 shows circuit arrangement to avoid ground loops. Here the load resistance R_L, input connections of op-amp are connected to one common point.

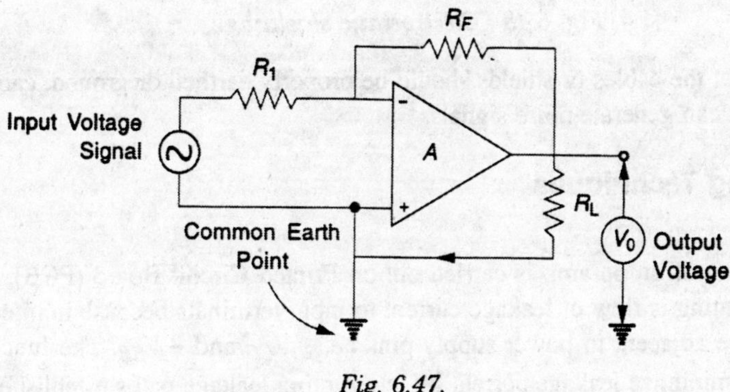

Fig. 6.47.

6.25 Shielding

Function of shield is to capture the electrostatic and inductive interference signals or charges. It also isolates the signal conductors from these interference signals. The signal conductor carries the signals from transducers to the input of op-amp. If these interference signals reaches to signal conductors it induces or produces noise signals in input signal. The captured charges or signals should be removed or grounded otherwise it may be coupled to signal conductors.

Fig. 6.48 shows electrostatic shielding. It consists of a shield which covers the signal conductors. It isolates signal conductors from the noise interference signals or charges. These charges gets drained off through connections. Fig. 6.48 shows how the electrostatic noise gets introduced into signal conductors and how it gets cancelled using earthed shielding. This system is applicable in minimizing inductive noise upto some extent.

For minimizing electromagnetic interference or magnetically induced noise, the shields of ferromagnetic materials are used *i.e.* magnetic shielding. The magnetic shielding provides high permeability path to reject or eliminate the interference magnetic flux. Thickness of magnetic field depends upon frequency and field strength.

Sometimes twisted conductors, wrapped foil shields and grounding drain wires may be used to reduce the electrostatic and inductive noise signals.

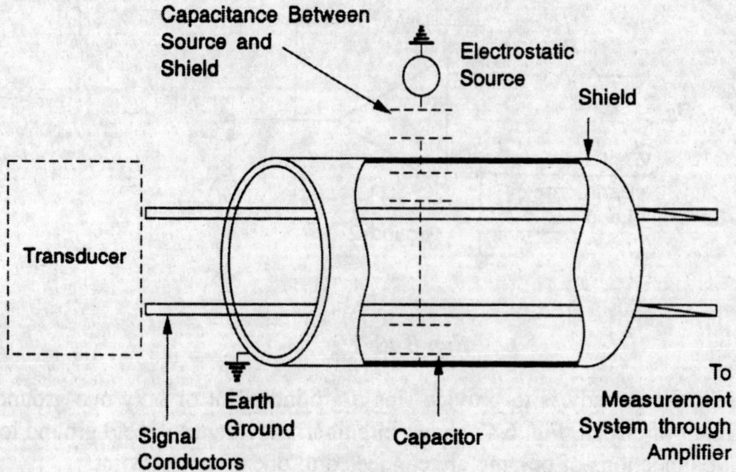

Fig. 6.48 : Electrostatic shielding.

Another aspect is that the cables or shields should be properly earthed or ground, cable should be free from flexing otherwise it can generate noise signals.

6.26 Input Guarding Techniques

The assembly or mounting of an op-amp is carried out on Printed Circuit Board (PCB). But the problem associated with this mounting is flow of leakage current to input terminals because in integrated circuits of op-amp the input pins are adjacent to power supply pins *i.e.* + V_{CC} and − V_{EE}. The function of **guard** or guarding technique is to minimize leakage current by interrupting leakage paths established.

If leakage current flows through the op-amp circuit configuration, it affects the input bias current of op-amp, hence guarding technique is very important aspect for op-amp. In guarding technique a guard consisting of a conducting ring is placed around the input pins of op-amp so as to prevent the input pins of op-amp from leakage current. It also isolates input pins from supply pins. Fig. 6.49(a) shows PCB layout board for *TO*–5 package with guard arrangements.

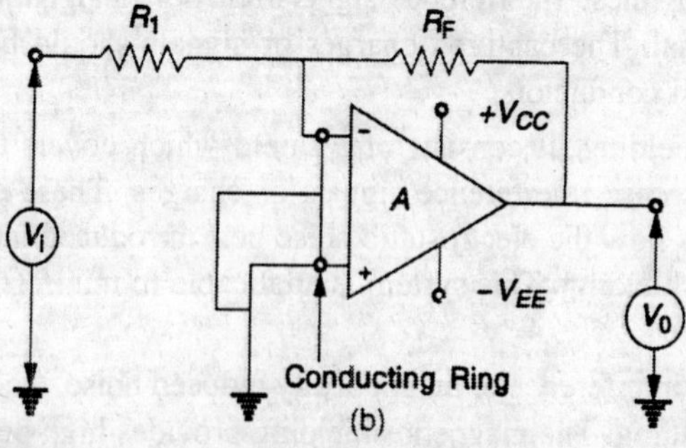

Fig. 6.49 : (a) and (b).

Another method of guarding is shown as below.

Fig. 6.49(b) and (c) shows op-amp configuration in inverting and integrator mode respectively. Here guard connection is provided to common feedback path. As the input terminals and guard are at equipotential

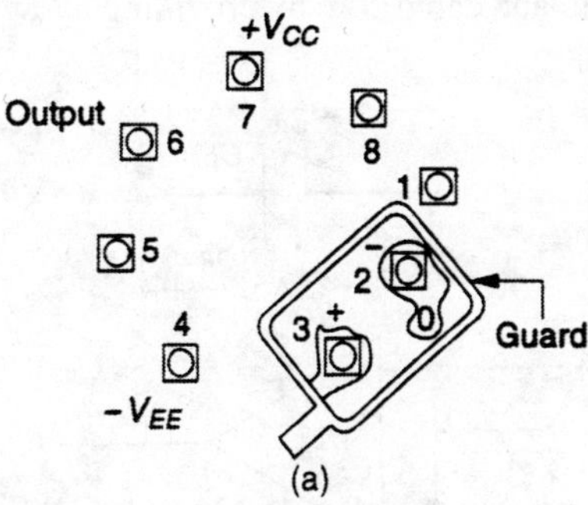

Fig. 6.49(c).

the guard ring will absorb leakage current flowing into input terminals of op-amp through the PCB board. Another aspect is that this ring acts as shielding by rejecting the noise signals and common mode signals. Sometimes the guard ring may be used on both sides of PCB for better results.

6.27 Power Isolation

After shielding another aspect is to reduce parasitic voltages. These parasitic voltages generally derives from utility power-line supply. Power isolation transformer is used to achieve protection from parasitic voltages. In following discussion power isolation is discussed.

Consider, the Fig. 6.50 in which the instrument is connected to the utility power line using 3-core cable. In 3-core cable one is neutral lead and two leads carry power current. The neutral lead is connected to ground at power distribution point such as step-down transformer. The safe guard is also grounded at the same point to which all metal housing are connected using third lead of utility power-line cable. This arrangement provides low-ohmic path to ground for currents caused due to insulation breakdown, accidental shorts etc.

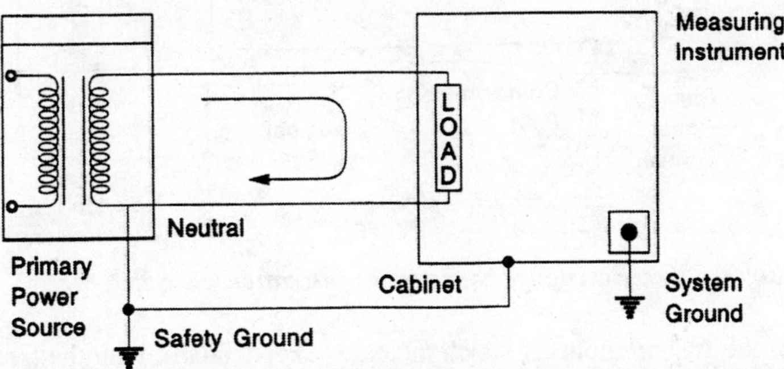

Fig. 6.50.

But in this arrangement a common-mode voltage can be formed between power ground and instrument ground when the two ground systems are connected. This formation of common-mode voltage developes large hum in the instrument. So for eliminating such common mode voltage (E_{cm}) power isolation transformer is used *i.e.* it provides shielding which is shown below.

Here, the shielding is provided around primary and secondary windings which are connected to one side of winding so that stray capacitances between windings and shields are eliminated at one end. A large capacitance is achieved between shield and cable core by providing a low impedance path to common mode generator E_{cm} as shown in Fig. 6.51.

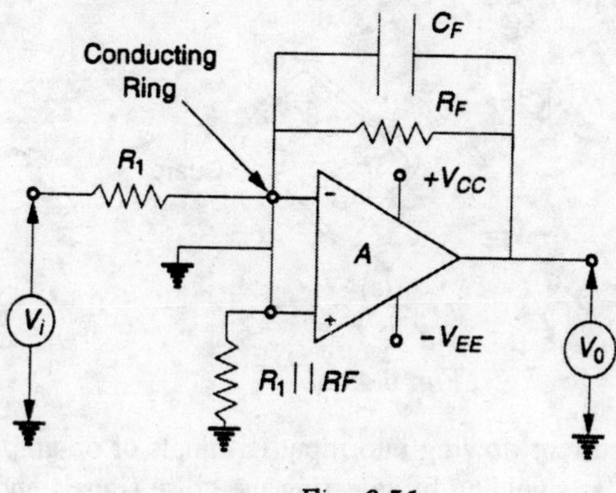

Fig. 6.51.

6.28 Filtering

This is also an important technique used for removing AC noise components. Due to circuit imbalances, normal mode noise pickups or fluctuations, conversion of common mode noise to normal-mode noise, etc. the noise signals get induced. For eliminating such noise signals filtering is used. Fig. 6.52(a) and (b) show filteration methods.

It consists of a resistance-capacitance filter which is connected between the signal source and amplifier unit as shown in Fig. 6.52(a).

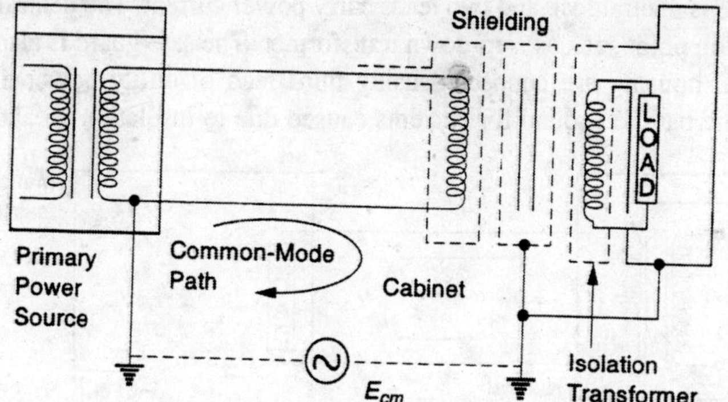

Fig. 6.52(a). Resistance-capacitance filteration method.

This circuit eliminates the AC components which increases exponentially with the frequency of noise signals. Here, the filtering action gets achieved by setting the time delay $T = RC$ between the change in signal at transducer and time of recognition of this change by measuring device. The time delay may be adjusted to one second or more depending upon higher frequency of the system.

For increasing the filtering efficiency, better filtered output quality and achieving better time delay control LC inductance-capacitance filter may be used but which is costly. This circuit is shown in Fig. 6.52(b).

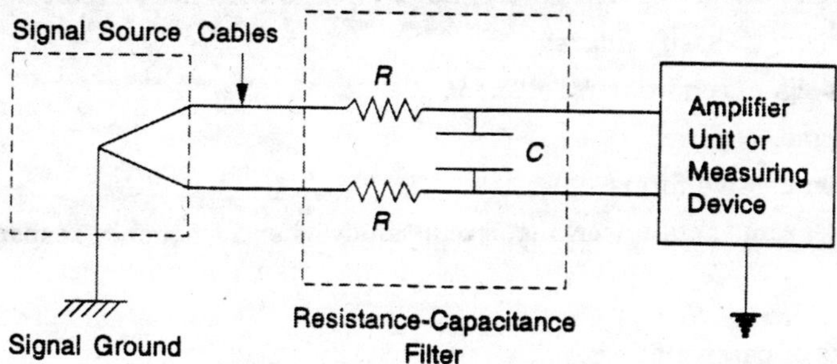

Fig. 6.52(b) Inductance-capacitance Filteration Method.

EVALUATION COMPONENT

1. Why signal conditioning elements are necessary in instrumentation systems ?
2. What are the applications of signal conditioning elements ?
3. Draw and explain the generalized block diagram of electronic measurement system.
4. With neat block diagram explain the working of DC signal conditioning.
5. How AC signal conditioning can be implemented ? Explain it in detailed manner.
6. Define the term OP-AMP. Explain the neat block diagram of it.
7. List the characteristics of ideal OP-AMP.
8. Why OP-AMP is used for designing or fabrication of signal conditioning elements.
9. Write a short note on :
   (i) Inverting amplifier
   (ii) Non-inverting amplifier
   (iii) Voltage follower
10. Explain with neat circuit diagram the working of summing amplifier. Also derive its gain equation.
11. Why logarithmic amplifier is used ? Explain its working.
12. Why transducer bridges are necessary ? Explain the function of transducer bridge with neat circuit diagram.
13. Why chopper stabilized amplifiers are used ? Explain its working.
14. What is the significance of instrumentation amplifier ? Mention the features of instrumentation amplifier.
15. Explain with neat circuit diagram the working of instrumentation amplifier. Also derive its equation for gain.
16. Define the following terms related to sample and hold amplifier :
   (i) Sample period
   (ii) Hold period
17. Explain with neat circuit diagram and waveforms the working of sample and hold amplifier.
18. Why the current to voltage and voltage to current converters are necessary in instrumentation system ?
19. Explain the working of :
   (i) Transconductance amplifier
   (ii) Transresistance amplifier

20. Explain the working of integrator and differentiator circuits with neat circuit diagrams.
21. Define the term "Filter". Classify filters.
22. How filter acts as a signal conditioning element ?
23. Explain the significance of filter.
24. Write a short note on passive filters.
25. With neat circuit diagram explain working, circuit analysis and frequency response of :
    (i) Low pass filter
    (ii) High pass filter
26. State the drawbacks of passive filters.
27. What are the advantage of active filters.
28. Define the term "Noise". List the types of noise ?
29. Elaborate the sources of noise signals.
30. What are the causes due to which noise signals get induced in the system ?
31. How the ground or earth loops gets formed ? Describe the method by which these loops get eliminated.
32. What is shielding ? Explain electrostatic shielding.
33. Why input guarding technique is necessary ? Explain input guarding technique in detailed manner.
34. Explain power isolation technique.
35. Write a short note on filtering technique.

Chapter 7

DATA ACQUISITION SYSTEMS

Learning Objectives

- To study significance of data acquisition system (DAS).
- To define features of the DAS.
- To study generalized DAS.
- To study signal conditioning elements used in DAS.
- To study operation of single channel and multichannel DAS.
- To define intelligent instrument.
- To define advantages of microprocessor based instrument.
- To study microprocessor based instrument system.
- To examine the operation of microcomputer based DAS.
- To define the term data logger.
- To study characteristics of data logger.
- To study block schematic of data logger.
- To develop microprocessor based long term temperature recorder.
- To describe need and operation of real time clock (RTC).

7.1 Introduction

In instrumentation systems, data acquisition system (DAS) plays an important role. It is widely used in industries, plants, biomedical measurements, process control etc. The function of data acquisition system is to collect the input data (in analog or digital form) efficiently, accurately and simultaneously to store or display the data.

Data is nothing but the signals obtained from the (1) output of transducers (e.g. thermocouple, thermistor, strain gauge etc.) or (2) Direct measurements of electrical parameters such as a.c. or d.c. voltage and current, frequency, resistance etc.

Data acquisition system consists of individual sensor or transducer associated with signal conditioning element, multiplexer (scanner) circuit, data conversion (signal conversion), data transmission, in the final stage storage element and display unit. Mostly analog data received from different ends is converted into digital form for accurate and fast processing, it also provides the facility of storage (temporary or permanent), display and transmission of signals without complex circuitry. Due to conversion of signals (analog to digital) microprocessors and microcomputers can easily be interfaced for measurement of complex parameters and controlling the processes.

Processing of signals is nothing but the operations like amplification, mathematical manipulations, information collection such as averages, statistics for calculating different parameters.

This chapter covers single channel and multichannel data acquisition systems. In single channel DAS, only one variable or input is sensed, conditioned and converted into digital form for display and storing. Whereas multichannel DAS accepts two or more inputs or channels. In this system individual sensor or transducer is provided for various physical inputs. Output of each channel is conditioned separately. Through multiplexer, sample and hold circuit every signal is transmitted to (A/D) converter. Now a days microprocessor based DAS is used.

Data logger is a system which records the variable readings or processed data of different instruments located at different areas of plant. Data logger uses microprocessor or computer for fast, accurate and precise measurements.

7.2 Significance of Data Acquisition Systems

DAS is used for collecting the input information or data in digital form accurately, rapidly and economically for processing, transmission, display and storage of the data. Proper signal conditioning elements should be provided to the DAS for better resolution and accuracy.

Data acquisition systems are classified as (1) Analog DAS (2) Digital DAS.

Analog DAS handles the analog signals as an information signal or data for measurement, processing and controlling. This system is used when the bandwidth of signals is large and inaccuracy is acceptable.

Digital DAS uses additional stages such as analog to digital converters. This is used when signal bandwidth is low and high degree of accuracy, low cost circuits are required.

Now a days, digital DAS is used because of its advantages such as easy storage, high degree of accuracy, better efficiency, precision, fast operating speed, easy to interface with microprocessors and computers, better measurement and controlling facilities.

Data acquisition system operates in suitable environmental conditions and in hostile conditions.

The suitable environmental conditions means minimum radio frequency interference and electromagnetic induction. This system mostly finds in laboratories, research and development fields. In such systems highly sensitive, accurate and calibrated instruments or elements are used for collecting the data.

In hostile conditions such as power plants, industrial process control, aircraft control. The elements of data acuisition system requires wide temperature handling capacity, excellent isolation and shielding, less distortion, proper data processing capabilities, good noise immunity. Such system uses highly sensitive, accurate, rugged and precise instruments.

Any data acuisition system should have following features :

(1) System should acquire the data quickly, accurately and precisely.

(2) It should record or monitor the complete on-line operations in the plant.

(3) System should be capable to process the complete data also to store or record the data for analysis, controlling and measurement purpose. Simultaneously the system should inform the condition of plant and processes.

(4) It should be flexible and can be expanded for further modifications or expansions.

In the above discussion various functions, constructional parameters, data handling phenomenon of data acquisition system is covered. Generalized representation and working of data acquisition is explained in the section 7.3.

7.3 Generalized Data Acquisition System

Fig. 7.1 shows block diagram of generalized data acquisition system. Function of each stage is explained as follows. Such systems are used in power plants, paper mills, chemical industries etc. Figure 7.1 shows transducer stages from 1 to n with proper signal conditioners 1 to *n* respectively.

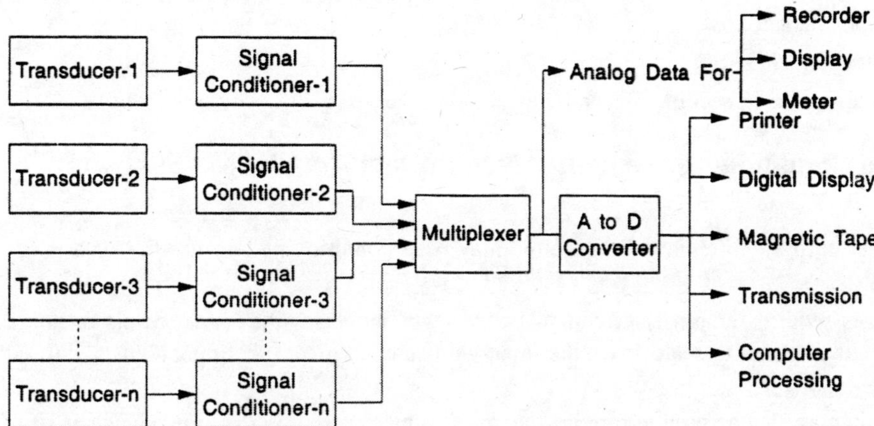

Fig. 7.1 : Generalized data acquisition system.

(1) Transducer/Sensor

Is the input stage of any data acquisition system. Physical quantities which are to be measured or controlled is applied to transducers. The function of transducer is to convert the physical quantities into electrical signals acceptable by the DAS.

(2) Signal Conditioning Elements (SCE)

Function of SCE is to modify or improve the output of transducers for recording or displaying and storing purposes. Most of the SCE are amplifiers, bridge circuit and instrumentation amplifiers, voltage to current and current to voltage converters, etc. as discussed in Chapter 6.

(3) Multiplexer

It is also called scanner. Multiplexer is used when several inputs or channels are to be handled. It receives multiple analog inputs (depending upon capacity) and connects sequentially the inputs to the measuring instrument or device located at the output stage. In instrumentation systems multiplexer circuit is used when distance between transmitter and receiver is large or several inputs are to be handled.

(4) Analog-to-Digital Converter (ADC)

This stage converts the analog signals into equivalent digital signals. Digital signals are easy to display or record using printer and can be easily interfaced with personal computer for further processing and controlling. This stage increases the speed, accuracy and efficiency of measurements, also minimizes the effects of interference signals, noise and hence enhances the performance and reliability.

(5) Output (recording) Stage

The output element or device is used to display or record the processed data proportional to the physical quantity applied to input stage *i.e.* transducers. This device may be analog or digital depending upon the derived output signal. For analog output signals the devices used are recorders, indicating instruments, oscilloscopes, analog meters etc.

Digital devices used are printer, visual display unit, floppy disc, magnetic tapes, punched cards, digital panel meters (DPM), display devices, digital storage oscilloscopes, microprocessors and personal computers etc.

Depending upon the number of channels handled, the data acquisition systems are classified as :

(1) Single channel DAS

(2) Multichannel DAS

These are discussed in sections 7.5 and 7.6

7.4 Signal Conditioning Elements Used in Data Acquisition Systems

Data (transducer output) which is to be acquired may need conditioning because this data is not from any sort of sources.

Some stages requires amplifier circuit to boost up or increases the signal levels on the other hand attenuator circuits are used to scale down the input gain so as to match the input signals with full range of converters.

Filter circuit may also be used to remove the noise signals present in the input signals if any. Filters are used at low level measurements, also provides high resolution for measurements.

Output signals or data from thermistors, R.T.D., flow meters, etc. Wheatstones bridge may be linearized by using linear approximation technique or smooth series linear approximation methods using low cost amplifiers.

The function of linearization is to produce the output linearly proportional to the input signal.

Digital signals can be linearized by storing the values in read only memory which is then used as a look up table or piece-wise linearization.

Rectification, logarithmic conversion, integration, differentiation and other types of processes are carried out for conditioning the signals in DAS if necessary. These devices are discussed in Chapter 6.

Two methods of signal conditioning are used in DAS :

(1) Ratiometric conversion

(2) Logarithmic compression

In ratiometric conversion consider the example of strain measurement using strain gauge with Wheatstones bridge. In this case the excitation voltage of bridge circuit is applied as a external reference voltage source to analog-to-digital converter. Here the conversion factor is proportional to reference voltage therefore the sensitivity of system depends upon excitation voltage.

7.5 Single Channel Data Acquisition System

Fig. 7.2 shows single channel DAS. It is used for handling only one input channel. As the input signal is conditioned or improved by a signal conditioning element it is fed to an analog-to-digital converter. This conversion is repetitive at free running internally determined rate. Digital data or information is fed to DPM or digital computer for analysis. External convert command is used to provide the conversion command externally from analog signals to digital signals and hold it.

The outputs are in digital code to indicate over-range, polarity and status of valid output digits.

Drawbacks of this systems are slow operation, for handling binary operated instruments the BCD coding has to be converted into binary form and as the conversion rate is free, interface register is necessary.

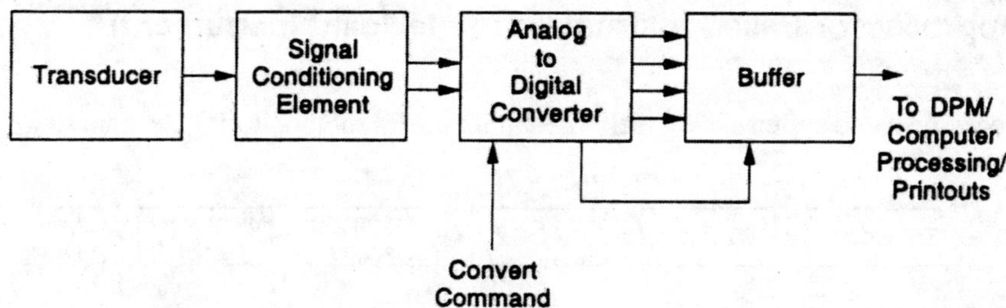

Fig. 7.2. Single channel data acquisition system.

7.6 Multichannel Data Acquisition System

It uses different sub-systems or channels which are time shared by using multiplexer arrangements.

Fig. 7.3 shows multichannel DAS. This system is applicable where several input channel are under measurement and controlling. It consists of individual transducer or sensor and signal conditioning elements for every channel respectively. Such conditioned signals are converted into digital signals by using analog-to-digital converter sequentially through multiplexer circuit. This system has 1 to *n* inputs with respective signal conditioning elements.

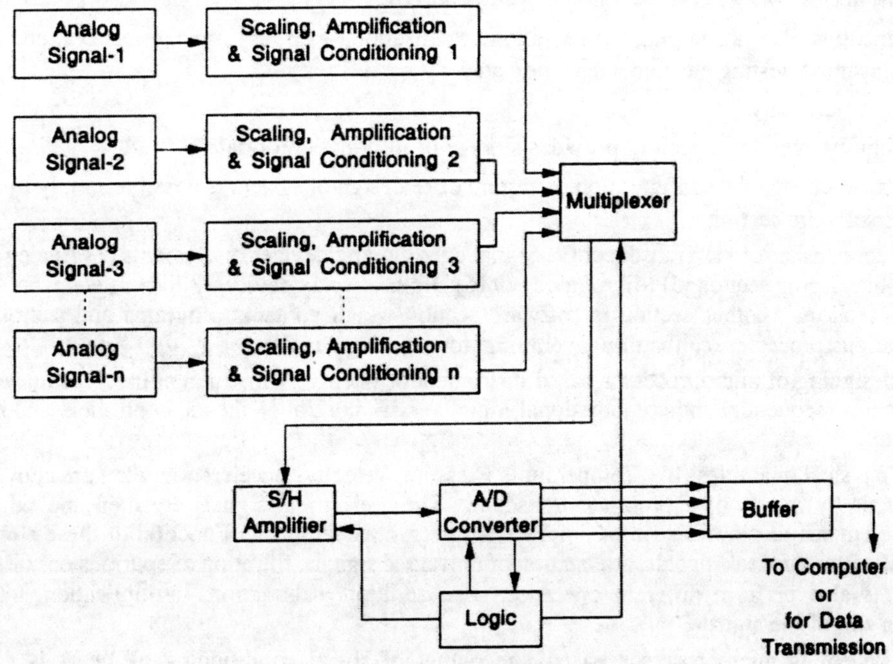

Fig. 7.3. Multichannel DAS.

The sample and hold circuit is provided to increase the speed with accurate conversion. This circuit is used to store or hold one signal while other is in A/D converter. This circuit increases the efficiency of system by utilizing time in such a manner that when multiplexer selects the data of one channel for conversion that particular data is stored or held in S/H circuit and converted into digital form. As the conversion gets completed S/H circuit returns to sample mode for storing or acquiring the data of next channel.

7.7 Microprocessor Based Instrument : (Intelligent Instrument)

The microprocessor-based systems are widely used in industries for monitoring the data as well as controlling the processes.

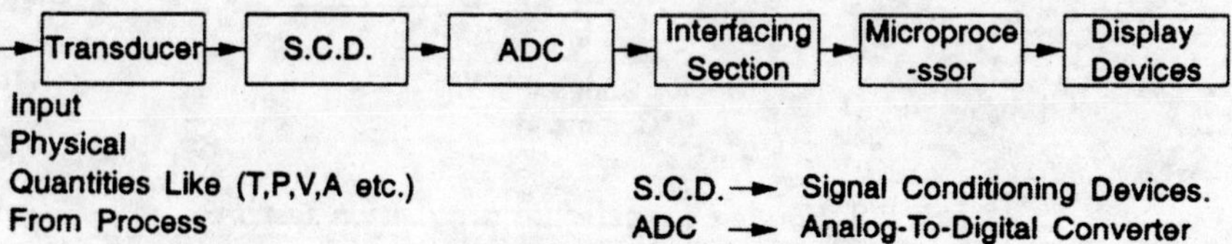

Fig. 7.4 :Block diagram of microprocessor based instrument.

Fig. 7.4 shows microprocessor based instrument system. As microprocessors are involved in measurement and instrumentation systems involvement of human being is completely eliminated or restricted. The instrument which operates on microprocessor systems or if the instrument is microprocessor based that instrument is called **Intelligent Instrument.**

Advantages of microprocessor based instruments :

(1) Speed of operation gets increased.

(2) Fast and automatic control action can be achieved.

(3) Functions like averaging, linearisation, totalization, gain adjustment, calibration, alarming, adjustment, testing etc. becomes automated.

(4) More flexibility.

(5) Memory interfacing facility provides storage of information or data.

(6) Easy to change the configuration of system or re-design of system is possible simply by changing the software section.

This system consists of (1) Transducer (2) Signal conditioning devices or elements (3) Analog to digital converter (4) Interfacing section (5) Microprocessor (6) Digital display devices. All these elements constitute the hardware section. Another section is software section which contains programs and subroutines for controlling various processes, calibration, averaging, totalization etc.

Overall designing of microprocessor based instrument or intelligent instrument involves linear circuits, digital electronics (sequential and combinational logic design), control elements, controllers and necessary software.

First the physical quantities (like Temperature, Pressure, Velocity, Acceleration, etc.) are converted into electrical signals by means of appropriate transducer. These electrical signals are then applied to signal conditioning elements or devices such as amplifiers, filters, attenuators etc. Function of these elements are such as amplification of weak signals, attenuation of unwanted signals, filtration of spurious or noise signals. These elements also perform different operations like addition, subtraction, multiplication, integration, differentiation etc. of the signals.

As this system is microprocessor based, the output of signal conditioning elements is applied to microprocessor so it is necessary to convert them into digital signals and hence analog to digital converter (ADC) stage is introduced.

Generally the ADC's used are ADC0800, ADC0808, ADC0809, ADC1210 and 1211 (ADC1210 and

1211 are 12-bit CMOS A to D converters). Accuracy of the system depends upon the parameters of analog to digital converters such as frequency, range, conversion speed, resolution, noise signal rejection, etc.

Various interfacing devices or chips (Integrated Circuits) such as Intel 8251 which is a Programmable Communication Interface (PCI). It is Universal Synchronous/Asynchronous Receiver/Transmitter (USART). This is compatible with microprocessors such as 8085 and 8086 and are used to transmit or receive the data. It's function is to accept the data in parallel format from microprocessor and converts this data into serial form. On the other hand it receives the serial data and convert it into parallel format.

Another interfacing chip is Intel 8253 which is a programmable counter/interval timer. It is used in timing and counting functions, to generate accurate time delay, generation of square wave signals of desired frequency etc.

Another interfacing elements are Intel 8259 which is programmable interrupt controller. Intel 8257 is a direct memory access (DMA) controller mainly used in data transfer.

Intel 8255 is programmable peripheral interface (PPI). These elements are used for interfacing the peripheral devices.

In block schematic next section is microprocessor. Desired microprocessors such as 8085, 8086, 80286, 80386 etc. can be used to perform different operations such as data averaging, linearisation, computation and calculation of data etc. Microprocessor is a device consisting of two sections. These sections are (1) Arithmetic and logic unit (ALU) (2) Timing and control unit. ALU performs arithmetic as well as logical operations. The timing and control generates timing and control signals which are necessary for the execution of instructions used in software programs.

Digital display devices or in-built display devices are used to display or indicate the measured quantity. These display devices may be LCD or LED which are interfaced with microprocessor.

7.8 Microcomputer Based DAS

Fig. 7.5 shows block diagram of microcomputer based data acquisition system.

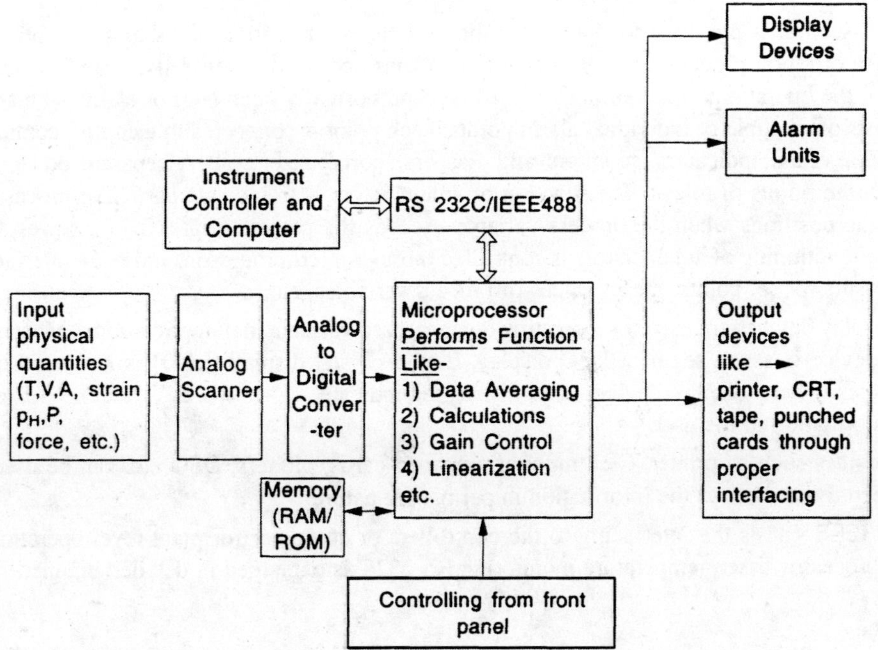

Figure 7.5 : Microcomputer based data acquisition system.

The basic difference between microprocessor and microcomputer is that : a microprocessor combined with memory and input/output devices forms a microcomputer *i.e.* microprocessor is the heart of microcomputer. The memories and input/output devices are interfaced with the microprocessor. In data acquisition system the designer has facility to select suitable memory configuration, input/output devices interfaced with microprocessor so as to perform particular task.

First the physical quantities (like temperature, velocity, pressure, acceleration, strain, pH, force etc.) gets converted into electrical signals by means of transducers. These electrical signals are applied to analog scanner. Function of analog scanner is to select the input signals in turn from the sequence (number of inputs) of inputs. If the signals are at low levels, they can be amplified up to + 5 Volt.

Also if the signals are not linearly proportional to measured parameter, these signals can be linearised by means of signal conditioning elements which is not shown in the block schematic. Modern scanners can scan the inputs at higher scanning rate *i.e.* up to 150 inputs/sec. The scanning rate should be matched with rate of change of input data and time required for the output devices to print or indicate the data.

After scanning of particular signal or the scanned signal is applied to analog to digital converters (ADC) where the analog input signal gets converted into digital signals. Output of ADC is interfaced with microprocessor by means of various interfacing chips like Intel 8255, Intel 8253, Intel 8251 etc.

Microprocessor performs different functions like data computations, calculations, data averaging functions, totalizing, linearisation, conversion, etc. Microprocessors such as Intel 8085, 8086, 80186, 80286, 80386 can be used for designing various systems depending upon the number of channels to be handled, data stream, operations to be performed. Necessary software programs, assembly language program's (ALP) can be developed so as to perform the above stated functions.

From the front panel, controlling of various operations and processing is possible. It facilitates the user to perform the operations like adjustment of amplifier gain, setting of linearisation factor, alarm setting, conversion operations of data converters like SOC (Start of conversion) and EOC (End of conversion). Input and output devices can also be controlled from front panel. All the controlling operations are controlled through this panel.

The alarm section is provided for indicating the various failures, indications of set points, hazardous situations in the different processes involved in the industries and plants. Generally, relay arrangements are used to control the operations (like normally closed NC and normally open NO) of alarm. The annunciator systems consists of multiple or individual alarm points (each point is connected to a control contact switch), a logic module, visual indicators, monitors and flashers, horn/buzzer/bell. A separate power supply is provided to all the points of relays. The function of annunciators is to indicate particular process variable, also indicate the positions when the process variable exceeds the present limits, failure stages, hazardous situations. The functioning of annunciators is controlled through microprocessors and relay interfacing using software programs for set points, preset values of process variables, etc.

Digital display devices are used for indicating the measured variable in numerical form. Most commonly used display device is seven segment LED display. Liquid crystal display (LCD) is also used to indicate process variables. These display devices also display or indicate +, –, "words", "letters" for conveying appropriate information to the user.

Output devices such as printer, CRT monitor, punched cards, plotters, DSO etc. can be used so as to display and record (hard copy) the information in permanent nature.

RS 232C/IEEE 488 is the interfacing to the controllers or computer for plant level operations. In the section microprocessor based temperature monitoring RS 232C is explained in detailed manner.

7.9 Data Logger

Now a days, in industries and process plants data loggers are becoming very popular so as to monitor, display, measure, store and control different process variables. Data is nothing but output from different transducers (analog or digital signals) and log means permanent storage of this data. The data logger handles digital information hence it requires fast, accurate analog to digital conversion techniques.

Heart of any data logger is microprocessor or computer based systems. In data loggers the output or recording devices may be printer, punched cards, plotters, teletype, magnetic tape etc.

Applications

Data loggers are widely used in power generation plants, petrochemical plants, oil refineries, industries (paper, sugar, plastic, rubber etc.), continuous process plants, engine testings, biomedical applications etc.

Characteristics of Data Logger

(1) More accuracy can be achieved.

(2) On-line monitoring of process variables is possible.

(3) Applicable or useful in hazardous or worst working conditions.

(4) System expansion is possible. This is the important characteristics of data logger *i.e.* number of input channel handling capacity can be increased without interrupting the working conditions. (More flexibility)

(5) Easy to use. That is the operator/user can operate all the functions in logical manner, simple operations, easy to handle and understand the system.

(6) Good reliability and linearity in the operations.

(7) Higher operating efficiency.

Fig. 7.6 shows the block diagram of data logger. Various sections of data logger are : (a) Input scanner, (b) Amplifiers, (c) Analog-to-digital converters, (d) Digital inputs to the microprocessor, (e) Programmer, (f) Microprocessor, (g) Memory, (h) Digital indicators, (i) Output devices.

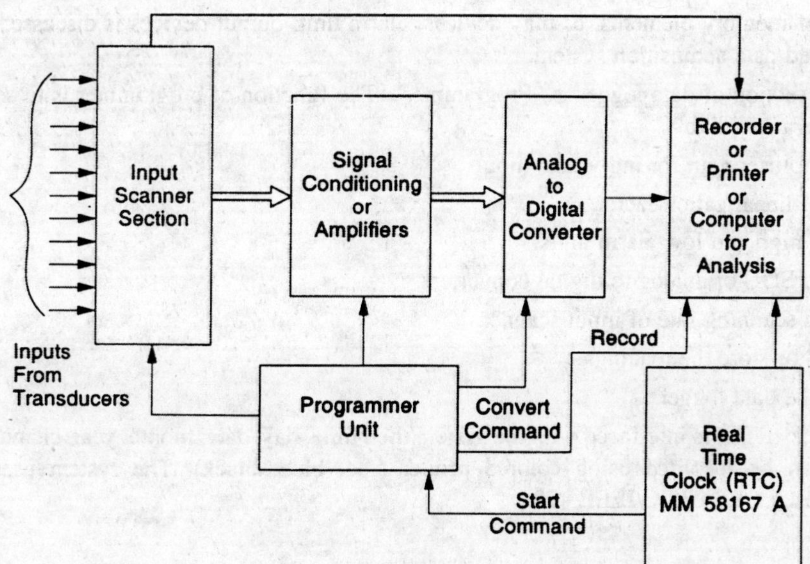

Fig. 7.6. Block diagram of data logger.

The function of input scanner is to select each input signal in turn. The scanner receives analog input signals from the transducers like thermocouples, R.T.D., thermistors, pressure transducers, strain gauges, LVDT, pneumatic transducers, potentiometers etc. Generally fixed scanning rate is provided so as to scan particular input channel for fixed duration of time intervals such as 30 to 60 minutes, every 10 minutes, every 5 minutes, few seconds. Typical scanning capacity of data logger is 128 to 150 channel inputs per second.

Now a days, the data loggers are integrated into total control system consists of the complete data acquisition system (DAS) and control capability. The scanning may be cyclic or continuously or intermittently. Sequence can be controlled or changed using elements such as push buttons, relays, clock frequency etc.

Amplifier stages are introduced so as to amplify the weak output signals of transducers. The gain of amplifiers can be adjusted through microprocessor using software control from programmer.

These amplified signals are converted into digital signals by means of A to D converters. The A to D converter should have fast conversion rate, good noise rejection capacity, great accuracy. Generally, dual slope or voltage to frequency type and integrating type A to D converter are used. Typically the range and resolution of ADC would be ± 40 mV (1 V resolution), ± 400 mV (10 V resolution), ± 4 V (100 V), ± 40 V (1 mV) with an input impedance of 200 M Ω.

The digital output signals from A to D converters and digital outputs from digital transducers such as rotary encoder is interfaced with the microprocessor. Microprocessor performs the functions or controls the operations like (1) data averaging (2) linearisation (3) alarm or annunciator settings (4) totalization (5) settings of set points (6) amplifier gain control (7) to control I/O devices (8) to provide control actions to the control elements (9) formatting of output signals (10) conversion of engineering units (11) interface control (12) Reset the system etc.

High level languages like BASIC, FORTRAN, C++ etc. are used to control all above mentioned operations. Large data loggers uses another computer on time sharing basis. These computers use various software packages, menu driven packages, system programs, subroutine subprograms, for handling the input channels, operations to be performed. The programming is possible by means of front panel control. These computers provide supervisory control and also minimizes the loading effect on mother board. The host computer can also be used to co-ordinate and minimize the load or burden on mother board.

The function of memory elements, display devices, alarm unit, output devices is discussed in section microcomputer based data acquisition system.

The important section of data logger is "Programmer". The function of programmer is to set or adjust the sequence of operations like :

(1) to set amplifier gain for individual input.

(2) setting of linearization factor.

(3) to adjust high and low alarm units.

(4) EOC and SOC of analog to digital converters.

(5) to set the scanning rate of input scanner.

(6) to record or store the readings.

(7) to reset the data logger.

If real time clock (RTC) is interfaced with the system then time, day, date, month, year, channel number, process variable can be recorded using colored printer (red, blue, black). The system uses standard communication interface RS 232 C/IEEE 488.

7.10 Microprocessor Based Long Term Temperature Recorder

Basically this is a long term microprocessor based data acquisition system used for recording the temperature data in deep waters. This system is useful in marine applications, sea water temperature measurements, to determine state of ecological health of water. It can be used in industrial and scientific applications by updating and modifying some system parameters. It is a submersible temperature recorder *i.e.* used to measure temperature of deep waters.

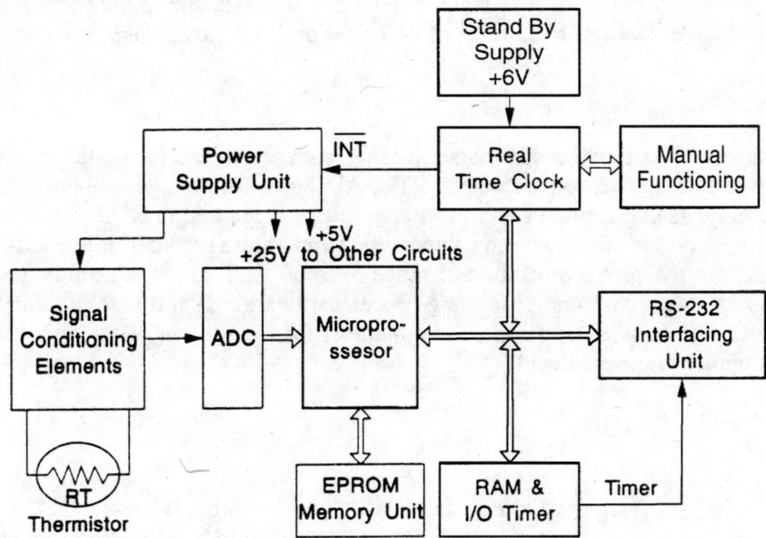

Fig. 7.7 : Diagram of microprocessor based long term temperature recorder.

Fig. 7.7 shows block diagram of the system. It consists of following sections.

(1) Temperature sensor

In this system thermistor is the temperature sensor. It can be changed according to the desired temperature range under measurement. Necessary signal conditioning stage such as amplifiers, voltage followers etc. (using OP-AMP) is provided to amplify the calibrated or converted output of thermistor.

Temperature is sensed by a thermistor operating in a resistive mode and supplied with constant current source. Thermistor provides resultant voltage proportional to change in resistance according to temperature under measurement. The thermistor can be placed in deep waters, remote areas by providing proper encapsulations and mountings.

(2) Analog to digital converter

This is a microprocessor based system, hence ADC is used to convert analog output voltage of thermistors into digital manner. Low cost 8-bit ADC can be used. Now a days different ADC's are available with greater accuracy and operating at + 5 V supply.

(3) Microprocessor and memory unit

Output of ADC is interfaced with microprocessor unit. The combination of microprocessor and memory unit interfaced with input/output devices is a microcomputer system.

Any 8-bit, 16-bit or other microprocessor can be used to control different operations, to carry out calculations and measurements, to monitor different temperature readings, transfer of data etc.

Separate memory unit is devoted to the system which contains random access memory (RAM) and ultraviolet erasable programmable read only memory (EPROM). These memories are required to store necessary software programmes of the system and output data. Generally 256 byte RAM is used for temporary storage. System needs approximately 12 kbytes memory to store one years recording. Separate UVEPROM is needed for monitor program.

(4) RS 232 Interface (Recommended Standard number 232 from the Electronic Industry Association (EIA))

It is a serial link or interface between data terminal equipment and data communication equipment using serial binary data interchange. Desired baud rates of 110, 300 or 1200 can be used for communication of data.

(5) Power Supply

Objective of power supply is to provide continuous biasing voltages to the integrated circuits (IC's) and different sections such as microprocessor, memories, RTC, ADC etc. DC voltage supplies or regulators of ±5V ± 12 V, ± 25 V can be used. Separate standby voltage source is applied to RTC.

This is the generalized system for long term submersible temperature recorder. It is a self-contained and low cost system, suitable to measure temperature of range 5° to 30°C. This system can be used to monitor the temperatures in hours, days, upto one year using proper hardware and software programs. Thermistor requires pressure tube of size 1 meter long and 15 cm diameter for installation. It can be useful to measure temperature of waters upto 100 meters depth.

7.11 Real Time Clock (RTC)

The MM 58167A is a 24 pin integrated circuit fabricated by low threshold metal gate CMOS circuit technology. It is developed by National Semiconductor Inc. RTC is a digital clock which provides exact time in hours and minute.

It consists of an addressable real time counter, 56 bits of RAM and two interrupt outputs.

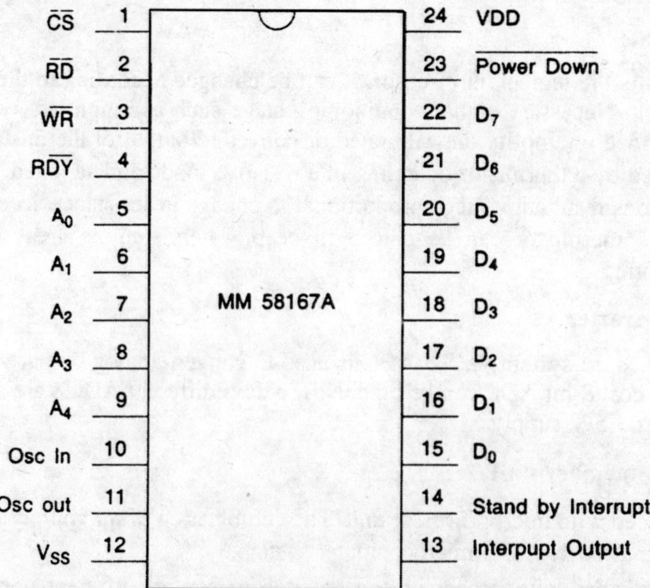

Fig. 7.8. Pin Diagram of MM 58167A.

Fig. 7.8 shows pin diagram of MM58167A microprocessor real time clock in dual in line package. It operates on the supply voltage of range 4·0 (minimum) and 5·5 V (maximum) when the outputs are enabled. Whereas in POWER DOWN mode the supply voltage is 2·0 (minimum) and 5·5 V (maximum). Voltage $V_{DD} - V_{SS}$ is 6·0 Volt.

MM58167A has following features :

(1) It is compatible with microprocessor as 8-bit data bus is available.

(2) Crystal oscillator having frequency 32,768 Hz.

(3) Random access memory of 56 bits with comparator for comparing the real time counter to the RAM data.

(4) Two interrupt outputs with 8 possible interrupt signals.

(5) Milliseconds through month counters.

(6) Calendar of four-year without leap year.

(7) Important features is POWER DOWN mode : This input provides the facility to disable the chip from rest of the system for standby low power operation. It disables all inputs and outputs except for one of the interrupts.

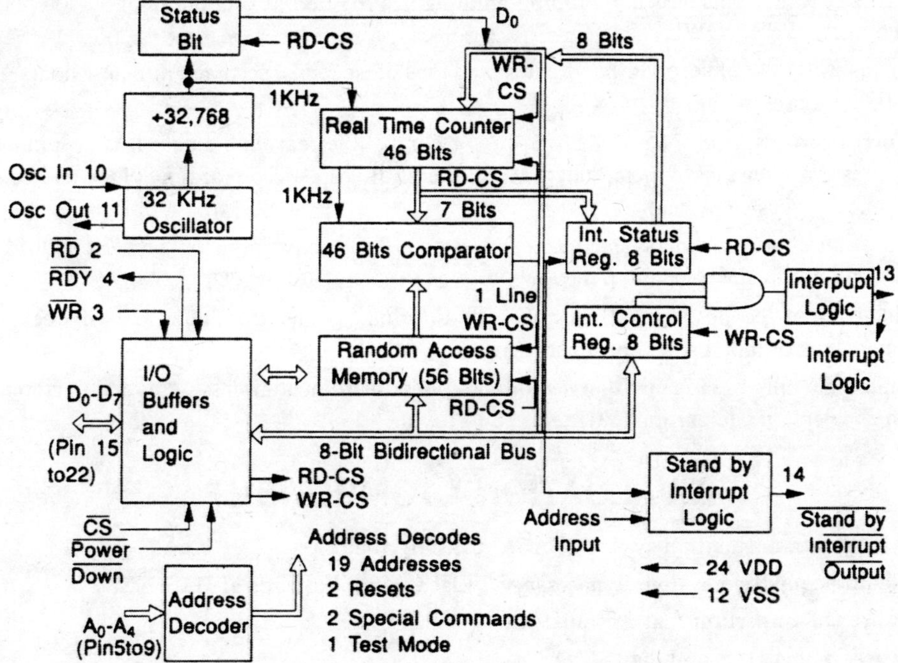

Fig. 7.9 : Block diagram of real time clock MM58167A.

Fig. 7.9 shows functional block diagram of RTC MM58167A. It consists of a real time counter of 46 bits. It is divided into 40 bit digits with two digits being accessed during any read or write cycle. The counter is a ripple counter and addressed by (D_0, D_1, D_2, D_3) and (D_4, D_5, D_6, D_7) *i.e.* two digits. It can be addressed from milliseconds to months.

A second section is random access memory (RAM) which is of 56 bits. The data in RAM gets compared with real time counter on digit basis. It is a 56 bits memory fabricated on chip and it can be used for any necessary storage of data during power down. RAM can also be used as an alarm latch for comparison to real time counter.

The comparator is nothing but a cascaded exclusive NOR gate arrangement. There are two interrupt signals which are INTERRUPT OUTPUT and STANDBY INTERRUPT. The INTERRUPT OUTPUT can be programmed to provide eight different output signals. These signals are 10 Hz, 1 Hz, once per month and when a RAM/real time counter comparison occurs.

Another interrupt is $\overline{\text{STANDBY INTERRUPT}}$ which occurs when enabled and when a RAM/real time counter comparison takes place.

A crystal oscillator is used to generate the necessary frequency signal. The oscillator circuit consists of a crystal, two capacitors and one resistor of 20 MΩ is connected externally. Generally, capacitor value ranges from 20 pF–25pF. The crystal which is used has frequency of 32,678 Hz. Between pin 10 and 11 *i.e.* OSC IN and OSC OUT a 20 MΩ resistor is connected to bias the internal inverter in the linear region.

The pins 2, 3 and 1 *i.e.* $\overline{\text{READ}}$, $\overline{\text{WRITE}}$, $\overline{\text{CHIP SELECT}}$ signals are active low inputs. $\overline{\text{READY}}$ signal is an open drain output.

The $\overline{\text{POWER DOWN}}$ pin 23 disables all inputs and outputs except $\overline{\text{STANDBY INTERRUPT}}$. When this pin is at logic O level the RTC will not respond to any external signals. During this mode STANDBY INTERRUPT gets turned on if programming is provided. If programming is done before logic 'O' is applied to $\overline{\text{POWER DOWN}}$

RTC counts 1/10,000 of seconds, hundredths and tens of seconds, seconds, minutes, hours, day of the week, day of the month, month.

For information : MM58174A is microprocessor-compatible real-time clock. It is 16-pin dual-in-line package which is a low-threshold metal-gate CMOS circuit. It works as a real time clock and calender in a bus oriented microprocessor system.

Important features are automatic leap year calculation, TTL compatible, low power standby operation, protection for read during data changing, internal pull-ups to safe guard the data.

It counts tenths of seconds, second, tens of seconds, minutes, tens of minutes, day of week, days, tens of days, months, tens of months, independent register.

This chip is generally used in microporcessor controlled instrumentation systems and in communication systems such as telephone, television, VCR etc.

EVALUATION COMPONENT

1. What is a data acquisition system (DAS) ? Classify them.
2. Why data acquisition system is necessary ? List the applications of DAS.
3. What are the important features of DAS ?
4. Compare analog DAS and digital DAS.
5. What is the significane of DAS ?
6. Draw neat block diagram of generalized DAS and explain it in detail.
7. What is the use of multiplexer in DAS?
8. What is ratiometric conversion? Explain it.
9. Write a short note on "logarithmic compression".
10. Explain with neat block diagram the working of :
    (a) Single channel DAS.
    (b) Multichannel DAS.

11. How digital to analog converters are useful in DAS?
12. Compare microprocessor and microcomputer.
13. How computer helps in DAS ?
14. Why digital DAS is advantageous over analog DAS ?
15. What is real time clock (RTC) ? What is the application of RTC ? Explain the neat block diagram of RTC.
16. Draw and explain the block diagram of microcomputer based DAS.
17. List the interfacing chips (IC's) necessary in microcomputer DAS.
18. What is the function of alarm unit in DAS.
19. How temperature can be monitored using RTC and microprocessor ?
20. What is data logger ? Compare it with DAS.
21. What are the characteristics of data logger.
22. Draw a neat block diagram of data logger and explain it in detailed manner.
23. How the data logger helps the operator or user ?
24. What is the function of programmer in data logger ?
25. Try to develop DAS for different physical quantities like pressure, liquid level, strain using proper transducers, necessary signal conditioning elements and other interfacing stages.
26. List the applications of data logger where it can be used.

Chapter **8**

RECORDERS AND DISPLAY DEVICES

Learning Objectives

- To study necessity of recorders.
- To study operation of strip chart recorders.
- To introduce with various marking mechanisms.
- To examine operation of various type of recorders.
- To study working of X-Y recorder.
- To study functioning of magnetic tape recording.
- To introduce digital recorders.
- To get aware in selecting parameters of recorders.
- To introduce digital display devices.
- To study working of light emitting diode. (LED)
- To study operating phenomenon of liquid crystal display. (LCD)
- To consider the working of analog displays and their applications.
- To examine the working of light indicators.
- To study advantages, disadvantages and applications of display devices.

8.1 Introduction

This chapter covers recorders, indicators, analog display and digital display devices. The function of all these devices is to indicate or monitor the output signals from transducers. Basic difference is that recorder provides permanent graphical representations of time-varying signals whereas the display devices provide visual indication.

In measurement and instrumentation systems, recorders and display devices plays vital role. These devices are connected or placed in the output stage. Generally, the output of different transducers or sensors are processed through proper signal conditioning elements and applied to recorders or display devices.

Recorders are classified into two types *i.e.* analog and digital recorders. There are again sub-types of recorders which are discussed in this chapter.

The display devices are analog display and digital display devices. Analog display devices are indicators (fixed scale and moving scale indicators). Light indicators like neon lamps, incandescent lamps are used for conveying different information like ON or OFF conditions of process, alarm conditions, normal and abnormal conditions in process.

Digital display devices are LED and LCD. These devices provide visual light necessary for indication. These devices indicate or display numeric and alphanumeric information. Nixie tube is also a display device which indicates numeric as well as alphanumeric information.

In this chapter, working principle, construction, operating ranges, advantages, disadvantages, features and selection parameters of recorders and display devices are discussed in detailed manner.

8.2 Necessity of Recorders

Recorders are becoming very popular devices in instrumentation systems.

These devices are useful in different plants, process industries, power stations, refinery plants etc. at instrumentation and control sections. Recorders are mainly applicable to record the signals and recorded information is used to control different parameters like pressure, temperature, velocity, acceleration, current. voltage, frequency, etc. involved in the industrial processes.

The term recorder is defined as it is an instrument or device used for recording the electrical or non-electrical parameters (as stated above) as a function of time or it can be defined as a device which provides graphical record of the variations in parameters which are under measurement.

Important aspect to record these quantities is to study and analyze the recorded signals for measurement and controlling the process under study. Recorder, records the electrical quantities directly but for recording non-electrical quantities, sensors or transducers are necessary for converting them into electrical signals. They are useful in research and development fields. Previously recorded data can be reproduced as a reference for comparison.

8.3 Classification of Recorders

Depending upon the type of system and output nature of system the recorders are mainly classified as :

(A) Analog recorders and (B) Digital recorders.

(A) Analog Recorders

These are used for recording the output of analog systems. They are again classified as :

(a) Graphical recorders.

(b) Oscillographic recorders.

(c) Magnetic tape recorders.

(B) Digital Recorders

They are used to record the output of digital systems. Digital recording devices are classified as :

(a) Return to zero recording (RZ).

(b) Non return to zero recording (NRZ).

(c) Return to Bias recording (RB).

Analog recorders :

(a) **Graphical recorders** : As the name indicates these recorder provides the graphical representation or waveform of varying voltage or current signals. Graph paper or chart with pen and ink (stylus) mechanism is used. Graphical recorders are classified in three types :

(1) Strip-chart recorders

(2) Circular chart recorder

(3) X-Y recorder

Different types of recorders are discussed in next sections.

8.4 Strip-chart Recorders

This type of recorder, records the variations in one or more variables with respect to time. The strip chart recorder consists of a long graph paper which moves vertically with mechanism necessary to move the paper at desired speed having speed selector arrangements. A pen or stylus is used to trace the waveforms. The pen or stylus moves horizontally in accordance with the variations in input signal *i.e.* the signals which are to be recorded. Driving mechanism is necessary to drive the stylus with respect to input signal.

Essential parts (assembly) which are necessary for strip chart recorder are discussed in preceding section.

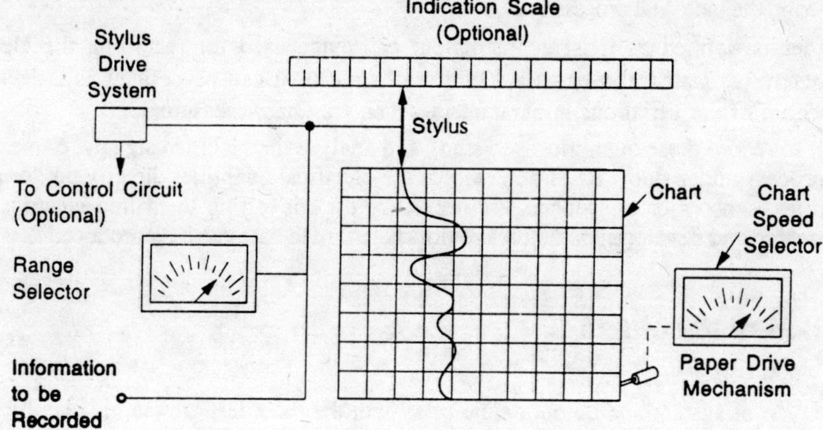

Fig. 8.1 : Strip chart recorder.

Fig. 8.1 shows the block diagram of strip chart recorder. It has three main parts :

(1) Paper Drive Mechanism

The paper or chart should be moved with uniform speed. Chart is usually moved by a synchronous motor with speed selector switch and gear train arrangements. The speed can be selected over the range from several inch/second to fraction of an inch/hour.

(2) Marking Mechanism

It includes the stylus or pen which moves over the chart. There are different types of stylus such as electric, pressure, light beam, electrostatic etc.. The stylus moves over the preprinted scaled circular or rectangular charts. Different stylus are as discussed below.

(a) Ink-filled stylus :

In this type, the stylus is filled with ink of different colours by using gravity or capillary actions. Ink reservoir is connected to the tip or point of pen. This is the most commonly used stylus because of low cost, wide range of operating speeds. Drawbacks of this stylus are (a) friction between the tip and graph paper

(b) when the stylus is at rest position, clog takes place also at higher speed ink splatters likewise the ink batches at low speed. (c) frequency response is limited upto few Hz.

(b) Inkless stylus :

(i) **Heated stylus** : This type of stylus is used in thermographic recorders. Such type of stylus moves over a special type of paper coated with white wax on a paper having black base. As the stylus gets heated it melts the wax and trace of waveform takes place in accordance with the input signal. Disadvantages of this system are high cost, recording gets affected due to the temperature produced in the process whose performance is under study.

(ii) **Optical stylus** : This technique uses photosensitive paper. The beam of light (mostly ultraviolet) traces waveforms on the paper. Advantages of this stylus are high recording speed can be achieved suitable to record higher frequency signals and good resolution. Drawback is high cost.

(iii) **Chopper stylus** : In this system a pressure sensitive paper with V-shaped pointer is used. The V-shaped pointer under chopper bar applies the pressure to the pen which makes the trace on paper. This system is suitable for recording slowly varying signals.

(iv) **Electric stylus** : This system consists of a current sensitive paper chart and the current flows to the paper through stylus. When the current flows through the stylus, a trace gets appeared on the paper. This system has higher recording speed, long life but the cost is high.

(v) **Electrostatic stylus** : In this method the graph paper or chart used is electrosensitive paper. This paper consists of a paper having coloured base like red, black etc. which is coated with thin powder of aluminium. Stylus is a thin wire of tungsten which move over the aluminium surface. The permanent record of the waveform is produced by making high voltage discharged at the potential of 35 V. Due to this discharge the aluminium powder gets removed and fine record gets produced on colored base of the graph paper or chart.

(3) Tracing System

The function of this section is to trace or produce the graphical trace (representations) of the quantities which are under analysis.

There are two tracing systems (1) curvilinear system and (2) rectilinear system. In curvilinear system the stylus moves through an arc and draws line across the chart. The line will be curved, time intervals will be on the curved segments. This tracing system is used in galvanometric type recorders.

Under rectilinear system the straight line gets recorded over the chart. In this system a line of constant time is perpendicular to the time axis. This system is used for thermal, self-balancing potentiometric systems.

8.5 Galvanometric Type Recorder

This type of recorder uses the principle of deflection in the moving coil instrument *i.e.* D'Arsonval galvanometer. When current flows through the coil torque gets developed resulting into deflection. Fig. 8.2 shows the schematic of galvanometric type recorder. The pointer of galvanometer is replaced by ink filled stylus or pen.

The coil is suspended inside the strong magnetic field with torsion spring and suspension arrangements. When the current proportional to quantity under measurement flows through the coil of galvanometer, deflection of coil takes place, it results to move the pointer. As the pointer carries ink filled pen or stylus with reservoir, it moves over the chart or graph paper. The pen is rigidly connected to the coil of galvanometer. Graph paper or chart is driven at constant speed by using electric motor and suitable roller arrangements.

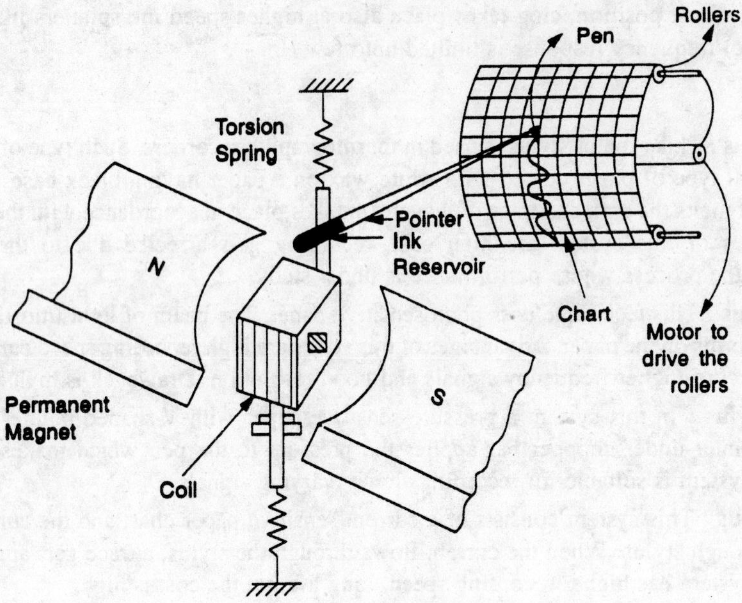

Fig. 8.2 : Galvanometric type recorder.

Spring arrangement is provided to exert the controlling torque when pointer with stylus arrangement comes to rest position. Gravity and capillary action provides the flow of ink to the pen with reservoir. If the amplitude of input signal is greater, more is the deflection. When the amplitude of input signal is smaller then amplifiers are used to increase the level of such signals.

This type of recorder is useful for recording of low frequency signals and has sensitivity of 0·4 mV/mm. Operates over the range of few mA to several mA and few mV to several mV.

Galvanometer type recorder is suffered by the problems like friction, small input impedance. If the assembly of galvanometer and chart or graph paper gets disturbed it results in tracing the distorted waveform.

8.6 Null Type Recorders

Basically null type recorder operates on comparison method. These recorder contains the measuring unit which gets disturbed or unbalanced by the output signal from the sensors or transducers. Unbalance of measuring unit produces the error voltage signal. This error voltage signal is used to make balance or NULL some device.

These recorders are classified as :

(1) Potentiometric recorder

(2) Bridge recorder

(3) LVDT recorder

In this book Potentiometric recorders are only covered.

(1) Potentiometric Recorders (Self Balancing)

This type of recorders are generally used for recording, analyzing the temperatures involved in different processes.

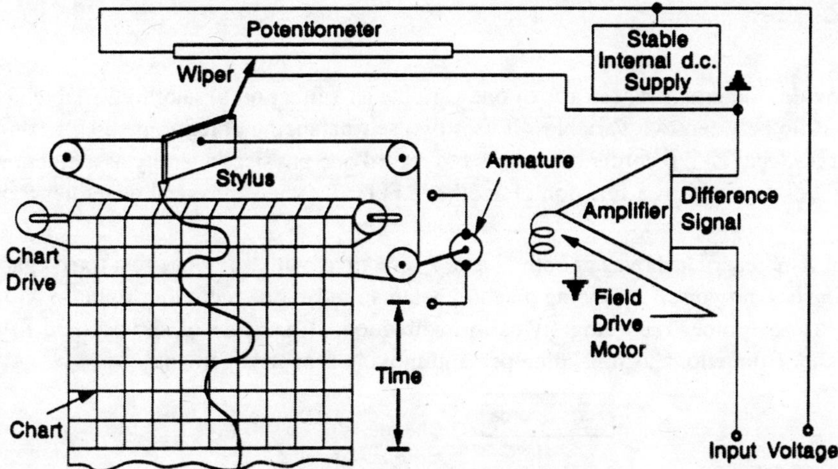

Fig. 8.3 : Potentiometric type recorder.

Fig. 8.3 shows the diagram of potentiometric recorder. It consists of a standard d.c. supply. The potentiometric voltage and input voltage signal is applied to the input of an amplifier. Comparison of input signal and reference signal is carried out by potentiometer arrangements. Function of an amplifier is to amplify the difference between these two voltages. Also the amplifier provides high input impedance which minimizes the loading effect.

The amplified difference signal is applied to field coil of d.c. motor. Wiper of the potentiometer is connected to the armature of d.c. motor. Wiper moves over the potentiometer in such a manner so as to reduce the error. When the error is becoming smaller the speed of motor slows down and motor stops where the error is zero, which results into NULL position. Stylus or pen is connected to the wiper of potentiometer so the pen moves in same direction as the wiper moves in particular direction *i.e.* pen moves in synchronization with the wiper of potentiometer to record the input waveform. A chart is moved at constant speed by means of a separate motor synchronized to power line frequency with gear train arrangements.

These recorders have high input impedance and sensitivity of 4 V/mm having error less than ± 0·25% with bandwidth of 0·8 Hz.

(2) Circular-chart Recorders

This type of recorder is used to record the waveforms of parameters like temperature, pressure, flow acceleration and liquid level etc. as a function of time or as a function versus angular displacement.

The construction of this recorder is as, it consists of a circular chart having concentric circles, the centre point is moved away from the centre. Chart is segmented into time lines by using arcs. This chart is mounted and clamped on a flat supporting also fastened to avoid the curling of paper. The chart rotates at geometric centre point, a motor with synchronous speed and centre spindle is connected to the chart for rotations. Speed of chart varies between one rotation in 15 minutes to one rotation in 30 days.

The pen or stylus used for recording should have light weight, flexible and smooth to avoid the friction between chart paper and pen during the twisting of the chart. Pens used for recording are '*V*' pen, fountain pen, reservoir pen and ball pen. Chart size is 10 inch diameter and for industrial applications 12 inch diameter is used.

Advantages of these recorders are the charts are flat and easy to handle, simple recording mechanism. Drawback of these recorders is that it is necessary to change the chart after completion of the process time period or process slot like 7 hr., 24 hr., etc. Another disadvantage is that it is difficult to read the time values near to centre.

8.7 X-Y Recorder

X-Y recorder provides the graphical records of one variable as a function of another variable *i.e.* it plots the graphical relationship between two variables. It uses two self balancing potentiometers. In connection with the appropriate transducer X-Y recorder can be used to record one physical quantity as a function of another. This recorder records an e.m.f. as a function of another e.m.f., in case of physical quantities transducers are essential.

The stylus of X-Y recorder moves in both X and Y directions with reference to a fixed chart of squared graph paper. It consists of two self balancing potentiometers to move one recording stylus in X-direction and another is used to move other recording stylus in Y-direction. It has pair of servosystem to provide the movement in X and Y directions using sliding pen and moving arm arrangements.

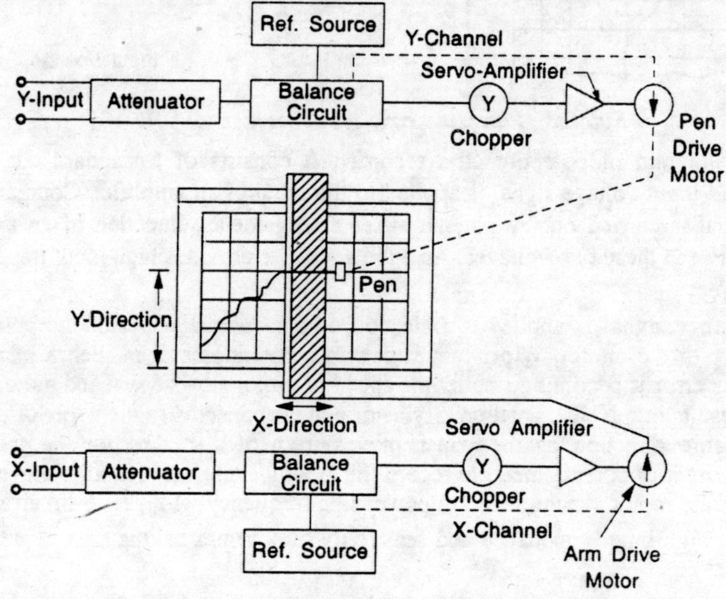

Fig. 8.4 : X—Y recorder.

Fig. 8.4 shows the block schematic of X-Y recorder. The input signals are applied at the X and Y channels through attenuator circuit. Attenuators are used to set the input signal range within 0 to 5 mV so that the recorder can operate satisfactorily within the specified range. At balance circuit the input voltage signal gets compared with the voltage of reference voltage source and error voltage gets developed. This d.c. error voltage is converted into a.c. signal using chopper circuit. Error voltage gets amplified by servo amplifier to drive the servomotor. Servo amplifiers are used because of the features like free from drift, inexpensive and higher gain. The servomotor has characteristics like high-power mechanical motion, better controllability, low friction, less power consumption and high torque.

The pen or arm mechanism operates in such direction to minimize the error and accordingly as the input signal varies the pen or arm mechanism maintains the balance and records the signals on graph paper. Input signal range for X-Y recorders is in between 0·25 mV/cm and 10 V/cm with ± 0·1% accuracy of full scale.

Applications of X-Y Recorders

Mainly this type of recorders are used to measure, analyze and study the following parameters :

(1) to plot the characteristics of devices like vacuum tubes, diodes, transistors, rectifiers etc.

(2) speed-torque characteristics of electric motors.

(3) to measure the physical quantities like temperature, pressure, velocity, acceleration etc.

(4) speed-load characteristics of governors.

(5) To plot stress-strain curves, hysteresis curves.

8.8 Magnetic Tape Recording

Magnetic tape recording is simple, most popular and cheapest method to record the information. In this type of recording the voltage signal gets recorded or stored, this recorded signals can be reproduced at any time in original electrical form. This system has good response to high frequency signals.

(A) Direct Recording Method

Fig. 8.5(a) and (b) shows construction of direct recording method.

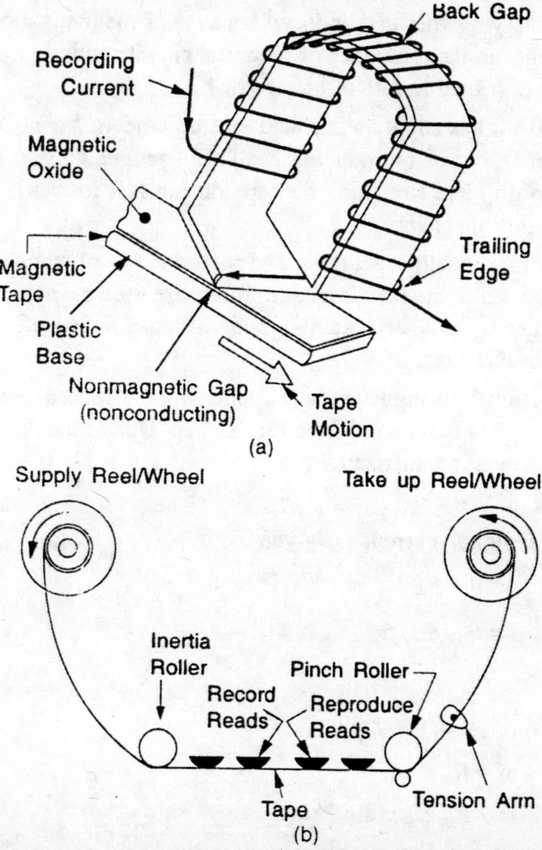

Fig. 8.5 (a) Recording head. (b) Tape transport mechanism.

It involves magnetic tape recording. The magnetic tape recorder consists of following parts :

(i) **Magnetic tape** : The information or voltage signals get recorded on flexible metallic tapes and wires. Most popular recording media is thin plastic base like polyvinyl chloride (PVC) or mylar ribbon coated with fine powder of magnetic iron oxide (Fe_2O_3) on one side only. Special type of binders are used for surface coating. Mostly mylar base is preferred because of its characteristics like flexibility, not affected by humidity and homogeneous coating of magnetic iron oxide is possible *i.e.* minimum surface defects (dropouts). Typical dimensions of magnetic tape are 12·7 mm wide and 25·4 μm thick.

(ii) **Recording head** : It is electromagnetic recording head which magnetizes the tape passing through it. Recording head consists of ring shaped magnetic core over which a coil is wound, this assembly is placed inside a metal case. A small air gap of size 5–15 μm is maintained in the head, magnetic tape is transferred from the air gap. Fig. 8.5(a) shows the recording head.

When current flows through the coil resulting to develop magnetic flux in the air gap. As the magnetic tape is transferred from air gap it magnetizes the iron oxide particles on the tape. Due to which residual flux remains on the tape. As the tape is passed through the recording head permanent magnetization pattern gets stored on the tape *i.e.* on the particles proportional to flux developed in the air gap.

(iii) **Reproducing head** : This head is similar to that of recording head. The head is electromagnetic with high permeability core, its function is to reproduce the magnetic recording. When the magnetic tape with recorded information signal is transferred from reproducing head, original electrical signal gets reproduced. At this head a varying flux gets induced because of magnetization pattern on recording tape. This head detects changes in the reluctance of magnetic circuit resulting to develop voltage in the winding. This voltage is proportional to rate of change of flux.

(iv) **Tape transport mechanism** : It has supply wheel and take up wheels. Function of this mechanism is to transfer the tape from recording or reproducing head with proper tension to the tape. It helps to move the tape without distorting and stretching the tape during fast forward and reverse operations. Fig. 8.5(b) shows tape transport mechanism

(v) **Conditioning devices** : For high quality recording and reproduction of signals, amplifiers and filters are used which are called as conditioning devices. Amplifiers are used to increase and maintain proper gain of the system with better impedance matching and without loading effect. To select particular frequency range signals filter circuits are used.

When the current i is proportional to input voltage applied to the record head, magnetic flux gets developed across the air gap. The tape is transferred from this air gap. Due to the developed flux, iron oxide on the tape retain the state of permanent magnetization.

The flux at air gap is $\phi = K\phi \cdot i$ ...(8.1)

If the input signal is sinusoidal signal, current i is given as ;

$$i = i_0 \sin \omega t$$

$$\therefore \quad i = i_0 \sin 2\pi f t \qquad \text{...(8.2)}$$

(ϕ is magnetic flux)

The intensity of magnetization is given by (m) ;

$$\therefore \quad m = K_m \, \phi$$

$$\therefore \quad m = K_m \, K_\phi \, i_0 \sin 2\pi f t \qquad \text{...(8.3)}$$

Suppose ; V is speed of tape (inches/sec.),

x is distance on the tape from air gap.

$$m = K_m \, K_\phi \, i_0 \sin \left(\frac{2\pi f x}{V} \right) \qquad \text{...(8.4)}$$

Here v/f is the wavelength of magnetization variations. Output voltage gets generated in the coil of reproduce head when the flux gets developed after passing the tape with recorded signal. Thus the output voltage gets developed due to rate of change of flux. This technique of direct recording is not suitable to record direct current signals because if such input signals are applied to the coil of reproduce head it will deliver zero output signals. Frequency range for direct recording is in between 100 Hz and 2 MHz.

Features of direct recording method :

(1) This technique has better frequency response of range 100 Hz to 2 MHz with tape speed of 3·05 m/s.

(2) Good bandwidth can be achieved.

(3) This system requires inexpensive and less complex circuitry such as amplifiers, filters etc.

(4) Signal to noise ratio is 22–30 db at 1% total harmonic distortion.

(5) The information on tape can be erased and reused to record the new information or data.

Drawbacks :

(1) Applicable when high bandwidth is required.

(2) If the coating of Fe_2O_3 is not homogeneous (dropout) the amplitude variations takes place resulting in insufficient amplitude stability *i.e.* it affects the quality of recording.

(3) In instrumentation systems it is essential to record the signals uniformly over the entire operating range. But this system is suitable to record speech and music which can sustain the amplitude variations, but it is not feasible in instrumentation and control systems.

8.9 Frequency Modulation (FM) Recording

The term frequency modulation is defined as the process of modulation in which the frequency of carrier signal is varied in accordance with the frequency of modulating signal or information signal by keeping amplitude as constant. This method is useful to record the d.c. input signals. Fig. 8.6 shows the block schematic of FM recording.

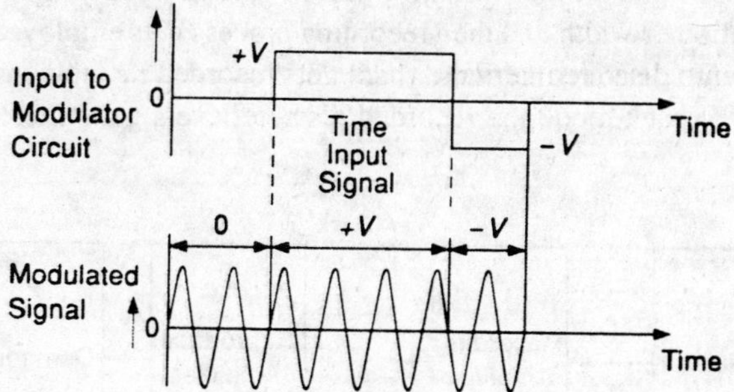

Fig. 8.6 : FM recording.

In FM recording technique the carrier frequency (432 kHz) from the carrier oscillator is modulated in accordance with the input signal which is to be recorded. This modulated signal get recorded on the tape as per the normal recording techniques.

If the input signal is of zero value then modulated signal is only the carrier frequency. When the input signal is positive going the carrier frequency signal gets deviated in one side according to the modulation index. Similarly if the input signal is negative going the carrier frequency signal gets deviated in opposite direction with respect to the modulation index. This modulated output is nothing but constant frequency for d.c. input and varying frequency for a.c. input. Such signals are applied to the tape for recording process. As the variations in the frequency is directly proportional to amplitude of input signal, these signals get recorded.

During playback process at reproduce head, its output is demodulated *i.e.* the carrier signal gets suppressed to reconstruct the original input (information) signal. Low pass filter is used to remove unwanted noise signals produced in the system if any. The signal to noise ratio of FM recording system varies in between 40 to 50 dB.

Advantages :

(1) This system is widely used to record the signals from d.c. voltages to several kHz.

(2) Used to record the output voltage signals of transducers which measures force, pressure, acceleration etc.

(3) Operation is not suffered by non homogeneous (dropout) coating of Fe_2O_3 on plastic tape.

Disadvantages :

Recording process may be damaged due to variations in the speed of tape.

This system may be suffered by complex circuitry involved in recording process, limited frequency response, it requires high speed of tape.

8.10 Pulse Duration Modulation (PDM) Recording

This system uses time-division multiplexing (TDM) technique to record the signals. In TDM the signals get sampled sequentially to fit into series of time slots in the transmission channel, mostly adjusted in a cycle. In instrumentation the signals are received from number of sensors or transducers. Pulse-width modulation is defined as the amplitude and starting time of the pulse is kept fixed and the width of each pulse is made proportional to the amplitude of the signal at that instant. In PDM it is necessary to convert the input sine wave into samples of uniform duration or width and then recording process gets employed. The instantaneous value of the signal gets converted into discrete intervals which gets recorded *i.e.* no continuous recording of the signal is carried out. During playback mode the recorded discrete levels gets filtered to reconstruct the original sine signals.

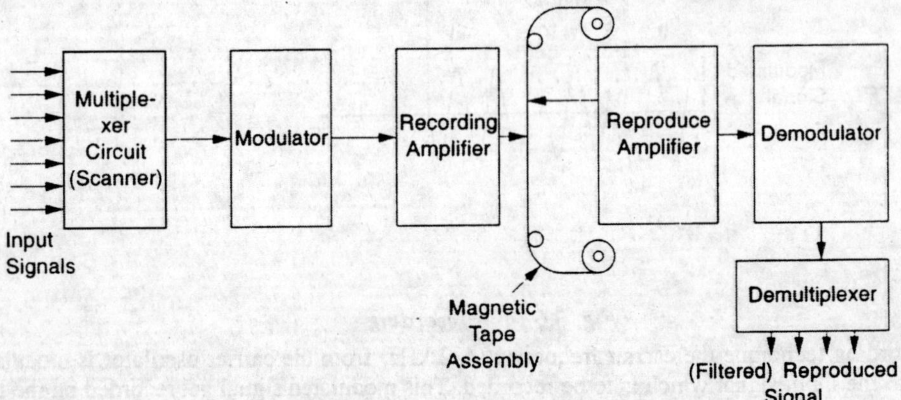

Fig. 8.7 : Block Diagram of PDM recording technique.

Fig. 8.7 shows the block diagram of PDM recording technique. The input signals or samples are scanned or multiplexed at the input stage which are modulated and recorded on the magnetic tape using usual procedure. In playback or reproduce mode the output of reproduce head is demodulated and filtered to reconstruct the original signal.

This technique is applicable for recording the information signals which are slowly changing and handle more number of channels. It has good S/N ratio and higher accuracy. PDM recording has limited frequency response and requires complex circuitry.

8.11 Digital Recorders

Digital signals have two values logic '0' and logic '1' (*i.e.* binary digits or bits). These logic levels of the signals or data can be stored on magnetic disk storage device or tape recorders.

Storage or writing of digital signals is carried out on a small surface of magnetic tape. Important aspects to be considered is the packing density should be high so that each bit to be recorded shound occupy little space which is economically advantages.

Another aspect is that the recording and reproduce mechanism should be reliable and precise.

The digital information which is to be recorded or stored on magnetic surface is supplied to the recording head *i.e.* write head. Logic levels '0' and '1' are recorded on appropriate tracks across the tape width. Recorded digital information can be reproduced by using "read" head.

Techniques used for digital recording are :

(a) Return to Zero (RZ) Technique

(b) Nonreturn to Zero (NRZ) Technique

(c) Return to Bias (RB) Technique

The recorded digital information can be reproduced by using "read" head with necessary reading system.

(A) Return to Zero (RZ) Technique

Fig. 8.8(a) and (b) shows block diagram and input/output waveforms of return to zero recording. In this recording positive pulse indicates logic '1' and negative pulse indicates logic '0'. Between the pulses input

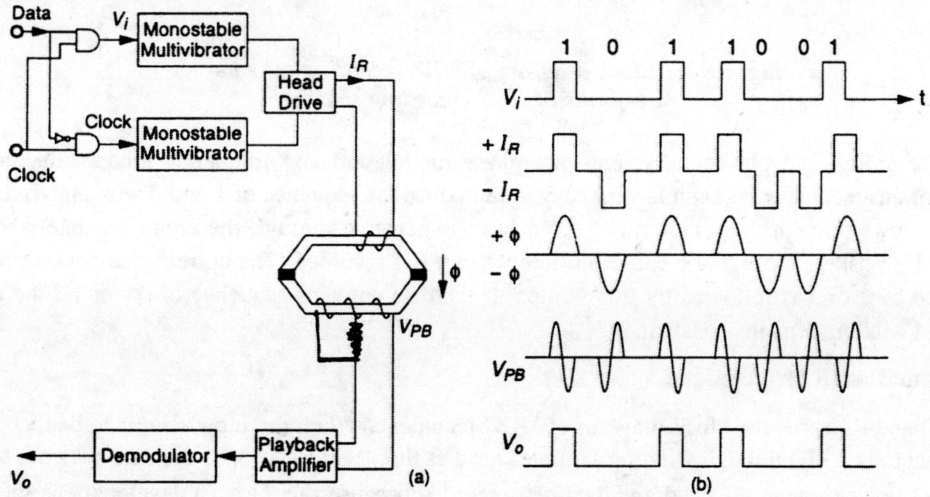

signal current returns to '0' hence this technique is called as "return to zero" (RZ) technique. The playback waveform V_{PB} consists of pulse doublets with phase difference of 180°.

When the recording process is under operation a positive current pulse is fed to recording head to record 1 present in the data similarly negative current pulse is fed for recording 0.

In the reproduce or playback mode demodulator stage is introduced to reconstruct the original signal which is recorded. In this system the magnetic field returns to zero flux when a 1 or 0 pulse is not present. Drawback of this system is that it is difficult to write over the previously recorded information.

(B) Nonreturn to Zero (NRZ) Technique

As its name indicates the input signal does not returns to zero.

Fig. 8.9 (a) and (b) shows the block diagram of NRZ technique with input and output waveforms. This system is not self clocking so a clock track must be recorded along with the input data which is to be recorded. In this technique the magnetic flux reversed only when a 1 is followed by a 0 and vice versa.

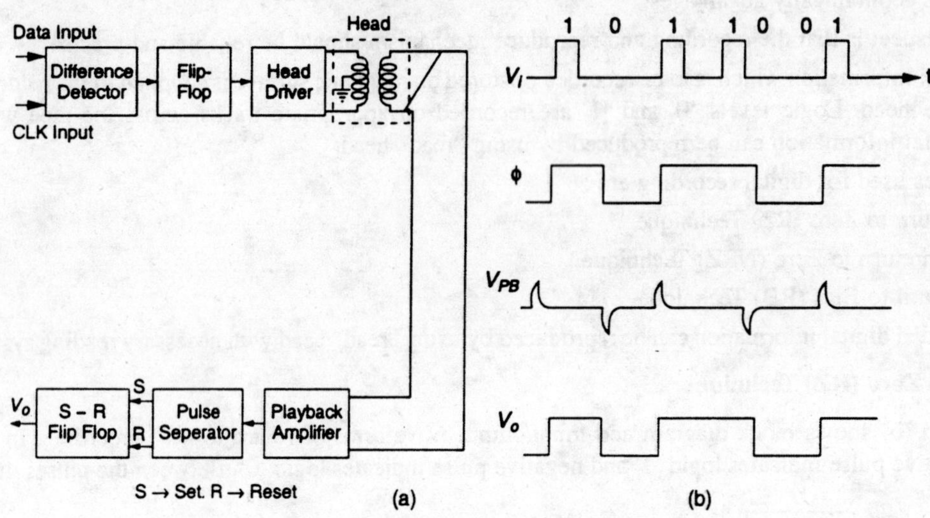

Fig. 8.9 (a) Block diagram of NRZ recording scheme
(b) Waveforms of NRZ recording scheme

Latest recording and playback system uses phase modulation and frequency modulation methods. Advantage of this technique is that it is very easy to reproduce the sequence of 1 and 0 with high frequency. The current flows through the winding of record head is negative through the entire bit time when 0 is recorded and is positive through the entire bit time when a 1 is recorded. The current changes only when a 0 is followed by 1 or 1 is followed by 0 is written. If the data contains sequence of 0's or 1's the current through head winding remains constant.

(C) Return to Bias (RB) Technique

Fig. 8.10 (a) and (b) show the block diagram of "RB" technique. When the input data is to be recorded or written, the current ($-I_R$) through winding saturates head at flux level ($-\phi_{sat}$), as the data switches to 1 the current ($+I_R$) flows through the head and data gets recorded because flux ($+\phi_{sat}$) develops. The saturation of magnetic material takes place over which the information gets recorded. '0' is represented by absence of pulse at clock time and '1' is indicated by positive pulse.

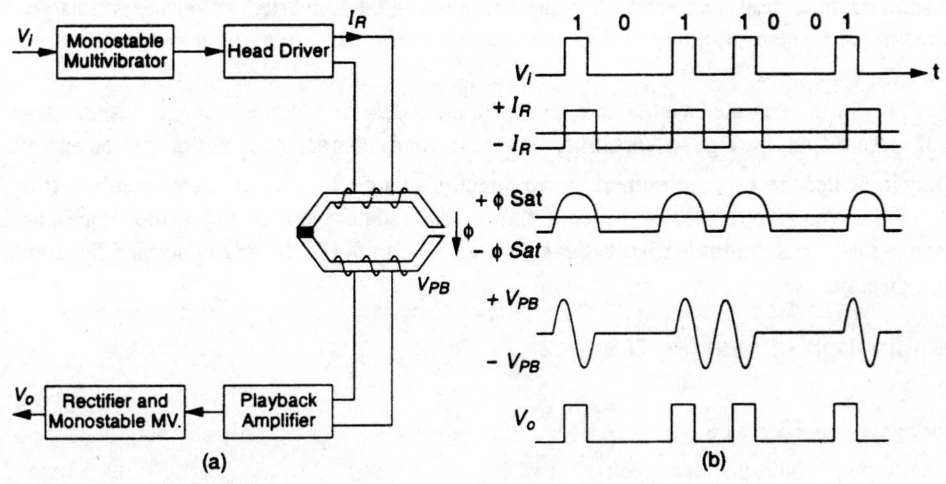

Fig. 8.10. (a) Block diagram of RB recording scheme
(b) Waveforms of RB recording scheme

At the playback or reproduce section playback amplifier is used to increase the voltage level. The playback voltage V_{PB} is proportional to rate of change of flux *i.e.* $d\phi/dt$. The magnetic flux changes at 1 so the flux changes due to which the read head reads the information so the clock pulse track is necessary.

The output voltage is rectified by means of rectifier and passed through monostable multivibrator to reconstruct the original signal which is recorded on tape.

Drawback of this system is to record the sequence of 0's because of saturation of head winding. It requires two flux transitions per bit to store the data hence packing density is reduced.

8.12 Selection of Recorders (Selection Parameters)

Following points should be considered while selecting recorder for a particular application

(1) Range of input voltage and current signals handled by the recorder.

(2) Frequency of the signals which are to be handled by the recorder.

(3) The recorder should have high input impedance (several thousand Ω).

(4) Type of writing mechanism required for recording such as ink pen, heated stylus, chopper stylus etc. according to the necessity of process or operation.

(5) Type of chart or graph paper *i.e.* wax coated paper, photosensitive paper etc.

(6) The recorder should have speed adjustment facilities.

(7) Accuracy and sensitivity of the recorders should be high.

8.13 Introduction of Digital Display Devices

Now a days in measurement and process control systems display devices are widely used to indicate the data or information in decimal digits. The evolution of analog to digital converters cover the wide spread use of display devices at moderate cost. Some of these devices are useful for displaying numeric and alphanumeric characters.

Digital display is an integrated system to convert the input data from some source to visual image. Major advantages of these displays are good versatility, low cost, sharp character resolution can be achieved, etc.

As display indicates the value in numeric form directly, a single display indicates numbers from 0 to 9. The number of digit corresponds to significant figure required to indicate the value. For indicating or displaying number of digits, multiple displays can be used. The multiple display is nothing but combination of single digit displays.

8.14 Classification of Display Devices

Display devices are classified as :

(1) Planner type (2) Non-planner type

In planner type displays the complete read-out characters are in the same plane. *e.g.* segmental displays.

The characters are displayed in different planes among the non-planner type displays. *e.g.* Nixie tube.

8.15 Light Emitting Diode (LED)

In instrumentation systems LED is most widely useful display device, it is also applicable in all type of electronic panel meters, instruments and products such as digital multimeters, watches, calculators, advertising displays etc. Most important application of LED is a light source for optical fiber communication.

Working Principle : It uses the phenomenon of electroluminescene *i.e.* illumination of light by the flow of current when voltage is applied to the material. Basically LED is a semiconductor P-N junction diode which emits light when current passes through it, if forward biased. In LED charge carrier recombination takes place at P-N junction when electrons from N-region recombines with the holes from P-region. Electrons from N-type semiconductor are in the higher conduction band whereas holes in P-type semiconductor are in the lower conduction band. The transition of electrons from conduction band to valance band makes emission of photons *i.e.* light and heat energy. Thus the emitted light comes but from the electron-hole recombination which occurs due to forward bias applied to the diode.

Due to forward bias the majority carriers electrons from N-region and holes from P-region cross the potential barrier and enter the opposite side. On the opposite side these are minority carriers resulting in increase in the minority carrier concentration, which increases the rate of combination. In reverse biased condition no charge carrier recombination takes place hence no emmision of light.

Construction : As discussed earlier, LED is P-N junction diode. The junction of this diode is heavily doped. Fig. 8.11 (a), (b) and (c) shows the construction of LED and symbol of LED.

The carrier recombination takes place in P-type, N-type layer is grown on a substrate and P-type is deposited on it by diffusion process. The P-type layer is kept uppermost. Metal connections are taken out from both sides by providing more upper surface of material for more radiations. The cathode connection is mostly on bottom side which is made of gold film to reflect the light.

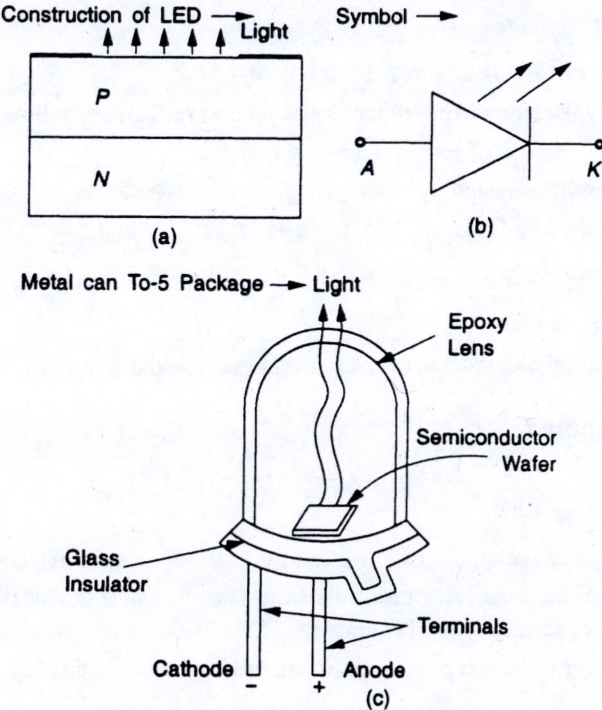

Fig 8.11 : Construction and symbol of LED.

Materials used for light emitting diode : Generally, materials used for manufacturing LED are gallium arsenide (GaAs), gallium arsenide phosphide (GaAsP), and gallium phosphide (GaP). These materials are widely used because they possess characteristics like maximum transition of electrons from conduction band to valance band, high refractive indices and high quantum output. Following table shows the typical LED materials with their respective colour.

Material	Colour
GaAs	Red
GaP (dopent N)	Green
GaP (dopent Zn, O)	Red
GaAs (dopent Si)	Infrared
GaAsP	Red/Green

Silicon and Germanium semiconductor materials are not used because most of the energy is given up in the form of heat so the emitted light is not acceptable. While selecting LED for particular application following points should be considered.

(1) Wavelength of the emitted light.

(2) Input power requirement.

(3) Location where the LED is to be mounted.

(4) Power requirements.

(5) Size of LED.

(6) Viewing angle.

(7) Package of LED.

Advantages of light emitting diode :

(1) Switching time of LED is less than 1 nanosecond.

(2) More versatility because of different colours such as red, green, yellow, infrared, amber etc.

(3) Good reliability because of longer operating lifetime.

(4) Operation cannot be affected by externally generated noise.

(5) Easily compatible with DTL and TTL family integrated circuits.

Limitations of light emitting diode :

(1) High power consumption.

(2) Can be damaged by overvoltage or overcurrent because these devices are sensitive.

8.16 Segmental Displays

(A) Segmental Display Using LED

In segmental displays single or individual LED is used for each segment of display.

Position of a digital display indicates numerals from 0 to 9; whereas alphanumeric display indicate numerals and alphabets with several symbol characters.

Seven segment display : These type of displays are used for numeric display. It consists of 7-segments

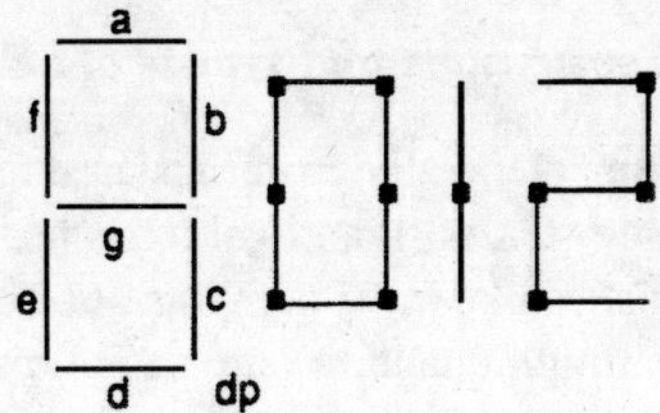

Fig. 8.12.

a, *b*, *c*, *d*, *e*, *f* and *g* with decimal point *dp*. By illuminating proper segments of display the numbers from 0 to 9 can be displayed as shown in Fig. 8.12. It is made up of seven rectangular light emitting diodes assigned by *a*, *b*, *c*, *d*, *e*, *f*, *g* and these LED's are called as segments. Now a days, LED's are available in single integrated package such as FND 500.

LED connections are provided by one common terminal and for corresponding segments other terminal is used.

(B) Fourteen-segment Display

These display devices are used for displaying both numerals and alphabets as shown in Fig. 8.13.

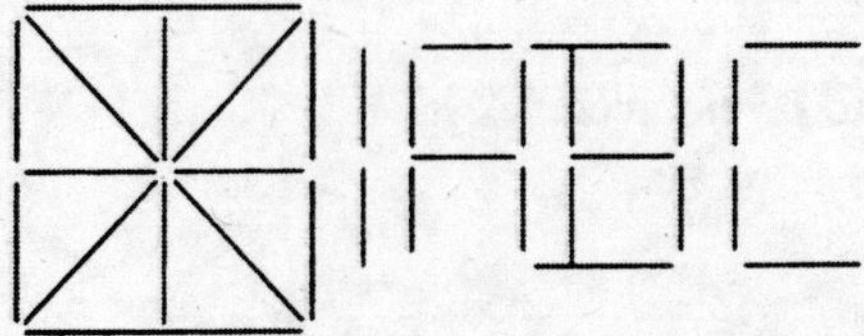

Fig. 8.13. Fourteen-segment display.

Common Anode (CA) connection :

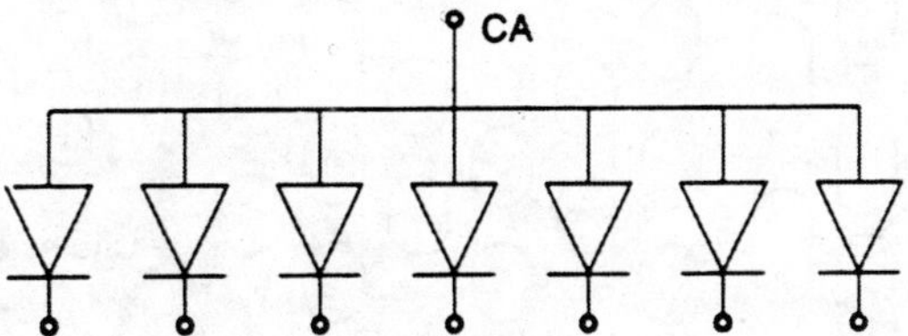

Fig. 8.14(a) Shows common Anode (CA) connection.

Common Cathode (CC) connection :

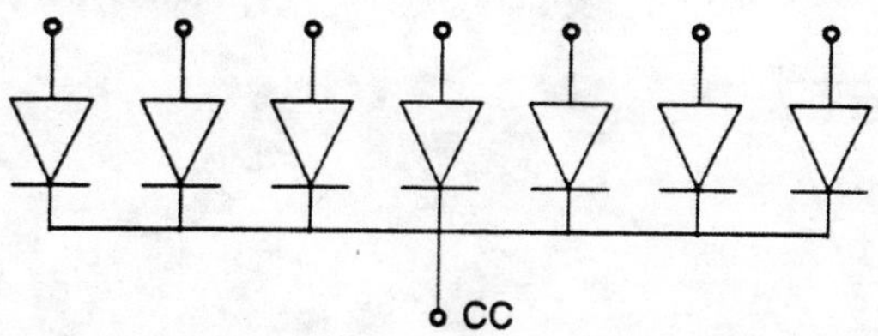

Fig. 8.14 (b) Shows common Cathode (CC) connection.

Dot matrices : In dot matrix type, arrays of LED displays are assembled in closed manner which provides better resolution and attractive performance due to their round shape. At each dot location one LED is placed.

8.17 Liquid Crystal Display (LCD)

This is also an electronic read-out and uses the characteristics of **Liquid Crystal** of materials. Like LED these devices are applicable for displaying numeric and alphanumeric characters in segment and dot matrix fashion.

Important advantage of these devices is that low power consumption of the order of few microwatt/cm. This is the passive device which modifies the light by scattering phenomenon. The intermediate stage between solid and liquid is called a **liquid crystalline stage**. In LCD, the electrical energy is not converted into visible light energy but it modifies the light so power requirement is in microwatt. Whereas in LED's, electical energy is converted in flourescent light and hence these devices require power in milliwatts.

The liquid crystals are simple organic compounds or mixtures of highly purified compounds. These crystal exhibits the optical properties like (1) birefringence (2) These crystals are sensitive to external excitation sources like electric and magnetic fields and heat. (3) Anisotropic nature of the dielectric constant allows the molecules to be assigned in an electric field. Fig. 8.16 shows construction of LCD.

In display technology the liquid crystal materials are classified as :

(1) Nematic crystal (2) Smectic crystal (3) Cholesteric crystal.

(1) **Nematic crystal** : In this type of crystal the rod like molecules are present and parallel to the one another separated by several micrometer distance as shown in Fig. 8.15(a). These molecules orients with respect to solid interface and perpendicular to the solid surface. The molecules placed are random in nature.

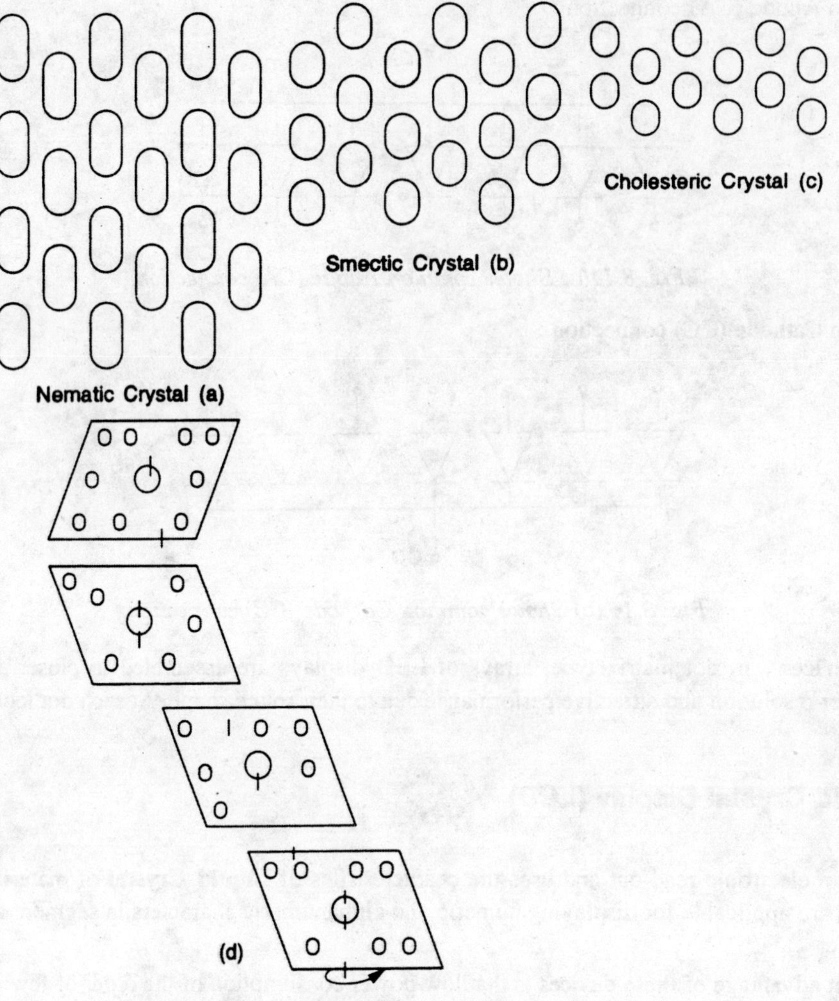

Fig. 8.15 : (a), (b), (c) and (d).

(2) **Smectic crystal** : In this crystal the molecules are aligned in layered structure *i.e.* the molecules form layers as shown in Fig. 8.15(b). The smectic crystal have slow response time to external fields.

(3) **Cholesteric crystal** : In cholesteric crystal the molecules are in complex structure. The molecules are not confined to materials derived from cholesterol but the axes of molecules twisted about the normal to the axis in a helical fashion. We can consider that the molecules are in sheets but the sheet is slightly rotated with respect to it's neighbours as shown in Fig. 8.15(c).

There are various modes of LCD operation such as dynamic scattering, twisted nematic (TN), phase angle and memory effects. Out of these most commonly used phenomenon is Twisted Nematic (TN) because the alignment of molecules is arranged parallel to the glass plates in a preferred direction. When the display crystal is observed from front to back the molecule forms helix. At front the molecules are placed vertically and at rare end they are placed horizontally. Molecules in center form 90 degree helix which is adjusted as viewing angle in construction.

Construction of LCD : It consists of two glass plates provided with electrodes, one side of this plate is transparent. The inner sides of glass plates are coated with Indium-Tin-Oxide (ITO) which may be transparent or reflective. For producing the different patterns of characters, segments, multisegment patterns the front edge of oxide coating is etched. After etching each segment is properly insulated from each other. In the hollow gap between the coated glass plates liquid crystal (LC) material is hermetically scaled as shown in Fig. 8.16. The liquid crystal may be Twisted Nematic (TN) or any other sealed material of few micro thick in size, generally it is 20 to 100 micrometer thick and uniformly aligned over the surface area. Sometimes the glass surface is coated with silicon oxide (SiO) layer for providing lining up to molecules with their directions parallel to glass surface.

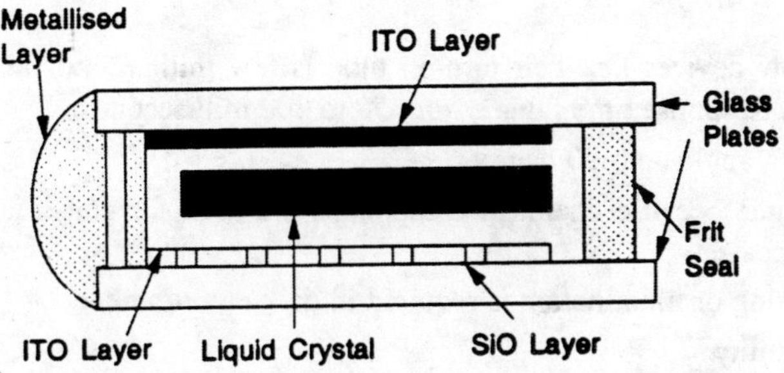

Fig. 8.16 : Construction of LCD.

The Fig. 8.17 shows effect of electric field on liquid crystal display (LCD). Due to electric field the helix of molecules will align with electric field, destroying the molecular helix. The polarized light entering from front of the liquid crystal will not be twisted but emerges from rear of display polarized in vertical plane. When vertically polarized light encounters the horizontally oriented real polarizer, the light will be attenuated. In this case if the reflector is placed no light will be reflected and display would appear dark.

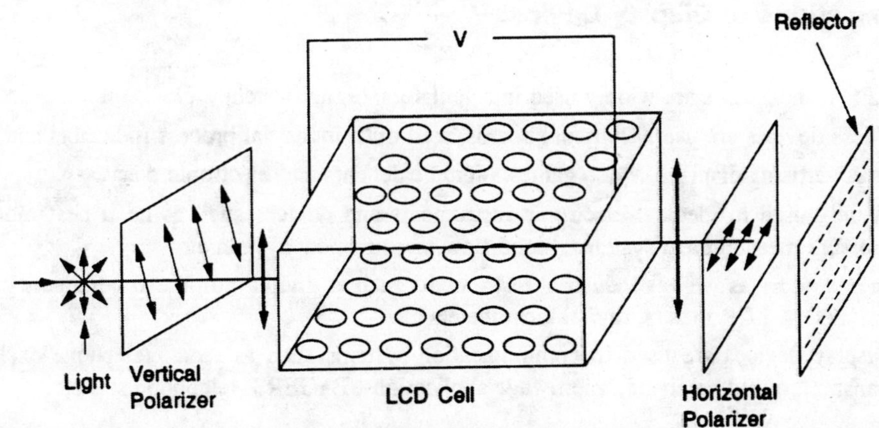

Fig. 8.17 : Effect of electric field on LCD.

Crystal compounds used or materials used for LCD :

These are simple organic compounds or mixture of highly purified compounds. The chemical compounds are :

(a) 4-Methoxy-4-n-Butyl-Azox-Benzene (MBAB)

(b) N-(p-Methoxy Benzylidene) p-n-Butylaniline (MBBA)

(c) Cholesterylerucate (CE)

(d) Cholesteryl Nonnoate (CN)

Advantages of LCD

(1) Low power consumption.

(2) They don't require external interface circuitry *i.e.* can be directly driven by IC.

(3) Low cost.

Disadvantages of LCD

(1) They are slow devices *i.e.* their turn-on time is few milliseconds and turn off time is tens of milliseconds. Response time ranges from 50 to 200 milliseconds.

(2) Small viewing angle upto 90 degrees.

(3) Limited lifetime because chemical compounds are used for construction which may be disintegrated or decayed.

(4) External lighting or illumination is required in dark environment.

(5) Limited reliability.

(6) Temperature range is limited *i.e.* 0 to 60 degrees.

These devices can be available or fabricated from few inches to a foot or larger. LCD arrays, dot matrix, segmental displays can also be fabricated.

Applications :

It is used as display device in commercial calculators, digital clocks, industrial panel control boards, panel of instruments, in defence instruments and equipments, multimeters, LCR meters, digital tachometers, in advertising display boards, industrial process controls, in biomedical instruments etc.

8.18 Applications of Display Devices

(1) LED's and LCD's are widely used in calculators, digital watches, clocks, microprocessor kits.

(2) These devices are useful in panel boards, read-outs, industrial process indicators and controllers.

(3) In advertising displays, video games, digital calendars and electronic diary.

(4) Widely used in medical electronic instruments and devices such as heart beat monitor, blood pressure measurement systems, blood flow measurement system etc.

(5) In laboratory as well as industrial instruments such as digital multimeters, frequency meters and voltmeters, LCR meters, digital tachometers etc.

(6) Display devices are used in communication systems such as radar, telephone exchanges, TV transmitters and receivers, microwave applications, PAGERS, telephone sets etc.

8.19 Indicators and Analog Displays

In instrumentation systems indicators and analog display devices play an important role. These devices are used for the measurement of a analog quantity or process variable. In general it consists of a graduated scale on which the digits are shown in a sequence over which the pointer slides or moves and indicates the digits of interest on scale.

In indicators scale and pointer are important parts. Every instrument uses any one movement from the various movements like D'Arsonval movement, moving iron, electrostatic, dynamometer, polarized iron. The D'Arsonval movement is applicable for measurement of DC and it can measure AC if rectifier stage is provided. The moving iron movement is applicable for the measurement of AC and DC variables. The electrostatic movement measures high value of AC or DC voltages. The dynamometer instrument measure AC or DC current, voltage and power parameters.

This discussion indicates that if any movement can be used the scale and pointer are necessary elements because it provides the accurate observations or scales. Indicator conveys the information or measurement of parameters using graduated scale over which digits are shown in sequence. Pointer is a movable reference used to indicate the digits on scale according to range and variable which is to be measured. Depending upon scale and pointer assembly the indicators are classified as :

(a) Fixed Scale Indicators

(b) Movable Scale Indicators

(A) Fixed Scale Indicators

In this type of indicators scale is fixed, whereas the pointer slides or moves over fixed scale. These are most commonly used indicators in instrumentation systems. The pointer consists of different coloured metal elements and the background colour is in contrast so as to get better readability. Sometimes transparent plastic may be used for pointers. The pointers may be coated with luminescent material so that good readability can be obtained in darkness. For good readability of observations (scale) the lighting source is introduced or placed near the scale. The lighting sources used may be neon lamps, fluorescent, incandescent lamps. Mostly instrument reads the scale from left to right or in clockwise manner. Fig. 8.18 & 8.19 shows various fixed scale indicators.

In measurements error can be occurred due to "PARALLAX" in instruments. This effect of parallax gets eliminated by using mirrored scales and knife edged pointer arrangements. Mirror is placed in the same plane of dial and observation or reading is taken by coinciding the pointer and mirror image of pointer.

Fig. 8.18 shows the dial type or gauge dial type indicator. It has a circular scale of 270 angular degrees. Generally it is used as a pressure gauge.

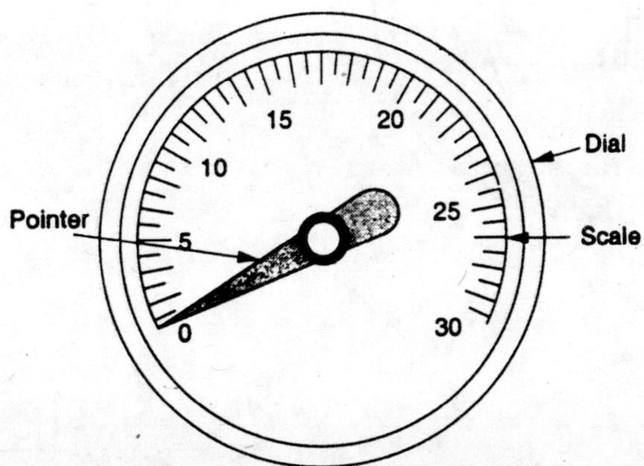

Fig. 8.18 : 270° gauge dial.

Fig. 8.19 shows multiple scale indicators. These indicator consists of different ranges on the scale for same variable. *e.g.* thermometer which indicates the temperature in two units like Farenheit and degree Celsius.

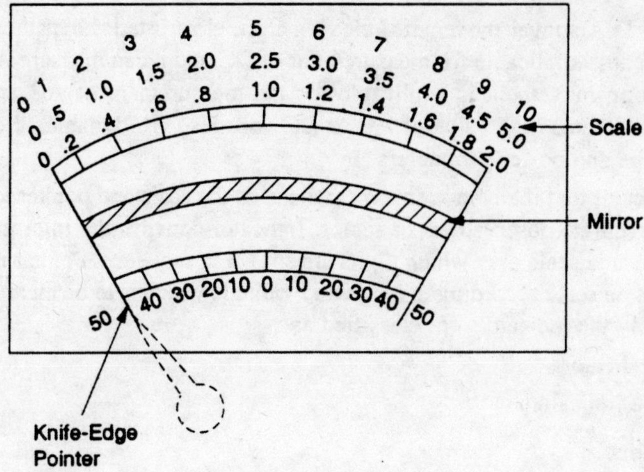

Fig. 8.19 : Multiple scale.

For multiple-scale meters the range selector switch is provided so as to indicate different ranges of same variable on selected range of scale.

(B) Movable Scale Indicators

This is the different method to that of fixed scale indicators. As the name implies the scale is moving or

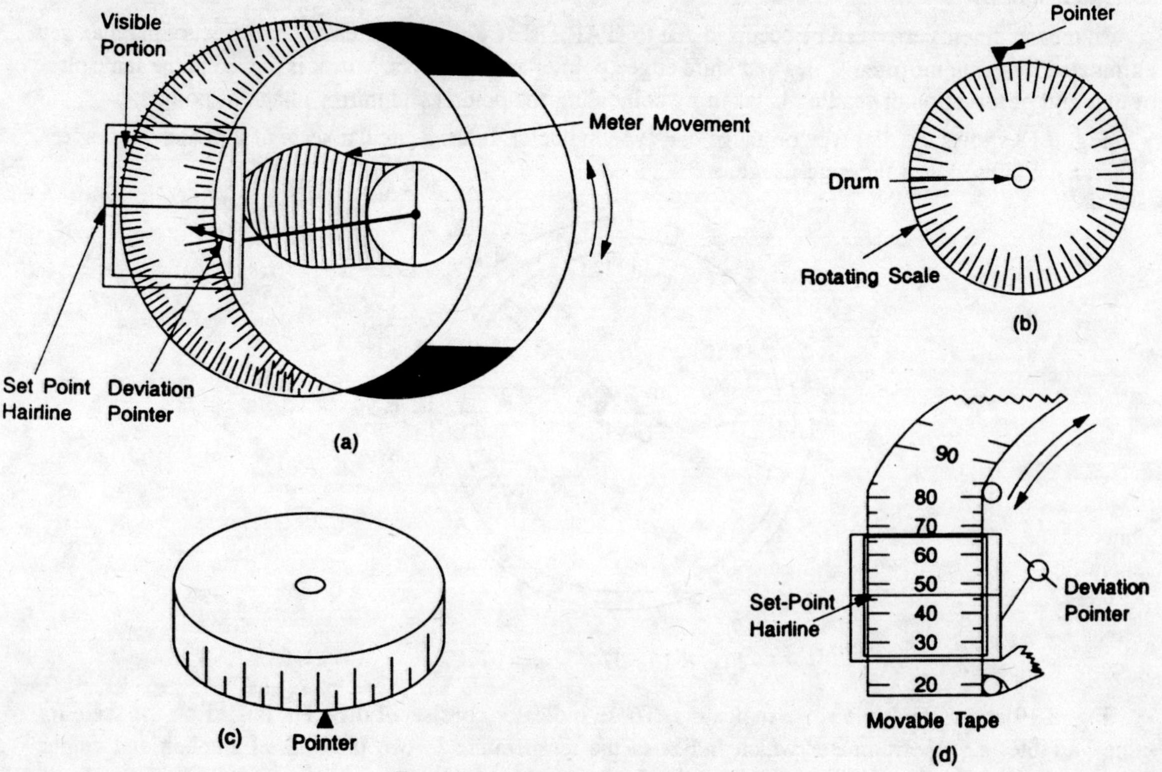

Fig. 8.20 : Movable scale indicators.

rotating whereas the pointer is stationary. Divisions and subdivisions are marked on a scale. This scale is moving on a rotating drum arrangements as shown in figure. It has deviation pointer and set point hairline. The visible window or portion is provided so as to read the scale. The scale consists of a large circular disc with graduations on its entire periphery, read at an index in 12 O'clock position. This system has great precision in reading the subdivisions.

Fig. 8.20(a), (b), (c) and (d) shows various constructional details of movable scale indicators.

8.20 Light Indicators

In instrumentation systems, various light indicators are used for conveying general information and representations to the operator or user. Generally light indicators are used for conveying the information such as status of the process as "ON" or "OFF", normal or abnormal conditions, alarm conditions, overflow, range, set point, failure etc.

Various indicators like coloured, flashing, non-flashing having various sizes, shapes, visibility are available. Light indicators are classified as neon lamp, incandescent lamp, solid state lamps. Following factors should be considered while selecting the light indicators :

(1) Size, shape and colour of the indicators.

(2) The light should be visible to the operator.

(3) The colour of indicators should be powerful so that it can be flashed in dark.

(4) Brightnes and visibility of the indicators should be excellent.

(5) Indicators should be located at proper angle of line of sight so that it can be visible clearly and sharply.

(6) Operating voltages and current required for light indicators.

(7) Necessary lamp holder and lens compatible with lamp should be provided.

(8) Sufficient space and area should be provided on the panel for installation of indicators.

(9) Light indicators should not be damaged by the parameters like temperature, humidity, dust, strong vibrations and shocks, etc.

There are three major types of light indicators as follows :

(a) Incandescent lamp

(b) Neon lamp

(c) Solid state lamp or Semiconductor devices : Light emitting diode (LED), LCD.

(A) Incandescent Lamp

Fig. 8.21(a) shows constructional details of incandescent lamp. It consists of a tungsten coil filament mounted on two support wires in an evacuated glass bulb. This assembly is fixed in a base through which electrical connection is possible. When the current flows through the coil it glows and emits the light and heat energy. Generally, the light emitted is white in colour and in visible band. These lamps operate on the range of 1 to 120 volts DC or AC and current from 10 mA to 10 Amp. It is low in cost, emits more light and is highly reliable. Specially designed incandescent lamps can withstand shocks and vibrations in operating conditions.

(B) Neon Lamp

Fig. 8.21(b) shows constructional details of neon lamp. It consists of two metal electrodes placed in a glass bulb filled with neon gas. This complete assembly is fixed in a base through which electrical connection is made.

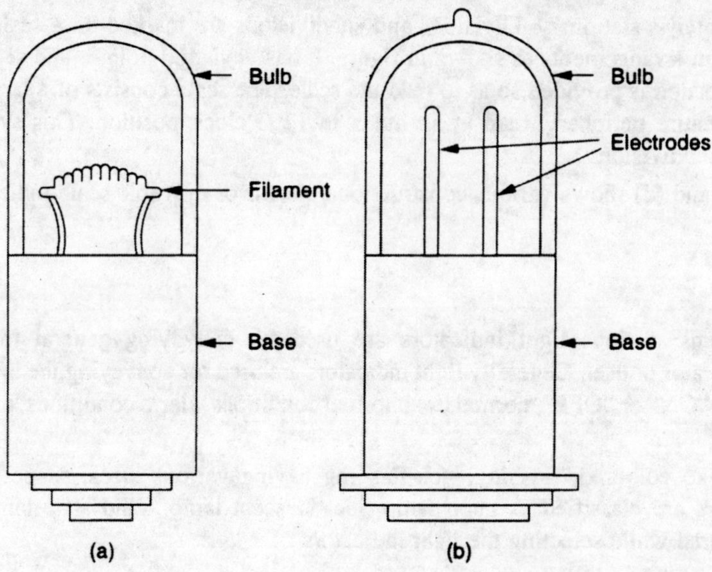

Fig. 8.21 : (a) Incandescent lamp, (b) Neon lamp.

When input voltage is applied to the metal electrodes, ionization of neon gas takes place, so current conduction takes place resulting to emit the light and heat energy. It emits orange colour light which is easily visible. Operating voltages of neon lamp is 120 or 240 AC voltages. As these lamps operate on higher voltages they require small range of current and power.

(C) Solid State Lamps

These lamps are nothing but semiconductor P-N junction light emitting diode. Its construction, working, advantages etc. are discussed in this chapter. Characteristics of these devices are longer life, efficient, available in different colours.

Components necessary for light indicators are lamp holder, lamp and lens.

8.21 Installation of Light Indicators

Following points to be considered for installation of light indicators.

(1) The light indicators should be installed with proper casing and cover so as to protect them from shocks, vibrations, temperature, dust, humidity etc.

(2) Watertight or waterproof design of indicators should be used in environment which contains moisture, humidity, and corrosive elements.

(3) Oiltight design should also be used.

(4) Various colours can be assigned so as to indicate different type of information such as :

(a) RED colour : Indicates critical information to the operator so as to take fast action.

(b) AMBER/YELLOW : Abnormal or caution condition gets informed to the operator.

(c) WHITE colour : Normal condition.

8.22 Gas Discharge Displays

The gas discharge displays are alphanumeric displays. These devices works on the phenomenon that is emission of light in a cold cathode gas filled tube under breakdown conditions.

Nixie tube is the cold cathode gas filled tube display or indicator. This tube consists of a glass envelope in which a common thin frame anode is used and positive voltage is applied to it. Also there are ten independent wire cathodes shaped numbers from 0 to 9. In a glass tube the gas filled is mixture of neon and argon. At bottom of the glass tube pins are provided for electrical connections.

When a negative voltage is applied to the selected cathodes, gas around the cathode gets discharged and emit a yellow-orange light. Due to this the selected character gets illuminated. For switching desired cathodes, transistorised driver circuit can be used.

Specially designed Nixie tube has additional cathodes provided so as to indicate (+) and (–) signs. Nixie tubes are also used for indicating alphanumeric characteristics for this additional (*e.g.* 15) cathodes are provided which constitutes the segments of character. These Nixie tubes are also applicable for indicating numeric values. Decimal point may be provided if desired. These tube requires DC voltage of 150 to 220 Volt and 1 to 5 mA current for selected cathodes.

Fig. 8.22(a) shows front view of Nixie tube.

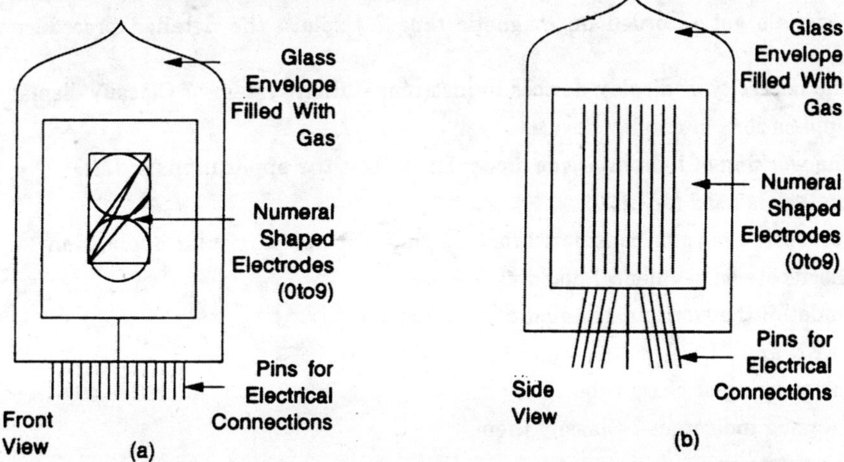

Fig. 8.22 : (a) Front view of Nixie tube, (b) Side view of Nixie tube.

Fig. 8.22(b) shows side view and Fig. 8.23 shows symbol of the Nixie tube.

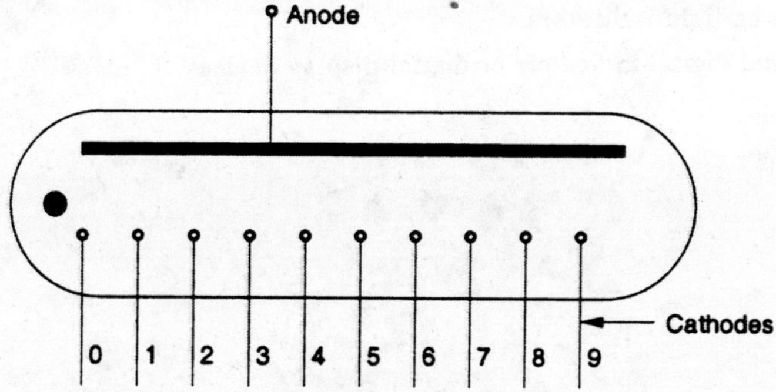

Fig. 8.23 : Symbol

EVALUATION COMPONENT

1. Define the term "Recorders". Classify them.
2. Why recorders are necessary in instrumentation systems.
3. What is the significance of recorders.
4. What are the essential parts of strip chart recorders ? Explain them.
5. Write a short note on "marking mechanisms" used in recorders.
6. What are the selection parameters to be considered for selecting a recorder for particular application ?
7. Explain the working of galvanometric type recorder.
8. Draw neat diagram of potentiometric type recorder and explain its working.
9. Explain in detail the working of (a) RZ (b) NRZ (c) Return to bias technique with neat diagram.
10. Write a short note on digital recording.
11. What are the applications of *X-Y* recorders ?
12. Explain frequency modulation recording method in detailed fashion.
13. How the signals get recorded on magnetic tape ? Explain the detailed procedure with neat diagram.
14. What is the necessity of display devices in instrumentation systems ? Classify display devices.
15. List the applications of display devices
16. Explain the working of light emitting diode (LED). List the applications of LED.
17. List the materials used for LED.
18. What parameters are to be considered for selecting LED for particular application ?
19. Write a short note on segmental displays.
20. Explain in detail the working of liquid crystal display.
21. Compare LED and LCD.
22. Explain the working of Nixie tube.
23. What are analog indicators ? Classify them.
24. Write a short note on :
    (i) Moving-scale indicators
    (ii) Moving-pointer indicators
25. What do you mean by light indicators ? Classify them.
26. Write a short note on light indicators.
27. Compare analog and digital indicators or digital display devices.

Chapter 9

CONTROLLERS

Learning Objectives

- To study the concept of Process Conrol.
- To study Automatic Controllers.
- To discuss analog and digital controllers.
- To elaborate various control modes like ON-OFF, proportional, derivative, integral, PI, PD, PID.
- To study electronic PI, PD, PID controllers.
- To discuss working of pneumatic controllers.
- To study hydraulic controllers.
- Introduction and concepts of fuzzy logic controllers.
- To study FLC's for different applications.

9.1 Introduction

The main objective of process control is to regulate or maintain a particular quantity at reference value (set point) by avoiding external interferences. Before development of automatic control system all the processes or functions were carried out by human beings. But disavantages of this controllers are slow operating speed, inaccuracy, improper control action, difficult to operate in hazardous working conditions etc.

Now a days computer based (digital) controllers, programmable logic controllers (PLC), Fuzzy Logic controllers (FLC) are becoming very popular in instrumentation systems and process control.

The function of controllers is to compare the measured value with set point and depending upon difference between these two parameters the correction signal or control action is generated for final control element.

This chapter covers different type of controllers like electronic, hydraulic, pneumatic, fuzzy logic etc. In section 9.5 different control modes with their analytical expressions is discussed. It also covers the advantages, disadvantages and applications of different controllers.

In the last section of this chapter recently developed Fuzzy Logic controllers (FLC) are discussed. This section covers basic block diagram and concepts of FLC. Applications of FLC in instrumentation fields like DC drive control and water heater, washing machine drier are discussed.

Controllers are widely useful in process industries like paper, sugar, plastic, textile etc., They are also used in power plants and in domestic appliances.

9.2 Process Control

Fig. 9.1 represents block diagram of systems used in process control, where

(1) Circle indicates the algebraic function such as addition or subtraction.

(2) Rectangular box indicates a dynamic function.

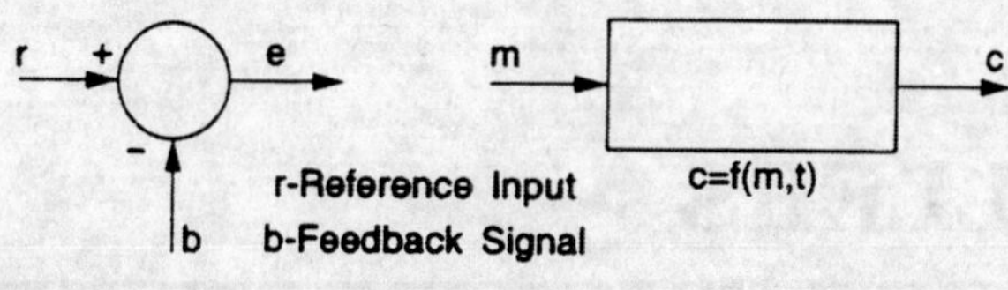

Fig. 9.1.

Here, c = Output or controlled variable.

b = Feedback signal

m = Input or manipulated variable.

The block diagram is used to study, analyse and represent the performance, behaviour of the system or process.

The process gain is ratio of change in output (dc) and change on the input that caused it (dm).

The process component changes its output in response to a change in its input. **Primary objective of process control is to regulate or maintain the desired quantity at fixed value**. Automatic control system uses feedback path which constitutes a closed loop action. In automatic control system no human aid is involved for controlling the process. It also maintains the controlled variable at the desired value or set point.

Automatic control systems are used to control the operations involved in industrial processes. It is useful in industries to control the variables such as temperature, pressure, liquid level, fluid flow, force etc. Widely this systems are used in chemical industries, sugar and paper industries, power plants, automobile industries, power mechanisms, food industries, aircraft's, missiles, ships etc.

Mostly automatic process control system is useful to perform tedious and critical operations in the process. Feedback is the important aspect in process control. This technique is the important aspect in process control. This technique is used to maintain or regulate the desired quantity by comparing it with the measured quantity (set-point). After comparing, the difference signal gets corrected by using corrective action. Feedback control is simple and accurate phenomenon for achieving desired quantity by eliminating the disturbances developed in the system.

9.3 Automatic Controllers

The automatic controller consist of measuring unit, input signal, actuating signal, controller and the output stage.

The term controller is defined as it generates a control for the final element depending upon the deviation. The function of automatic controller is to compare the actual value with the desired value and to determine the deviation, it also provides counter action for maintaining smallest deviation.

The method of producing counter action is called control action or mode of control. The control actions can be produced by using electrical, mechanical, hydraulic, electronic, pneumatic, automatic controllers etc.

The important functions of automatic controllers are to receive the measured value of variable which is to be controlled. It compares the measured variable with set point. These controllers calculate the magnitudes of deviation between measured value and set point. The calculation of deviation provides control signal to reduce the deviation up to zero value.

For selection of type of controller for a particular application, points to be considered are nature of industry or plant and operations involved, accuracy, security, economical conditions, size and weight of controllers etc. The device which maintains a process variable value at set point is called **controller**. The controller should have dynamic characteristics such as flexibility, wide operating range etc.

The automatic controllers should follow the functions like :

(1) to receive the actual measured value of the variable which is to be controlled.

(2) to compare the measured value with set point.

(3) to determine the deviation or error.

(4) to generate the output control signal in such manner to reduce the deviation to zero or small value.

9.4 Electronic Controllers

The electronic controllers are classified as :

(1) Analog electronic controllers.

(2) Digital electronic controllers. (Digital controllers)

Analog controllers are simple, less expensive and do not require analog-to-digital and digital-to-analog converters. It is easy to interface analog controllers with other control components of the system. But analog controllers are suffered by the errors such as hysteresis, non-linearity, drift etc. The essential requirements for analog controllers are input section, output section, control unit, switching section, power supply and display unit.

(A) Basic Block Diagram of Analog Controllers

Fig. 9.2 shows basic block diagram of analog controllers. The control section may consists of automatic controller or manual controller. The output of controller can be selected by switching arrangement. The deviation signal gets generated at input section by comparing the process variable and set point signals. By providing switching mechanism, automatic or manual mode of control can be selected.

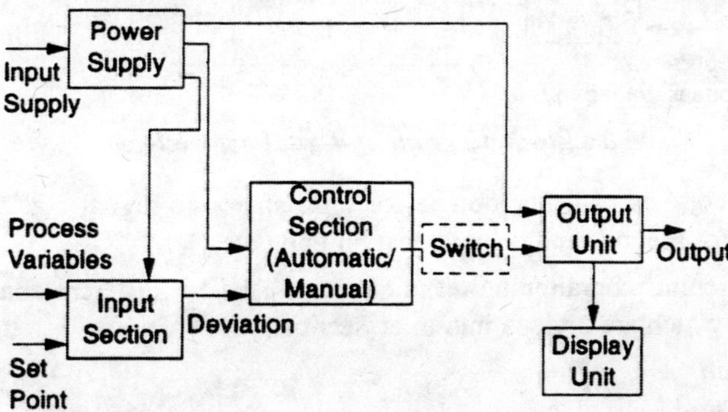

Fig. 9.2 : Basic block diagram of analog controller.

The display section is provided to indicate and display the values of set point, process variable, deviation and output of controller. Power supply is necessary to excite the instrumentation system and circuitry involved for controlling the process. Power supply also contains rectifier unit for converting the a.c. signals into d.c. supply.

In analog controllers the various circuits used are integrator, differentiator, amplifiers etc. to generate required control signals. Set point in the system can be generated internally or externally from the remote locations.

(B) Digital Controllers

Now-a-days digital controllers are widely used in the industries for controlling different processes. The advantages of digital controllers are :

(1) More accuracy.

(2) Precise control action with more stability.

(3) Not suffered by the problems such as drift, distortions, stray effects in the components which may be possible in analog controllers.

(4) Operating speed of digital controllers is fast.

(5) It is possible to change the configuration or redesign the system by changing the software or algorithms without changing the hardware and involved parts of the system.

(6) It is possible to transmit the data over long distances to other devices and can be displayed accurately also in detailed manner. It provides information to the operator regarding different parameters involved in the processes.

Fig. 9.3 shows block diagram of digital controllers. Digital controllers have two important parts :

(1) Hardware section (2) Software section.

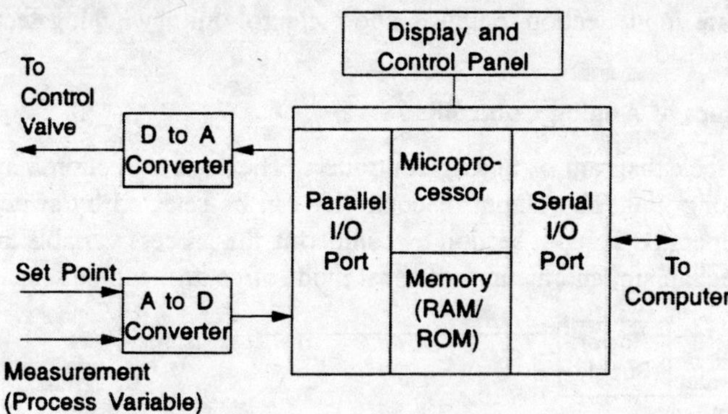

Fig. 9.3 : Block diagram of digital controllers.

The hardware section consists of a microprocessor unit, analog-to-digital and digital-to-analog converters, memory units, control sections and communication ports etc.

In software section different programming techniques are used for different control functions. It has various computer programs which are divided into three sections :

(1) Execution program.

(2) Application program.

(3) System support software.

The execution program is used to execute the application program to control complete hardware section involved in the system, to control and supervise the input and output operations, to store the signals into memory.

Application programs are used for converting the input data involved in different processes into engineering units, to control the correction, computations, optimization, to manage logging and console of the system.

System support software program involves assemblers, compilers, loader, linker, debugger etc.

Block diagram of digital controllers : For digital control of the system single processor is necessary to execute different programs and control loops. It is possible to develop digital control algorithms in machine language or assembler etc. Now-a-days the manufacturers are providing standard algorithms with the processor which are called as **Blocks.**

Initially, the analog signals (set point and process variables) are converted into digital signals with better accuracy and high resolution. In memory section the programs necessary for controlling the various operations are stored. The microprocessor is used to calculate and analyze the set point.

9.5 Control Modes

The function of controller is to compare the measured value of process variable to its set point. The difference is error signal '*e*' it generates a correction signal to final control element so as to eliminate the error. Following are the basic control actions :

(a) Two Position or ON-OFF Control Mode.

(b) Proportional Control Mode.

(c) Integral Control Mode.

(d) Derivative Control Mode.

(e) Proportional-Plus-Integral (PI) Control Mode.

(f) Proportional-Plus-Derivative (PD) Control Mode.

(g) Proportional-plus-Integral-plus-Derivative (PID) Control Mode.

These controls are as discussed below :

(A) Two Position or ON-OFF Control

This is the old method of control action. As the name indicates ON-OFF, this control action has two fixed positions *i.e.* ON means position 1 and 0 it indicates OFF position. These two positions are "ON" and "OFF" positions of the process.

Consider an example of room heating system, which turns ON the heating system when room temperature is low and turns OFF when room temperature is at desired set point or desired level.

Fig. 9.3 (a) represents the block schematic of ON-OFF controller. The output signal '*m*' depends upon the error signal '*e*'.

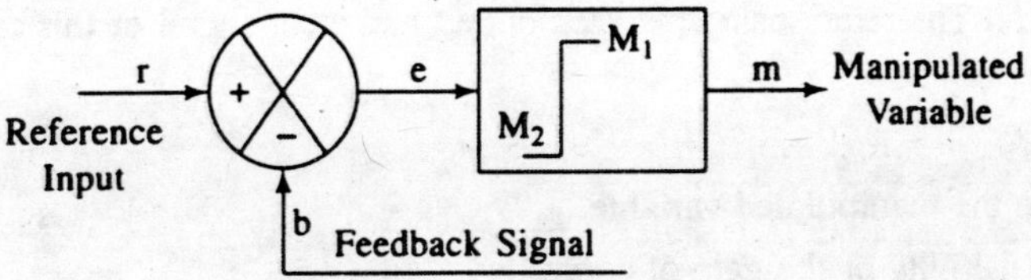

Fig. 9.3 (a) Block diagram of ON-OFF controller.

If the error signal *e* is positive the output is maximum and if error signal *e* is negative the output is minimum. The control action is represented as :

$$m = M_1; e > 0. \qquad ...(9.1)$$
$$= M_2; e < 0.$$

where ; M_1 and M_2 are constants.

The controller should get "ON" position when the measured value or variable is below the set point and the manipulated variable has maximum value M_1 *i.e.* when $e > 0$. If the measured value or variable is above the set point and manipulated variable has minimum value M_2 *i.e.* at $e < 0$; controller is in the "OFF" position.

In the operation of "ON-OFF" controller the differential gap is developed due to mechanical friction, electrical contacts, mechanical arcing, lost motion etc.

If sinusoidal input is applied to two position controller then the output response is as shown in Fig. 9.3(b).

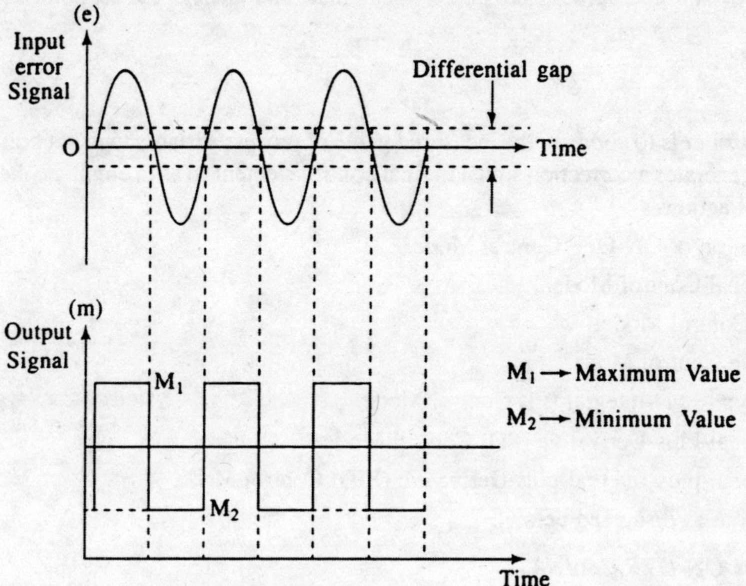

Fig. 9.3 (b) Response of two position controller to Sinusoidal input.

It is necessary to eliminate the differential gap. The two position controllers are inexpensive. These controllers are widely useful in industrial and domestic control systems.

(B) Proportional Control Action

In proportional control action there is a continuous linear relationship between the input and output of the controller. The output of the controllers is proportional to error *i.e.* the correction developed by this control action is proportional to error. The relationship between output and error signal of this controller is as :

$$m \propto e$$
$$\therefore \quad m = Kp \cdot e \qquad ...(9.2)$$

where ; m = output signal to the manipulated variable.

Kp = proportional sensitivity or the gain of controller.

e = error signal or deviation from set point.

The proportional sensitivity Kp is the parameter used to adjust the gain of the controller. Fig. 9.3(c) shows the proportional controller.

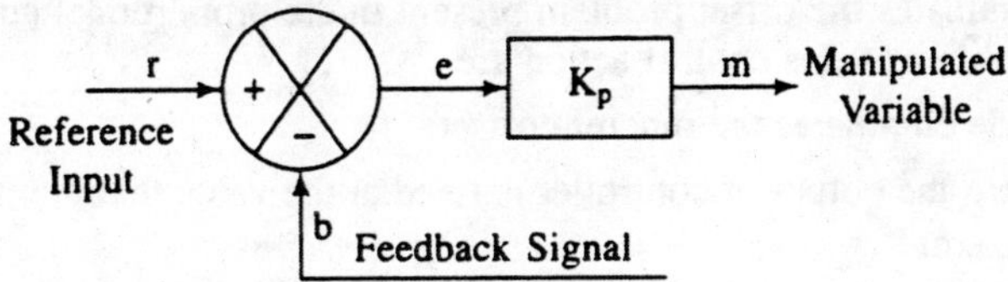

Fig. 9.3 (c) Block diagram of proportional Controller.

The proportional sensitivity *Kp* is expressed in terms of percentage proportional band (PB) and is given by :

$$Kp = 100/PB$$

so the relationship is expressed as :

$$m = (100/PB)\, e + b \qquad ...(9.3)$$

where ; PB = Proportional band = 100/*Kp*

b = Live zero or bias of the output (controller output with no error).

From above equation it is observed that if there is no error the output of the controller is equal to bias.

The drawback of proportional control action is that it does not maintain the controlled variable at desired set point.

(C) Integral Control Mode

The drawback of proportional controller is offset, this error is eliminated in integral control mode. This mode is also called reset mode because after a load change, it returns the controlled variable to set point and eliminates the offset. The relationship between output and error signal of the controller is expressed as :

$$m = 1/T_i \int_0^t edt + b \qquad ...(9.4)$$

where ; T_i = Integral time setting of the controller.

The integral control mode is defined as a device whose output is time integral of the deviation. This mode is used in different flow control systems and valve position control.

(D) Derivative Control Mode

The relationship between output and error signal of the controller is given as :

$$m = T_d\ de/dt \qquad ...(9.5)$$

where ; T_d is the time derivative. It provides high sensitivity. It is always used in combination with proportional or proportional-plus-integral controller because it operates on rate of change of error and not the error itself.

(E) Proportional-plus-Integral Mode (P-I Mode)

This control action is the combination of proportional and integral control actions. Its output is proportional to a linear combination of input error and time integral of input error. The analytical equation of this controller mode is expressed as ;

$$m = Kp\left[e + \frac{1}{T_i}\int_0^t edt\right] + b.$$

$$\therefore \qquad m = Kpe + Kp\,.\,\frac{1}{T_i}\int_0^t edt + b \qquad ...(9.6)$$

This control action eliminates the offset problem present in the proportional controller. It is also called "Reset" control action. Features of this control action are ;

(i) The integral mode eliminates the inherent offset.

(ii) When error is zero the output of controller is fixed at the value that the integral term had even if error reaches to zero.

(iii) When error is non zero the proportional controller provides correction and the integral time (T_i) varies (*i.e.* either increases or decreases) the accumulated value depending upon the error and type of action (*i.e.* direct or reverse).

(iv) The integral control action can be adjusted by varying the integral time (T_i).

(v) This control action is useful where large load variations are present.

(F) Proportional-plus-Derivative Mode (P-D Mode)

As name indicates it is the combination of proportional and derivative control actions. It is widely used in industrial applications. Its control action is defined mathematically as ;

$$m = Kp\left(e + T_d \frac{de}{dt}\right) + b$$

$$\therefore \quad m = Kpe + KpT_d \frac{de}{dt} + b \qquad \text{...(9.7)}$$

This control action sometimes fails to eliminate the offset of proportional controllers. It is also called **Rate** control action. The controller output varies with respect to rate of change of actuating error signal.

(G) Proportional-plus-Integral-plus-Derivative Mode (P-I-D Mode)

This controller combines the proportional, integral and derivative control modes. The analytical expression for this controller is

$$m = Kp\left(e + \frac{1}{T_i}\int_0^t edt + T_d \frac{de}{dt}\right) + b \qquad \text{...(9.8)}$$

This controller eliminates the problem of offset in proportional controller. It is fast and powerful controller but its designing is complex.

Depending upon actuating media the controllers are classified as (i) Electronic Controllers. (ii) Pneumatic Controllers. (iii) Hydraulic Controllers.

Which are discussed in next sections of this chapter.

9.6 Electronic Controllers

In analog (electronic) controllers the active element used is OP-AMP (operational amplifier) because of its important characteristics and widely useful applications. Other circuit elements are also used with OP-AMP for designing the controllers. The controller accepts analog signal which may be electric current or voltage. In following section different electronic control modes are discussed.

(A) Two-position Control

Two-position control is widely used in household appliances and devices such as air conditioners, water heaters and are also used in industrial control. These are inexpensive controls. In this type of control if the controlled variable increases or decreases than certain fixed value (set point) then the electrical signal operates (drives) a correcting unit. The correcting unit may be a relay or a solenoid or a motor controlled valve.

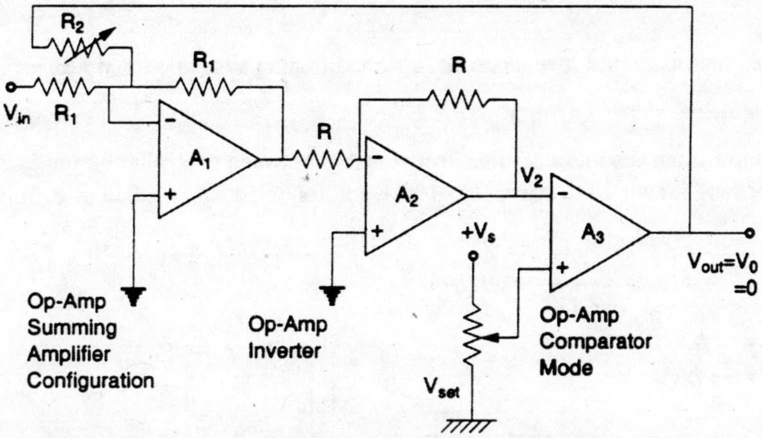

Fig. 9.4 (a) Two position control.

Fig. 9.4 (a) shows two position or ON-OFF controller which consists of adder (summing amplifier) and comparator configurations of op-amp.

Here, V_{in} is the input voltage signal of the controller. The applied voltage levels are supposed as V_{ON} is the "ON" or High voltage and V_{OFF} "OFF" or LOW voltage. V_2 is the output of inverter A_2. At the input of A_3 the voltage levels V_2 and V_{set} gets compared. The output of comparator A_3 is V_{out}. The comparator changes its output level V_{out} when the voltage V_2 is equal to set point voltage V_{set}. The design values of resistances of inverter circuit lies in the range of 1 to 100 kΩ.

Circuit Analysis :

Case I : When the input voltage level is V_{ON} (HIGH).

$$V_{ON} = V_{set} \quad \text{...(9.9)}$$

Case II : When the input voltage level is V_{OFF} (LOW).

$$V_{OFF} = V_{set} - R_1/R_2\, V_0 \quad \text{...(9.10)}$$

Response of two-position controller : Fig. 9.4 (b) shows response of two-position or ON-OFF controller. It is the relationship between input and output voltage levels. Here N.Z. is the neutral zone or differential gap in the response. Width of nautral zone is in between V_{OFF} and V_{ON} which can be set or adjusted by changing R_2.

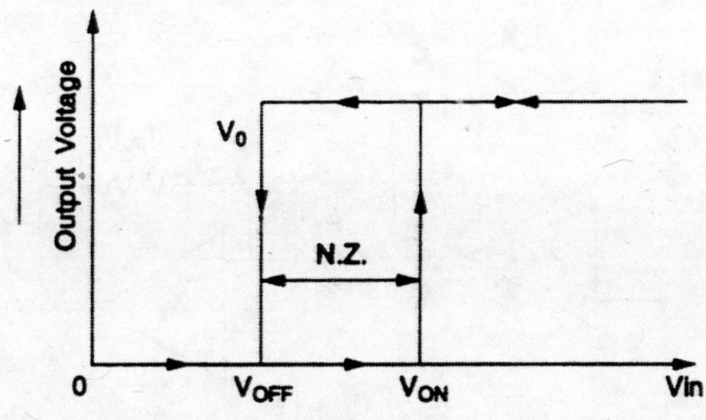

Fig. 9.4 (b)

Applications :

In temperature controls, liquid level controls, air conditioning systems, room heaters etc.

(B) Proportional Control

In proportional control mode smooth and linear relationship between controller output and the error can be achieved. The error range required to cover 0 to 100% controller output is called as **proportional band.**

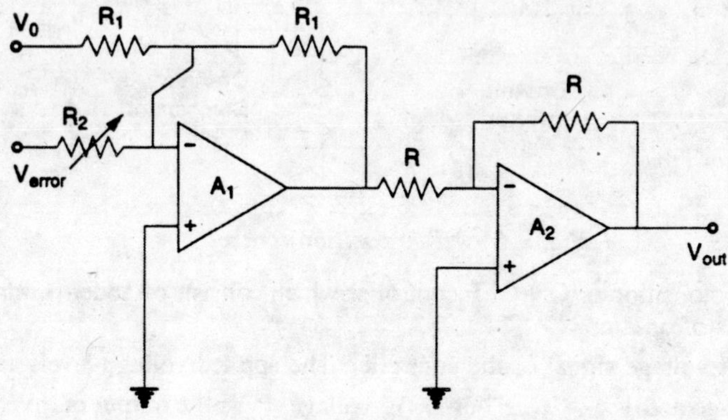

Fig. 9.5 : Proportional control.

Fig. 9.5 shows electronic proportional controller. It consists of summing amplifier configuration (whose input signals are error voltage V_{error} and V_0) and inverter circuit. The equation of output voltage V_{out} of the controller is specified as :

$$V_{out} = G_p\, V_{error} + V_0 \qquad \text{...(9.11)}$$

where ; V_{out} = Output voltage

V_{error} = Error voltage

G_p = Gain R_1/R_2

V_0 = Output voltage with zero error

(C) Integral Controller

This mode is also called as reset action. Fig. 9.6 shows integral controller using op-amp. It consists of op-amp A_1 in integrator configuration and A_2 is inverter. The output equation of this controller is :

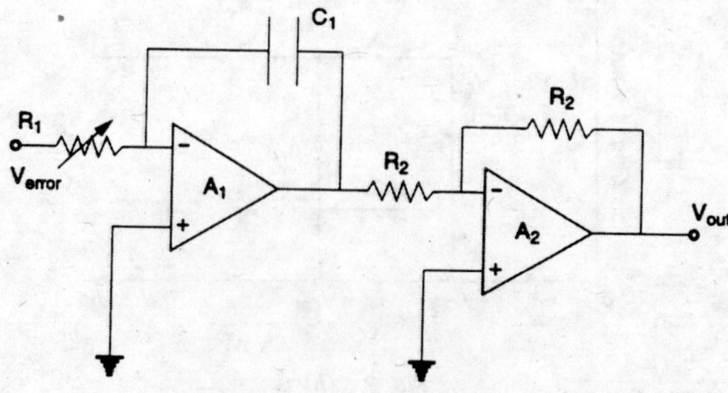

Fig. 9.6 : Integral controller.

$$m = \frac{1}{T_i}\int_0^t e\,dt + b \qquad \text{...(9.12)}$$

From this equation the output voltage of controller is :

$$V_{out} = G_I \int_0^t V_{error}\,dt + V_{out}\,(0)$$

$$V_{out} = 1/R_1C_1 \int_0^t V_{error}\,dt + V_{out}\,(0) \qquad \text{...(9.13)}$$

where ; $G_I = 1/R_1C_1$ = Integral gain (integral time in sec.)

V_{error} = Error voltage

V_{out} (0) = Initial input voltage

The values of R_1 and C_1 determines or used for adjusting the integration time. This integration time constant determines the rate at which controller output increases when error is constant.

(D) Derivative Controller

This controller is also called as rate or anticipatory control. Derivative controller is not used alone because if the error is zero or constant the controller has no output. The output expression for this controller is :

$$m = T_d\, de/dt \qquad \text{...(9.14)}$$

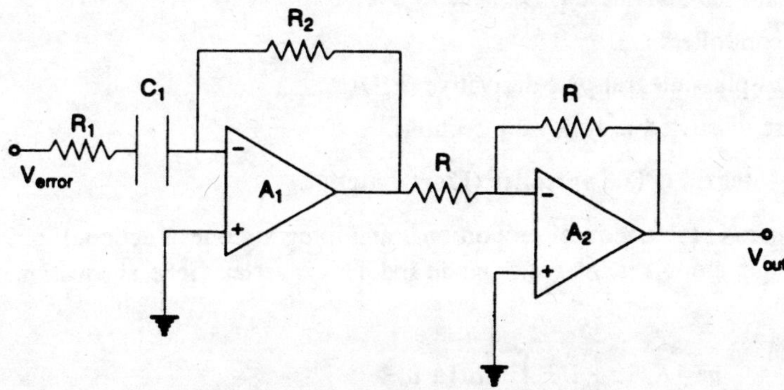

Fig. 9.7 : Derivative controller.

The op-amp derivative controller can be implemented as shown in Fig. 9.7. It consists of op-amp A_1 in differentiator configuration and A_2 is inverter. R_1 is the input resistance which provides stability against randomly varying signals. Output voltage equation of this circuit is :

$$V_{out} = G_D \cdot \frac{d\,V_{error}}{dt} \qquad \text{...(9.15)}$$

where ; V_{out} = Output voltage

V_{error} = Error voltage

G_D = Derivative time constant = R_2C_1 (sec.)

Above equation indicates that if R_1 is not used in the circuit its gain increases with increasing frequency due to which the oscillations at the output increases. If R_1 is added the gain of circuit gets controlled by the ratio R_2/R_1 at higher frequency thus R_1 provides stability at higher frequencies.

Applications :

Pure derivative mode is seldom (rarely) used in process control. It is used or implemented in combination (composite controller mode) with other controllers because it provides more advantages. In next section composite controllers are discussed.

9.7 Composite Mode Controllers

(1) In process industries various parameters are to be controlled at desired fixed value or set point. But sometimes it becomes very difficult and tedious job to find accurate (control action) controllers.

(2) Sometimes (single) individual controller has some limitations or disadvantages.

To avoid above stated problems or limitations of individual controllers, combination of the controllers is implemented which are called as composite mode electronic controllers. In these type of controllers advantages and applications of individual controllers are used whereas limitations of the controllers are eliminated.

The composite mode controllers are :

(1) Two mode controllers :
   (a) Proportional-plus-integral (PI)
   (b) Proportional-plus-derivative (PD)

(2) Three mode controllers :
   (a) Proportional-plus-integral-plus-derivative (PID)

These controllers are discussed in following sections.

(A) Proportional-Plus-Integral (PI) Controller (Electronic)

This type of controller uses combination of proportional and integral control action. Fig. 9.8 shows circuit diagram of PI controller. Op-amp A_1 is integrator mode and A_2 is inverter. General equation of this controller is :

$$m = K_c\,[e + 1/T_i \int_0^t e\,dt] + b \qquad ...(9.16)$$

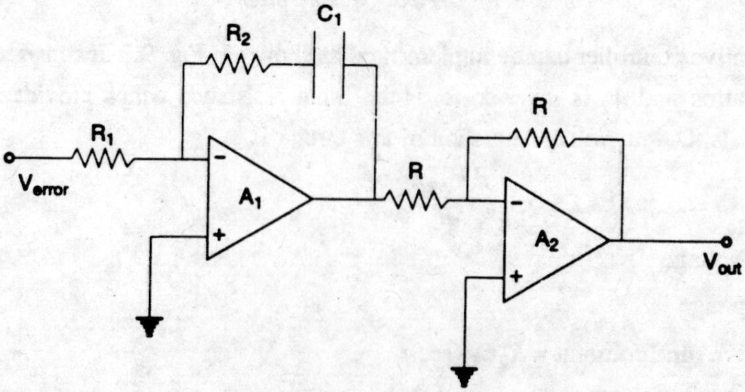

Fig. 9.8 : P–I controller.

The relationship between input and output is given by :

$$V_{out} = (R_2/R_1)\ V_{error} + 1/R_1C_1 \int_0^t V_{error}\,dt \qquad ...(9.17)$$

According to controller action the gain of proportional and integral controller is as follows :

$$V_{out} = (R_2/R_1)\ V_{error} + (R_2/R_1)\ 1/R_2C_1 \int_0^t V_{error} + V_{out}(0)$$

$$V_{out} = R_2/R_1\ [V_{error} + 1/R_2C_1 \int_0^t V_{error}\,dt] + V_{out}(0) \qquad ...(9.18)$$

The proportional band of this controller is adjusted by proportional gain $G_p = R_2/R_1$ and integration gain $G_I = 1/R_2C_1$.

(B) Proportional-Plus-Derivative (PD) Controller (Electronic)

Fig. 9.9 shows PD controller. It is a combination of proportional and derivative control action. Its general equation is :

$$m = K_c\ [e + T_d\ de/dt] + b \qquad ...(9.19)$$

Using op-amp, this control action can be implemented as shown in Fig. 9.9.

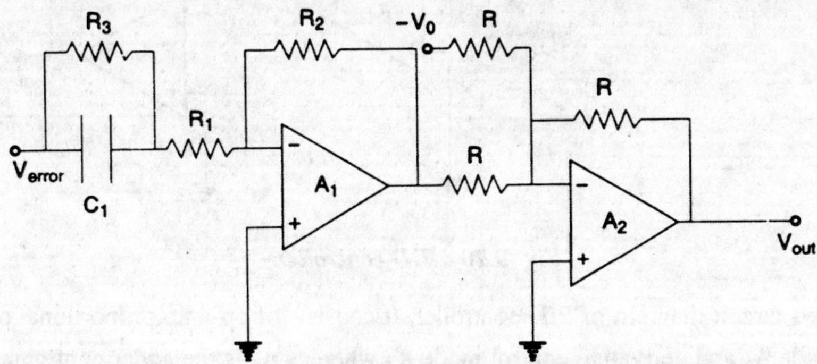

Fig. 9.9 : P–D controller.

The output voltage equation is :

$$V_{out} = \left(\frac{R_2}{R_1 + R_3}\right) V_{error} + \left(\frac{R_2}{R_1 + R_3}\right) R_3C_1 \frac{d\,V_{error}}{dt} + V_0 \qquad ...(9.20)$$

where ; $G_P = \left(\frac{R_2}{R_1 + R_3}\right)$ = Proportional gain constant

$G_D = R_3C_1$ = Derivative gain constant

(C) Proportional-Plus-Integral-Plus-Derivative (PID) Controller (Electronic)

This controller is combination of proportional, integral and derivative control actions. It is a powerful control and has fast response, noise free operation. In general the output equation for PID controller is specified by :

$$m = K_c\ [e + 1/T_i \int_0^t e\,dt + T_d\ de/dt] + b \qquad ...(9.21)$$

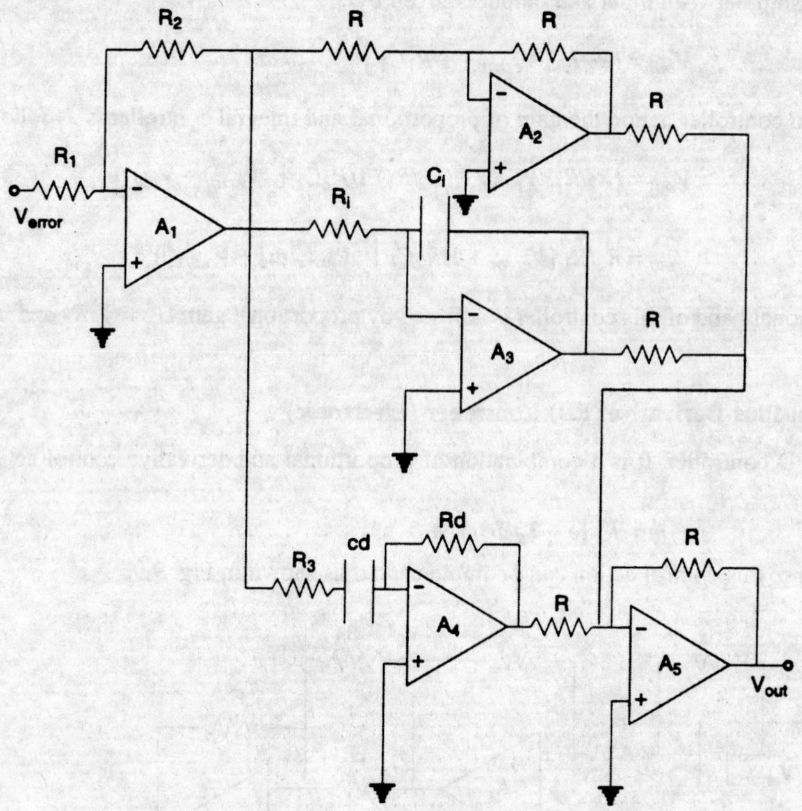

Fig. 9.10 : PID controller.

Fig. 9.10 shows circuit diagram of PID controller. It consists of op-amp proportional control mode A_1, integrál control mode A_3 and derivative control mode A_4 whereas A_5 is the adder configuration so as to get complete PID action.

Therefore ; the output voltage equation of this controller is specified as :

$$- V_{out} = R_2/R_1 \, V_{error} + (R_2/R_1) \frac{1}{R_i C_i} \int_0^t V_{error} \, dt + \left(\frac{R_2}{R_1}\right) R_d C_d \frac{d V_{error}}{dt} + V_{out}(0)$$

$$- V_{out} = R_2/R_1 \left[V_{error} + \frac{1}{R_i C_i} \int_0^t V_{error} \, dt + R_d C_d \frac{d V_{error}}{dt} \right] + V_{out}(0) \quad ...(9.22)$$

where ; $G_P = R_2/R_1$ = Proportional gain constant

$G_I = \frac{1}{R_i C_i}$ = Integral gain constant

$G_D = R_d C_d$ = Derivative gain constant

9.8 Pneumatic Controllers

Pneumatic controller system uses the pressure of gas or air in a pipe. These systems are widely used in industrial applications, process control, industrial instrumentation, power plants, and domestic services.

These controllers are simple, inexpensive, reliable, safe and also requires less maintenance. As stated above in pneumatic system pressure of gas or air in a pipe is used. Consider a pipe arrangement of any length and if pressure of gas is increased from one end then this increased pressure propagates down until the pressure is increased to new value. Generally, the pneumatic signals are carried out with dry air or gas lies in the pressure range of 3 to 15 psi or 20 to 100 kPa.

The basic pneumatic controller system is nozzle-flapper system which converts pressure into mechanical displacement or displacement into pressure signal. In industrial process control the pneumatic valve requires high power for operation so it becomes necessary to amplify the pneumatic power. The device used for pneumatic power amplification is called as pneumatic relay or Booster. It is connected with nozzle-flapper arrangement.

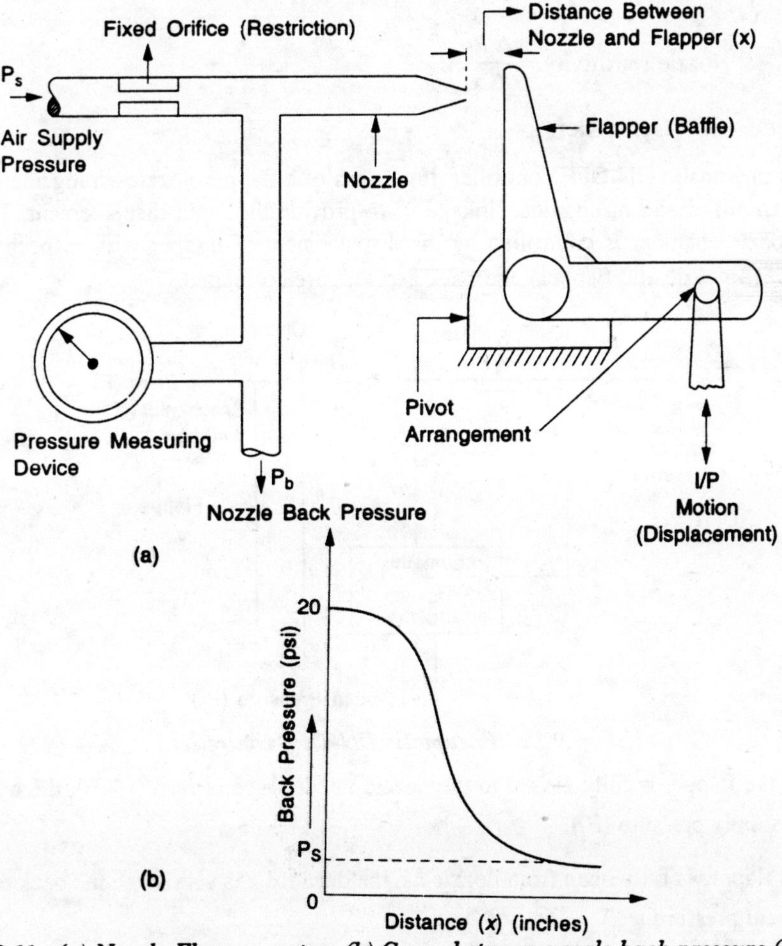

Fig. 9.11 : (a) Nozzle-Flapper system (b) Curve between nozzle back pressure (Pb) and distance between nozzle and flapper (x)

Fig. 9.11(a) shows basic arrangement of nozzle-flapper system. It is also called baffle-nozzle system.

This system consists of a nozzle supplied with air pressure which is restricted by a fixed orifice plate. Near to nozzle the flapper is placed or located which is moved or displaced by means of pivot arrangement. Generally the air supply pressure is maintained at about 20 (above atmospheric (gauge) pressure in pounds per square inch) using orifice plates. Nozzle diameter is 0·025 inches. The diameter of orifice is 0·010 inches. Nozzle diameter is larger than orifice diameter for more accuracy.

When input motion or displacement is applied to flapper the distance '*x*' between nozzle-flapper gets changed. As flapper approaches to nozzle the distance '*x*' is reduced resulting in increase of the signal pressure (P_s) which is also called back pressure. If the flapper and nozzle surfaces gets closer to each other the signal pressure becomes equal to input air pressure supply (P_s). As the flapper is moved away from nozzle the pressure gets set at maximum value by considering maximum pressure leakage through nozzle.

Fig. 9.11(b) shows curve between Nozzle Back Pressure (P_b) and distance between Nozzle-Flapper (*x*). It shows the relationship between back pressure and distance '*x*'.

The slope of curve at any point is called as "Nozzle Sensitivity". In central region nozzle sensitivity will be high.

$$\text{Nozzle sensitivity} = \frac{\Delta P_b}{\Delta x} \qquad \text{...(9.23)}$$

(A) Pneumatic ON-OFF Controller

Fig. 9.12 shows pneumatic ON-OFF controller. It consists of a flapper-nozzle arrangement with pneumatic relay or power amplifier and mechanical linkage so to provide the input displacement. The back pressure P_b in flapper nozzle chamber is controlled by the displacement of flapper with respect to nozzle *i.e.* the distance *x* (between nozzle and flapper) and back pressure are co-related.

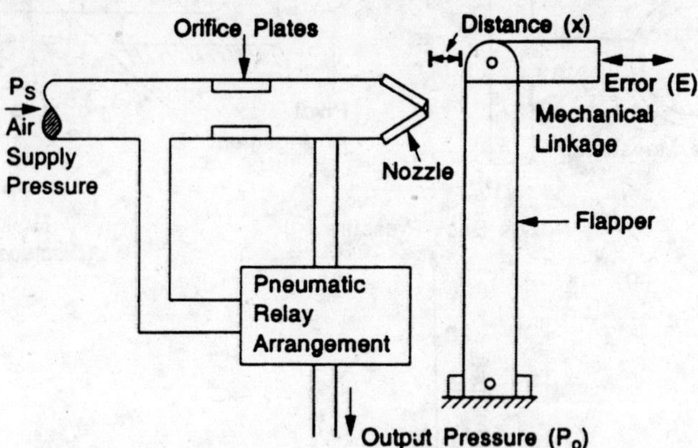

Fig. 9.12 : Pneumatic ON–OFF controller.

Mode I : If the flapper is fully closed to the nozzle *i.e.* distance is zero ($x = 0$), the back pressure (P_b) will be equal to supply pressure (P_s).

Mode II : If flapper is fully open from nozzle *i.e.* the distance *x* is very large the back pressure (P_b) will be equal to ambient pressure (P_a).

Here, as the back pressure (P_b) increases the output pressure (P_0) decreases because in this configuration the relay is reverse acting.

Following equation describes the ON-OFF control action.

$$-P_b = P_0 = -K \cdot x \qquad \text{...(9.24)}$$

where ; K = Sensitivity.

This equation indicates that (1) If the value of x is positive the output pressure P_0 will be maximum. (2) If x is negative the output pressure P_0 will be minimum. Therefore ;

$$P_0 = \text{Max if } x < 0$$

$$= \text{Min if } x > 0 \qquad \text{...(9.25)}$$

This type of controllers are used in temperature controls, humidostats.

(B) Pneumatic Proportional Controller

Fig. 9.13 shows diagram of pneumatic proportional controller. In this controller the constructional elements are flapper-nozzle arrangement which is first stage amplifier and pneumatic relay assembly is the second stage amplifier. Bellows and spring arrangement is connected in the bottom of flapper. In feedback path the bellows and spring is placed so as to reduce the actual movement of flapper. The flapper is extended and mechanically linked to which the input signal 'E' is applied. This input signal controls the position of flapper above nozzle. In this system pneumatic relay is used in reverse action.

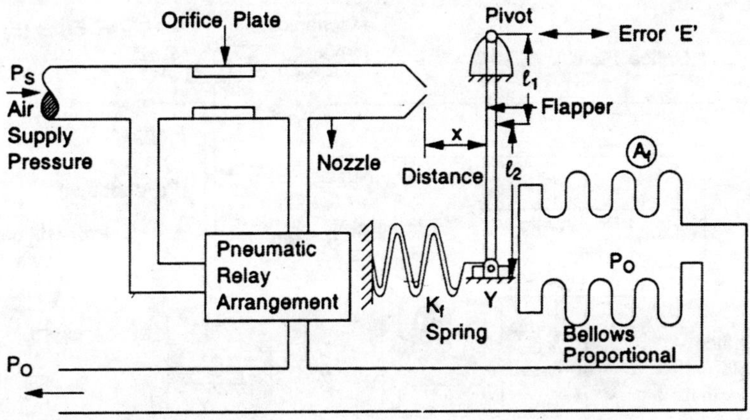

Fig. 9.13 : Pneumatic proportional controller.

Here 'x' is the distance between nozzle and flapper, 'Y' is the position of bellows and spring. The relative position x of flapper and nozzle is determined by Y and input signal E. The output equation of this controller is specified as :

$$P_0 = K_P\, e \qquad \text{...(9.26)}$$

$$P_0 = \frac{l_2 K_f}{l_1 A_f} e \qquad \text{...(9.27)}$$

where ; l_1 = Input lever arm (m)

l_2 = Feedback lever arm (m)

P_0 = Output pressure

K_f = Spring constant

A_f = Feedback bellow area

$$K_P = \frac{l_2 K_f}{l_1 A_f} = \text{Gain of proportional controller.}$$

In this type of controller the gain can be varied by adjusting l_1 and l_2 due to which the position of flapper gets varied at which the error signal is applied.

(C) Pneumatic proportional-Plus-Integral control mode (P-I Mode)

This is the combination of proportional and integral controller. It is also called as positive feedback restriction. This control action is specified as :

$$m = K_c\left(\frac{1}{T_i}\int_0^t edt + e\right) + b \qquad \text{...(9.28)}$$

Fig. 9.14 shows construction of pneumatic PI controller. In this type of controller the integral (reset) bellow is added in the system. This integral bellow is also called positive feedback bellows which is connected opposite to the proportional bellows and in addition of an adjustable restriction orifice plate which provides integral control action and calibrated in time units. The pneumatic relay is in reverse action.

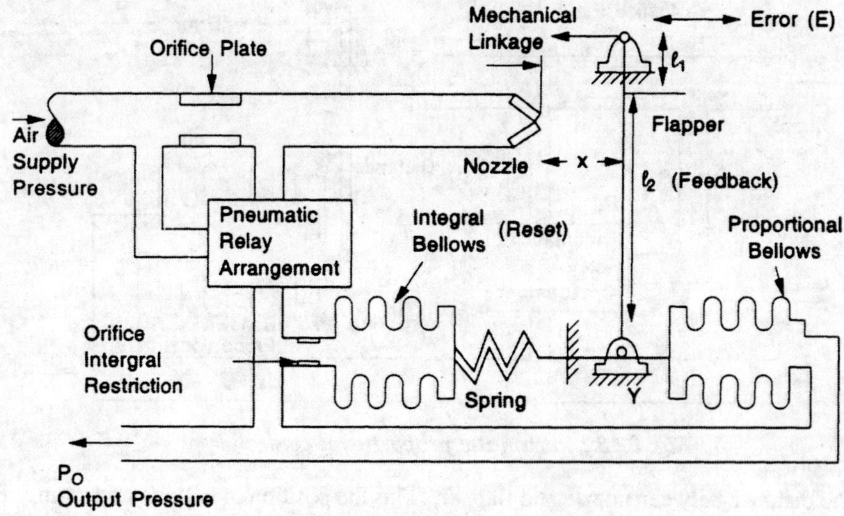

Fig. 9.14 : Pneumatic P–I controller.

When the input error (E) of the controller changes the output pressure (P_0) changes. The output pressure P_0 is applied to integral restriction orifice plates and to the feedback bellows. Pressure in the integral bellows changes slowly as the air is passed through the integral restriction to counteract gradually the pressure in feedback bellows due to which the gain of controller is increased. The integral action rate depends upon rate of air flow through the integral restriction orifice plates. In this controller the integral action increases with increase in the rate of air flow through integral restriction.

Case I : In this system gain of controller increases when integral restriction is completely open. In this case the system acts as two step controller.

Case II : When integral restriction is closed completely the system acts as a proportional controller.

(D) Proportional-Plus-Derivative Controller (P-D Mode)

This is the combination of proportional and derivative control action modes. This is also called as negative feedback restriction. Fig. 9.15 shows diagram of pneumatic proportional-plus-derivative controller. It consists of general constructional assembly such as flapper-nozzle arrangement, pneumatic relay. In this type of controller the derivative action is achieved by placing derivative restriction (orifice plates) with the feedback or proportional bellows as shown in Fig. 9.15.

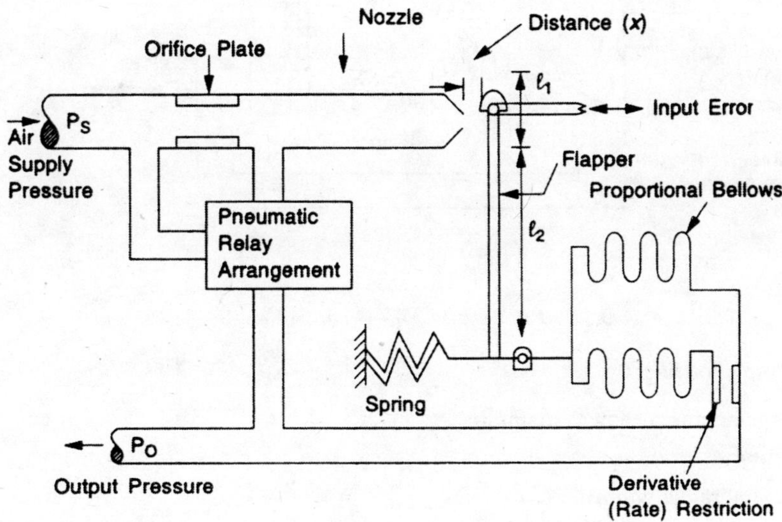

Fig. 9.15 : P–D controller.

Case I : When the input error signal (*E*) increases the flapper moves towards the nozzle due to which nozzle back pressure increases. In this case suppose the derivative restriction is absent the pressure on proportional bellows gets increased. So it replaces the flapper and limits the output pressure to a value proportional to derivative.

Case II : Now if the derivative restriction is replaced in system, due to which the pressure in feedback path will be less than the back pressure in nozzle, the flapper approaches to nozzle than it would if the derivative restriction were not present.

If the derivative restriction is fully opened the controller produces simple proportional control action but when the derivative restriction is fully closed it behaves as ON-OFF control action.

(E) Proportional-Plus-Integral-Plus-Derivative Control (P-I-D Mode)

In this controller, proportional, integral and derivative control actions are used. Fig. 9.16 shows diagram of pneumatic PID controller. It consists of flapper-nozzle assembly, pneumatic relay and proportional bellows, integral bellows and derivative restriction. The proportional gain, reset time, rate of the controllers can be set by adjusting bellows separation and restriction size.

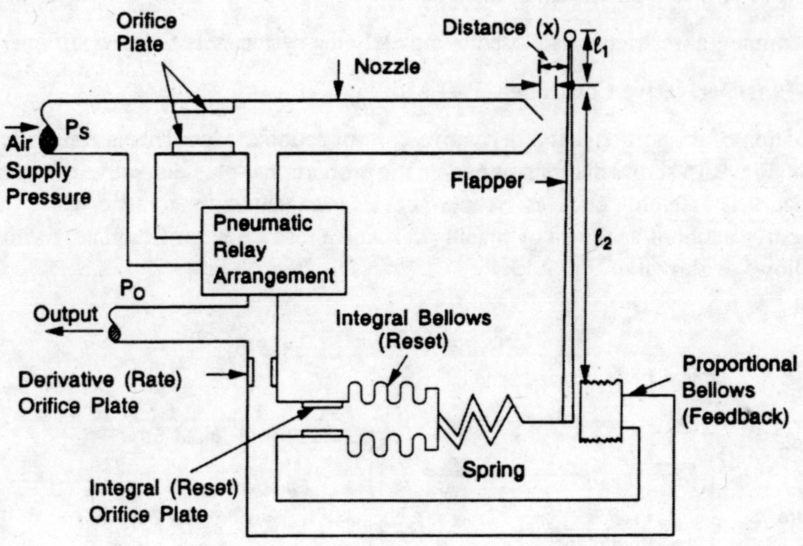

Fig. 9.16 : Pneumatic P–I–D controller.

Advantages of pneumatic controllers :

(1) Simple arrangement and easy to maintain.

(2) More reliability.

(3) Cheaper than hydraulic controllers.

(4) No problem of short circuits, fire hazards.

Disadvantages of pneumatic controllers :

(1) Leakage of tubings.

(2) Blocking of orifice plates, nozzle and relay may affect the control action.

(3) It is difficult to install the system in a common place.

(4) It is necessary to maintain dry air supply at constant pressure.

Precautions for pneumatic controllers : In pneumatic system it is necessary to maintain the clean and dry air at constant pressure. The air should be free from moisture, humidity, corrosive fluids, oils etc. otherwise it can cause troubles. Necessary filters, compressor arrangement, chemical or electric dryers etc. should be used for eliminating the above stated parameters.

Chemical dryers such as silica gel, activated alumina, calcium chloride, glycol and other drying agents should be used. Porous stone filters, dehydrators should be installed for removing oil and moisture percentage present in the air supply if any.

9.9 Hydraulic Controllers

In these type of controllers oil pressure signal is used. These type of controllers use liquid control medium to provide an output signal which is a function of input error signal. The oil may be petroleum oil or fire resistant liquids are used. This oil provides high pressure and auxiliary power. Sometimes high pressure water media may be used. In hydraulic controllers more controlling power and correct control action can be achieved because they use high pressure oil such as auxiliary power source. So these controllers are used where high power and fast operating response is required.

Necessary constructional elements for hydraulic controllers are :

(1) Hydraulic Relay or Amplifier

Its function is to provide small mechanical displacement for controlling the output of hydraulic power supply so to drive a work load. There are three type of relays used in hydraulic controllers.

(a) Jet-pipe valve

(b) Double throttle or flapper valve

(c) Four-way or spool valve.

(2) Error Detector

It produces actuating error signal necessary for controllers.

(3) Sensor or Transducer

This produces output signal proportional to measured value.

In following section different hydraulic controller modes are discussed.

(1) Hydraulic Integral Controller

Fig. 9.17 shows hydraulic integral controller which uses jet-pipe valve as a hydraulic relay for flow regulation *i.e.* to control the flow of fluid and maintain it at set point using a control valve. It consists of a diaphragm

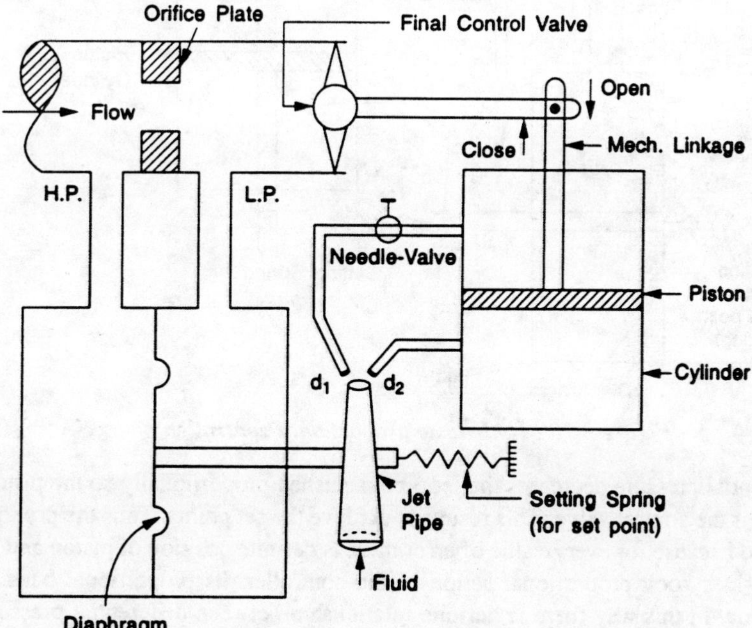

Fig. 9.17 : Hydraulic integral controller.

connected to the two differential pressure tappings (high pressure and low pressure) with the main line. The differential pressure is developed by orifice plates and applied to diaphragm due to which contradiction and expansion of diaphragm takes place. The diaphragm is linked to a jet pipe with nozzle, also a spring is provided for setting. Piston of the cylinder is connected to main control valve which opens or closes the main line with respect to movement of piston.

Case I : When the fluid flow is at set point *i.e.* the force exerted by setting spring is equal to force proportional to measured value then the jet pipe remains in the mid or centre position. Due to this the piston remains stationary or unmoved and flow rate remains constant.

Case II : Suppose if the flow rate is changed then the differential pressure at diaphragm increases or decreases resulting to move the jet pipe to right or left from mid position. Because of this pressure at one end of orifice plate is different than the pressure at other end. This differential pressure causes to move the piston which in turn opens or closes the control valve and reduces the error. The rate of movement of piston is proportional to deflection of jet pipe from mid position.

(2) Hydraulic Proportional Controller

Fig. 9.18 shows diagram of hydraulic proportional controller. It consists of feedback mechanism by coupling the control valve to controller setting spring at the bottom of cylinder.

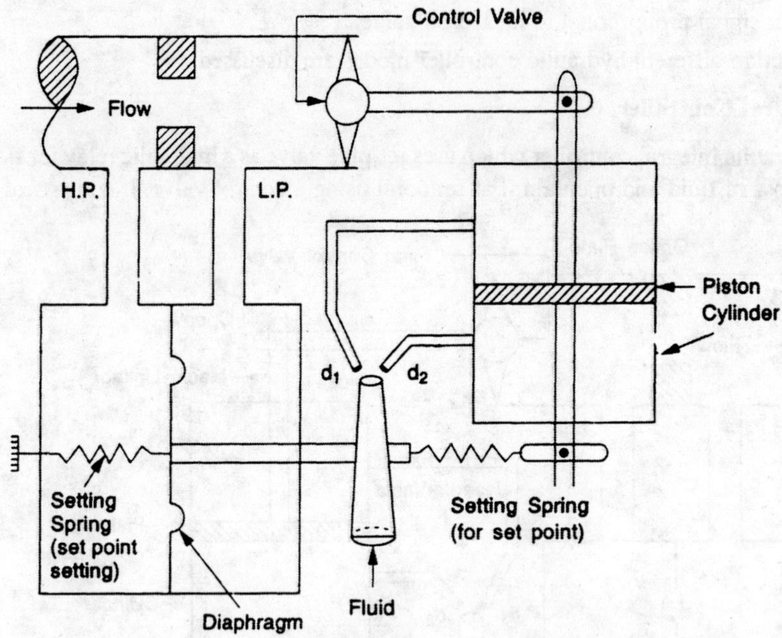

Figure 9.18 : Hydraulic proportional controller.

When the differential pressure decreases the jet pipe is pushed proportionally so the piston is moved to new position and opens the control valve. This results to achieve the set points. Thus the proportional control action can be achieved because for every value of error there is definite position of piston and control valve.. Feedback spring provides good proportional action of the controller. Its proportional band can be set by adjusting spring tension. In this way there is definite relationship between differential pressure and control valve.

Advantages of hydraulic controller :

(1) High power gain can be achieved.

(2) Fast response.

(3) They require small size of power units

(4) Simple arrangement.

(5) Low inertia or torque ratio.

(6) High amplification can be achieved using minimum amplification or relay stages.

(7) These controllers can easily be protected from overloads.

(8) Long operating life.

Drawbacks of hydraulic controller :

(1) Proper sealing and connections are necessary so as to avoid the leakage of hydraulic fluids or liquids.

(2) Because of leakage accuracy may be decreased.

(3) These controller requires more space for installation.

(4) They require periodic maintenance.

(5) Requires clean fluids so their filtration is necessary.

(6) As the control medium is circulated, they requires two lines to achieve desired control action.

(7) Accuracy of the system depends upon viscosity of fluids which are to be used, gain of relays or amplifiers, ambient temperature variations and corrosive effects.

Applications :

(1) Widely used in pressure regulation systems, flow control.

(2) Hydraulic controllers are widely used in power plants, oil refineries, petrochemical industries, steel plants etc.

9.10 Fuzzy Logic Controllers (FLC)

Introduction and Concepts of FLC

Fuzzy Logic was first proposed in 1965 by Professor L.A. Zadeh of University of California at Berkeley. The fuzzy logic techniques are widely useful in industrial process controls, industrial instrumentation systems. The successful applications of FLC are in cement kilns, paper machines. These type of controllers are very useful when the processes are too complex for the analysis of information sources available which are interpreted qualitatively and uncertainly. Another applications of FLC are in different fields such as operation research, robotics, computer science, artificial intelligence, medicine, economics, pattern recognition, control systems, chemical engineering, transducer and sensor controls, machine controls etc.

Fuzzy logic is based on set theory which allows for degree of truth or falsehood. The term logic deals with true or false. As in binary logic which defines the levels *i.e.* logic 0-false, logic 1-true means there is certainty or assertion (confirmation) whereas in fuzzy logic the assertion can be partly true or partly false.

Consider an example that "It will rain today" in this statement there is certainty. Means this statement can be true or false *i.e.* this statement has truth values only 1 or 0 so this statement is crisp and has certainty. On the other hand if the statement is "I think it will rain today". In this statement there is very less certainty about 0·8 rather than 1, means in this statement no fixed logic can be obtained. Such type of statement is called fuzzy logic *i.e.* partly true or partly false.

Actually, the fuzzy logic is not "Fuzzy" *i.e.* uncertain but the fuzzy controller has set of rules and by using these rules desired final control action is calculated. Every rule in the set is a linguistic expression of the control action related to specific set of process conditions. The rules are like IF ..... THEN its format is IF (condition) THEN (action). In this rule the condition may be "AND" and "OR" (also called as connectives). These rules are IF-THEN statements of the form IF {set of inputs} THEN {set of outputs}. The IF side of rule contains one or more conditions called antecedents THEN side contains one or more actions (outputs) called consequences.

In the fuzzy rules there are fuzzy sets and there are membership functions in the fuzzy sets. The membership function determines the type of fuzzy sets. The membership function determines the type and degree of control action required. The membership functions are represented by membership curves having shapes like triangular, *S*, *Z*, bell-shape and trapezoidal. In this section some applications of fuzzy logic controllers in instrumentation and control systems are discussed by assuming that the reader is well familiarize with FLC.

Block Diagram of Fuzzy Logic Controller

Fig. 9.19 shows block diagram of Fuzzy Logic Controller (FLC). It shows the design methodologies for a fuzzy logic control system. This system consists of different sections which are discussed below.

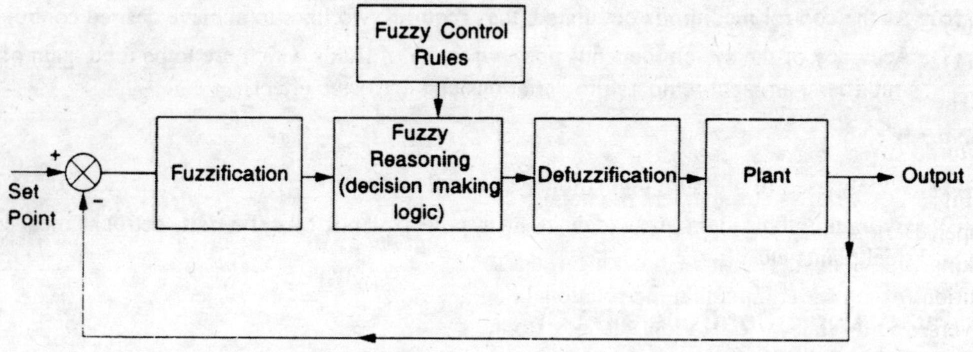

Fig. 9.19 : Block diagram of fuzzy logic controller.

(1) **Fuzzification interface** : The function of this unit is to define fuzzy subsets over a universe of discourse of a variable and to assign suitable linguistic labels to the subset, that represents the meaning of a suitable variable.

(2) **Fuzzy control rules** : These rule defines the relationship between process input and control output variable. These rule characterises the control goals and control policy of domain experts using set of linguistic rules. A database which defines the membership functions of fuzzy sets is used in fuzzy rules and fuzzy data manipulation in FLC.

(3) **Decision making logic** : It is also known as fuzzy reasoning. It provides the simulation of decision making based on fuzzy concepts and fuzzy control using fuzzy implications and rules.

(4) **Defuzification interface** : It converts the range of values of output variables into corresponding universe of discourse. It calculates the crisp value of control action at the output. Most widely used method of generating the control action is center of gravity (COG).

9.11 Applications of Fuzzy Logic Controllers

Different applications of fuzzy logic controller are discussed as follows :

(A) IGBT (Insulated Gate Bipolar Transistors) Chopper Fed DC Drive Control

System Configuration : (FLC with drive arrangement)

Fig. 9.20(a) shows block diagram of the scheme. This system is divided into three sections as follows :

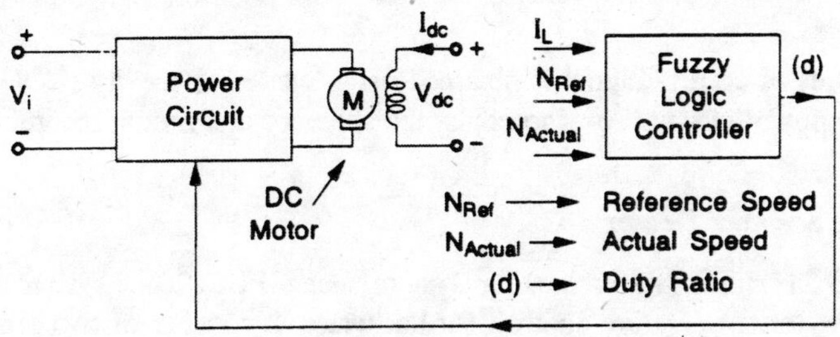

Fig. 9.20(a): Block diagram.

(1) Power circuit.

(2) Controller circuit.

(3) Software section.

These sections are discussed below.

(1) Power circuit

Fig. 9.20(b) shows arrangement of chopper circuit and DC motor. Chopper circuit is two-quadrant IGBT chopper. DC motor is fed by the chopper circuit. This circuit is used for (i) motor mode (ii) regenerative breaking mode. In regenerative breaking mode most of the breaking energy is returned to supply. The condition for regeneration is that the rotational e.m.f. must be more than applied voltage so that the current is reversed and mode of operation changes from motoring to regenerating. In this operation some of the energy is recoverable so regenerative breaking is widely used in drives. In the circuit T_1 is $IGBT_1$ and T_2 is $IGBT_2$. The function of chopper circuit is as ON/OFF switch (thyristor/power device) that connects and disconnects the load from power supply and produces chopped load voltage from a constant input supply. An IGBT is a voltage controlled device. Advantages of IGBT are (a) High input impedance (b) Low ON state voltage drop (c) Lower switching and conducting losses. (d) Peak current capability (e) Ease of gate drive (f) High operating speed.

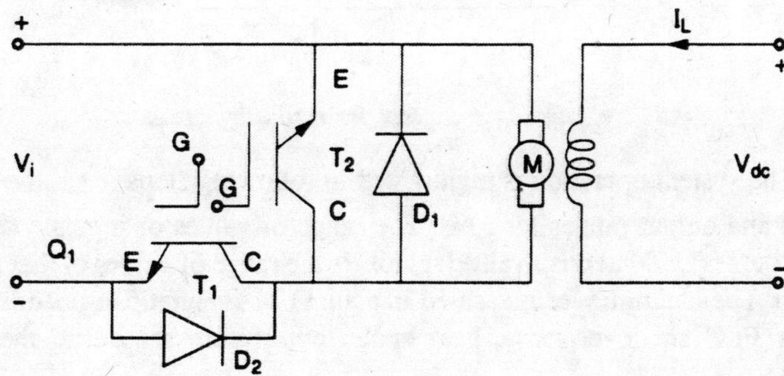

Fig. 9.20(b) : Chopper circuit.

(2) Fuzzy logic controller (design of rule-base and fuzzification)

The important function of FLC is to control the duty ratio (*d*) of chopper so as to achieve constant speed. But the duty ratio depends upon the parameters (a) Change of input voltage (*dV*) and rate of change of input voltage (*dV/dt*) (b) Change of speed *dN* and rate of change of speed (*dN/dt*) (c) Load current (I_L). For these parameters different membership functions can be defined in the fuzzy sets.

For the membership functions of *dN*, *dN/dt* and duty ratio (*d*) triangular membership curves can be used. Using software simulation like MATLAB or any other softwares, the result can be obtained.

(3) Defuzzification

In this step the crisp values of control signal is obtained using center of gravity (COG) method. This provides the control signal to power circuit. So as to adjust the speed of d.c. motor according to duty ratio (*d*).

(B) Fuzzy Logic Control for a Water Heater

System description : Fig. 9.21 shows water heating system using FLC. This system provides good performance and control (in water temperature control) for hot water. It consists of two sensors which are water temperature sensor and water level indicator in the tank. Water is stored in a tank with the help of water inlet at the top and hot water outlet at the bottom of tank. The outputs of these sensor is provided to the inputs of fuzzy logic controller (FLC). Output of FLC is applied to knob of heater because this knob controls the heating element power. Depending on the amount of heat of the heater element, water gets heated. In this system automatic water flow control is provided which determines how much the cold water flows into the tank from main supply.

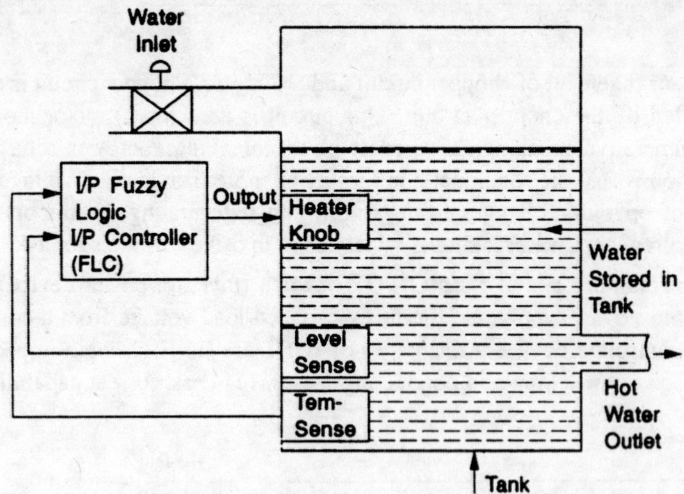

Fig. 9.21 : Fuzzy logic control for a water heater.

System operation : The system operation is highlighted in following steps :

(1) To define input and output ranges for FLC. The range of values of inputs and outputs is called universe of discourse (UoD) at crisp value (A UoD is the range of values of that a process variable may take place). The UoD may be measured in actual or computational manner. UoD for inputs and outputs for FLC are level sense, heat knob, temperature sense and their maximum and minimum values.

(2) **Fuzzification of inputs** : The input signals to the FLC are level sense and temperature sense. In fuzzification triangular membership functions can be used. During this process the fuzzified values across the universe of discourse are distributed. Also the fuzzy membership functions for level sensor, temperature sensor can be prepared.

(3) To set up fuzzy membership functions for the output signal which is heater knob.

(4) **To develop fuzzy rule base** : Upto this point the membership functions for input and output signals are developed now the next task is to set the rules for operation of FLC. Generally, these rules are in IF-THEN format or these rule can be developed using neural networks or by human experts (heuristics).

(5) **Defuzzification** : In this method heater knob is controlled by obtaining crisp values. In fuzzy rules there are fuzzy sets which assigns different membership (values) variables to all the outputs which are to be applied to the process. So it is necessary to get the crisp value of the control action that can be applied. In short the outputs have been fuzzified so it is necessary to obtain single crisp value which is called diffuzification. There are different methods available for obtaining the crisp value. Most commonly used method is center of gravity (COG) of fuzzy interface to get final control action.

Steps involved in COG method :

(1) Determination of the centroid point on *X*-axis for each output membership function.

(2) Calculation of area of the memberships.

(3) Defuzzified output (crisp value control action) is obtained by using weighted average of *X*-axis centroid points and computed areas.

(C) Fuzzy Logic Control for Washing Machine Drier

One of the important feature of fuzzy logic is that it provides better control action for complex processes which operates over wide range of conditions. Another application of FLC is to control the drier current of a washing machine. In a washing machine the drier current is controlled by fuzzy logic methods. The phenomenon used in washing machine for drying the clothes is drier current, this current can be controlled by following parameters:

(1) Wetness of the clothes.

(2) Quantity and nature of the clothes *i.e.* load.

Above discussion indicates that both the parameters *i.e.* wetness and load of the clothes is to be considered for designing a FLC for drier of a washing machine. If heavy canvas cloth is considered then more drier current is required for drying irrespective of the quantity as compared to the cotton clothes.

Steps involved in designing of FLC for drier :

(1) To define the input and output ranges of the parameters for the FLC of drier. These parameters are wetness of the clothes, dry moist, less moist, more moist cloth. (*i.e.* to identify I/O process variables and UoD).

(2) **Fuzzification of inputs** : Initial task is to define fuzzy subsets of variables and assigning linguistic labels to the fuzzy subset. These labels are such as dry, less moist, more moist and wet (for cloth dryness).

(3) To define the membership functions which acts on input variables, usually from sensor data. The selection of membership function may be triangular or trapazoidal.

(4) **To develop fuzzy rule base** : In this step the task is to set the rules for smooth operation of controller. IF-THEN statements may be used for developing the fuzzy rules. The rules are like

IF {wetness is dry} AND {load is less} THEN current of drier is zero.

The amount of current required for drier depends upon degree of wetness and load of the clothes so the fuzzy rules may be like :

IF {wetness is more moisture} AND {load is minimum} THEN {current is maximum}.

During preparation of fuzzy base rules consider the cases like wetness is dry, wetness is less moisture, wetness is more moisture with load may be minimum, more, heavy and to find out current requirement *i.e.* zero, more maximum and less.

(5) **Defuzzification** : This method is carried out so as to get the crisp values of the control action that can be applied to the process. In fuzzy subset different membership functions are assigned so as to achieve single crisp control action (output signal) and the defuzzification process is used. Most widely used method is center of gravity (COG). In heater control method COG is discussed.

9.12 Parameters to be Considered while Selecting Controller Mode

In process control systems it is very difficult or impossible to design universal controller which is suitable for all processes. It is very critical and complicated way to design, assemble and install proper controller for any process. Following important parameters are suggested for selecting proper controller.

(1) Accuracy of the controller.
(2) Quality of the controller and its control action.
(3) Reliability and efficiency of the controller.
(4) Cost of controller.
(5) Response and stablity of controller.
(7) Safety factor.
(8) Erection and maintenance of controller.
(9) It should be mechanically rigid and strong.
(10) It should not be affected by drift, hysteresis, stray electric and magnetic fields.

EVALUATION COMPONENT

1. **Write a short note on "Process Control".**
2. **Explain Automatic controllers.**
3. **With neat block diagram explain ;**
   (i) Analog controllers
   (ii) Digital controllers.
4. **What are proportional, integral, derivative control actions ? Explain them.**
5. **What are the advantages of electronic controllers ?**
6. **Compare**
   (i) Analog and Digital controllers.
   (ii) Pneumatic and Hydraulic controllers.
7. **Explain, with a neat circuit diagram the working of Electronic PID controller.**
8. **Why composite mode controllers are used ?**
9. **Explain the working of pneumatic ON-OFF controller with neat diagram.**
10. **What are the precautions necessary for pneumatic controllers ?**
11. **Explain hydraulic integral controller.**

12. List applications of hydraulic controllers.
13. With neat block diagram explain Fuzzy Logic Controllers.
14. How the FLC can be implemented for DC drive control ?
15. Compare PI, PD and PID controllers.
16. Draw PI controller using OP-AMP. Explain it.
17. Explain following control modes.
    (a) ON-OFF
    (b) Proportional + Derivative
18. State the criterion or parameters for selecting a controller for typical application.
19. Suggest a Fuzzy logic controller system for
    (a) Liquid Level control.
    (b) Flow rate control.

INDEX

A

B

C

D

P

Q

R

S

T

U

V

W

X

Z

REFERENCES

1. D. PATRANABIS "Principles of Industrial Instrumentation" Second Edition, Tata McGraw-Hill Publishing Company Limited, New Delhi.
2. ERNEST O. DOEBELIN "Measurement Systems. Application and Design" Fourth Edition, McGraw-Hill International Editions, Mechanical Engineering Series.
3. GOLDING and WIDDIES "Electrical Measurements and Measuring Instruments" Fifth Edition, Wheeler Publishing.
4. C.S. SHARMA, G.R. SHARMA, V.S. MANI "Instrumentation Devices and Systems" Seventh Reprint 1989, Tata McGraw-Hill Publishing Company Limited.
5. BARRY E. JONES " Instrumentation Measurement and Feedback" Twelth Reprint 1994, Tata McGraw-Hill Publishing Company Limited.
6. S.K. SINGH "Industrial Instrumentation and Control", Fifth Reprint 1992, Tata McGraw-Hill Publishing Company Limited.
7. BERNARD M. OLIVER, JOHN M. CAGE "Electronic Measurements and Instrumentation" McGraw-Hill International Editions, Electrical and Electronic Engineering Series.
8. H.S. KALSI, "Electronic Instrumentation" Fifth Reprint 1998, Tata McGraw-Hill Publishing Company Limited, Technical Education Series.
9. A.K. SAWHNEY, PUNNET SAWHNEY, "A Course in Electrial and Electronics Measurement and Instrumenttion" Dhanpat Rai and Sons.
10. W.D. COOPER and A.D. HELFRICK, "Electronic Instrumentation and Measurement Techniques" Third Edition 1985, Printice-Hall of India.
11. SOL. D. PRINSKY, "Electronic Instrumentation" Printice-Hall of India.
12. G.B. CLAYTON, "Operational Amplifiers" Second Edition, English Language Book Society/Butterworths.
13. RAMAKANT A. GAYAKWAD, "OP-AMPS and Linear Integrated Circuits" Third Edition. Printice-Hall of India Pvt. Limited.
14. D. ROY CHAUDHARI, SHAIL JAIN, "Linear Integrated Circuits" New Age International Pvt. Limited Publishers.
15. I.J. NAGRATH, M. GOPAL, "Control Systems Engineering" Second Edition, Wiley Eastern Limited.
16. BELA G. LIPTAK, "Process Measurement and Analysis" Instrument Engineers Handbook, Third Edition, Butterworth-Heinemann Ltd.
17. DONALD P. ECKMAN, "Automatic Process Control" Wiley Eastern Limited.
18. F.G. SHINSKEY, "Process Control Systems Application, Desin and Tuning" Third Editioin, McGraw-Hill International Editions, Chemical Engineering Series.
19. CURTIS JOHNSON "Process Control Instrumentation Technology" Fourth Edition, Prentice-Hall of India Pvt. Limited.

20. B.C. NAKRA, K.K. CHAUDHARI, "Instrumentation Measurement and Analysis" Tata McGraw-Hill Publishing Company Limited, New Delhi.
21. BELA G. LIPTAK, "Process Control" Instrument Engineers Handbook, Third Edition, Butterworth-Heinemann Ltd.
22. A.J. BOUWENS, "Digital Instrumentation", McGraw-Hill Book Company, International Student Edition.
23. Dr. DEEPAK PATRANABIS, Dr. SUDEEP GHOSH, Sri. KALYAN MAJUMDAR, Sri. KALLOL BHATTACHARYA, "Transducer Engineering", January 1993, (Code No. 266) Update for Working Professionals, All India Council for Technical Education, Continuing Education Programme.
24. Logic Databook, Volume 1, National Semiconductor Corporation.
25. DOUGLAS V. HALL, "Microprocessors and Interfacing Programming and Hardware", Tata McGraw-Hill Publishing Company Limited, New Delhi.
26. M.D. SINGH, K.B. KHANCHANDANI, "Power Electronics", Tata McGraw-Hill Publishing Company Limited, New Delhi.
27. LESLIE CROMWELL, FRED J. WEIBELL, ERICH A PFEIFFER, "Biomedical Instrumentation and Measurements", Second Edition, Prentice–Hall of India Private Limited, New Delhi.
28. R.S. KHANDPUR, "Hand Book of Biomedical Instrumentation", Tata Mcgraw-Hill Publishing Company Limited, New Delhi.

Article Papers

1. K.C. NG and C.D. THOMAS, "A Long Term Submersible Temperature Recorder", TENCON'84, Trends in Electronics, IEEE International Conference on Consumer and Industrial Electronics and Applications (April 17-19, 1984) Singapore, pp 160-162.
2. V.R. SINGH and E.S.R. GOPAL, "Progress of Instrumentation During Past Fifth Yearrs in India", IETE Technical Review, Vol. 15, No. 5, September-October 1998, pp 305-318.
3. R.P. BAJPAI, "Microelectronics Instrumentation at CSIO", IETE Technical Review Vol. 15, No. 1 and 2, Januarry-April 1998, pp 13-14.
4. Products and Processes, "Fibre Optic Sensors" Technorama-August 1999 (A Supplement to the Journal of the Institution of Engineers (India)), pp 32.
5. H.O. MODI, V. JAGADEESH KUMAR and P. SANIKARAN, "An Optical Fibre Sensor for Non-Sinusoidal Current", IETE Journal of Research, January-February 1997, pp 79-83, Vol. 1, No. 1.
6. P. SESHADRI, V.K. SHRIRAM and C. RAJA RAO, "An ECG Simulator", Practical Design, IETE Technical Review, March-April 1996, pp 119-123, Vol. 13, No. 2.
7. K.N.S. ACHARYA, A.C. METI, S.K. KADKOL and S.A. SAVANUR, "Determination of Response of Thermocouples", IETE Technical Review, Vol. 14, No. 6, November-December 1997, pp 479-482.
8. Y.P. SINGH, N. AHMAD, "Fuzzy Linguistic Control Systems for Process Control Applications", Journal of the IETE, November-December 1996, Vol. 42, No. 6 pp 363-376.
9. K.B. KHANCHANDANI, "A Fuzzy Logic Controller for IGBT Based Chopper Fed DC Drive Control", "Proceedings International Conferene on Cognitive systems", ICCS'98, Dec. 13-15, 1998 New Delhi, India pp 298-306.
10. C.L. PUTTASWAMY, BHIMSINGH and B.P. SINGH, "Fuzzy Based Sliding Mode Control of Permanent Magnet Brushless DC Motor", Journal of IETE, July-August 1995, Vol. 41, No. 4 pp 245-252.
11. S. SELVAM, "Fuzzy Logic Control for the Drier of a Washing Machine", IETE Technical Review, May-June 1998, Vol. 15, No. 3, pp 217-219.
12. K.B.M. NAMBUDIRIPAD, DEVDAS MENON, "Application of Fuzzy Logic in Student Evaluation", The Indian Journal of Technical Education, Vol. 21, No. 4, Oct.-Dec. 1998, pp 1-7.
13. ST CHOW and CH CHEW, "Radio Telemetry system for Vibration Measurement of Rotating Machinery", TENCON' 84, Trends In Electronics, IEEE International conference on consumer and Industrial Electronics and Applications (April 17-19, 1984) Singapore pp 3-5